Information Systems Engineering and Management

Volume 31

The book series "Information Systems Engineering and Management" (ISEM) publishes innovative and original works in the various areas of planning, development, implementation, and management of information systems and technologies by enterprises, citizens, and society for the improvement of the socio-economic environment.

The series is multidisciplinary, focusing on technological, organizational, and social domains of information systems engineering and management. Manuscripts published in this book series focus on relevant problems and research in the planning, analysis, design, implementation, exploration, and management of all types of information systems and technologies. The series contains monographs, lecture notes, edited volumes, pedagogical and technical books as well as proceedings volumes.

Some topics/keywords to be considered in the ISEM book series are, but not limited to: Information Systems Planning; Information Systems Development; Exploration of Information Systems; Management of Information Systems; Blockchain Technology; Cloud Computing; Artificial Intelligence (AI) and Machine Learning; Big Data Analytics; Multimedia Systems; Computer Networks, Mobility and Pervasive Systems; IT Security, Ethics and Privacy; Cybersecurity; Digital Platforms and Services; Requirements Engineering; Software Engineering; Process and Knowledge Engineering; Security and Privacy Engineering, Autonomous Robotics; Human-Computer Interaction; Marketing and Information; Tourism and Information; Finance and Value; Decisions and Risk; Innovation and Projects; Strategy and People.

Indexed by Google Scholar. All books published in the series are submitted for consideration in the Web of Science.

For book or proceedings proposals please contact Alvaro Rocha (amrrocha@gmail.com).

Kenneth David Strang · Narasimha Rao Vajjhala
Editors

International Program and Project Management—Best Practices in Selected Industries

Springer

Editors
Kenneth David Strang
W3-Research-and-University
of the Cumberlands
Williamsburg, KY, USA

Narasimha Rao Vajjhala
American University in Bulgaria
Blagoevgrad, Bulgaria

ISSN 3004-958X ISSN 3004-9598 (electronic)
Information Systems Engineering and Management
ISBN 978-3-031-80274-4 ISBN 978-3-031-80275-1 (eBook)
https://doi.org/10.1007/978-3-031-80275-1

This Springer imprint is published by the registered company Springer Nature Switzerland AG
The registered company address is: Gewerbestrasse 11, 6330 Cham, Switzerland

Foreword

In an era where globalization and technological advancements are transforming industries at an unprecedented pace, the role of project and program management has never been more crucial. International projects, often spanning multiple countries, cultures, and regulatory environments, present unique challenges that require innovative solutions and strategic foresight. As we continue to navigate this rapidly changing landscape, it becomes essential for organizations to adopt best practices that not only optimize efficiency but also align with broader organizational goals and societal needs. *International Program and Project Management—Best Practices in Selected Industries* is a timely and valuable contribution to the body of knowledge in this field, offering insights from a diverse array of industries and project environments.

This book's strength lies in its interdisciplinary approach, drawing from industries such as financial technology, biopharmaceuticals, software engineering, and public administration. The authors have meticulously crafted case studies that provide practical applications of project management frameworks, demonstrating how universal methodologies like ISO 21500 and agile can be adapted to meet industry-specific challenges. Each chapter builds on the foundational principles of project management while addressing the nuances that make each sector unique. This blend of theory and practice offers readers a comprehensive guide to managing complex international projects, equipping them with the tools needed to succeed in a variety of environments. One of the standout features of this book is its focus on the integration of emerging technologies in project management. In particular, the chapter on artificial intelligence (AI) projects addresses the ethical, social, and legal challenges associated with deploying AI systems. As AI continues to reshape industries, project managers must be adept at navigating the risks and complexities inherent in these technologies. The authors offer a robust framework for managing AI projects that emphasizes transparency, stakeholder engagement, and ethical responsibility, which will be invaluable for professionals working in AI-driven environments.

The exploration of human resource management (HRM) within project contexts is another significant contribution of this book. The chapter on HRM in the software engineering industry, for example, highlights the importance of team development, leadership, and performance evaluation in project success. By contrasting traditional organizational HRM with project-specific HRM practices, the authors provide a clear understanding of how project managers can effectively manage their teams to ensure optimal performance. This focus on the human element of project management is a vital reminder that, despite the technological advancements, people remain at the heart of successful project execution.

Additionally, the book examines stakeholder management in complex R&D projects, offering a novel approach to balancing short-term efficiency with long-term value creation. In sectors like biopharmaceuticals, where innovation and collaboration are paramount, the ability to manage stakeholder relationships effectively can determine the success or failure of a project. By applying complexity theory, the authors offer a framework that helps project managers navigate the uncertainty and ambiguity that often characterize collaborative R&D projects, ensuring that stakeholder needs are met while maintaining strategic focus. The book also sheds light on the growing importance of agility in project management, particularly in the context of Business Intelligence and Analytics (BI&A) projects. These projects, characterized by their exploratory nature and rapidly changing requirements, demand a level of flexibility and adaptability that traditional project management approaches often struggle to provide. By using a Complex Adaptive Systems (CAS) lens, the authors demonstrate how project teams can self-organize, adapt, and thrive in high-uncertainty environments. This chapter serves as a critical resource for project managers working in data-driven industries, where agility and iterative processes are key to success.

Public administration, another critical area explored in this book, is undergoing a transformation with the adoption of digital technologies. The chapter on Nigeria's public sector highlights how emerging trends like automation and data analytics are revolutionizing governance and project management practices. This chapter offers valuable lessons for public administrators around the world, demonstrating how the integration of technology can improve transparency, efficiency, and citizen engagement in governmental projects. As governments face increasing pressure to deliver results in a fast-paced, globalized world, the insights provided here are particularly relevant.

Finally, this book concludes with a focus on risk management, particularly in highly complex and uncertain project environments. The chapter on risk management presents advanced techniques, such as nonlinear Monte Carlo modeling and dynamic adaptive methodologies, that address the limitations of traditional risk management approaches. These techniques provide project managers with the ability to anticipate and mitigate cascading risks, ensuring that projects are not derailed by unforeseen challenges. As projects become more complex, these risk management strategies will be essential tools for project managers looking to deliver successful outcomes.

International Program and Project Management—Best Practices in Selected Industries is a comprehensive resource that offers valuable insights into the challenges and opportunities of managing international projects. Whether you are a seasoned project manager or new to the field, this book provides a wealth of knowledge that will help you navigate the complexities of today's globalized business environment. By drawing on real-world examples and integrating cutting-edge methodologies, this book equips readers with the skills and strategies needed to thrive in the ever-evolving world of international project management.

Dr. Mathias Fonkam
Associate Professor
College of Information Sciences
and Technology
Pennsylvania State University
Pennsylvania, USA

Preface

The field of international program and project management has grown increasingly complex and vital in today's globalized world. Organizations across industries are recognizing the need to manage large-scale, multi-national initiatives with precision and adaptability. This book, *International Program and Project Management—Best Practices in Selected Industries*, seeks to provide a comprehensive exploration of the best practices and methodologies that drive success in this dynamic environment. By drawing on real-world case studies from diverse industries such as technology, healthcare, finance, and construction, we aim to equip project managers, executives, and students with actionable insights and strategic frameworks that can be applied across borders and industries.

Our focus on selected industries highlights the importance of contextualized approaches to program and project management, recognizing that different sectors face unique challenges and opportunities. Through this book, readers will gain an understanding of how successful organizations navigate cultural differences, regulatory landscapes, and emerging technologies while adhering to best practices in project governance, risk management, and stakeholder engagement. Whether you are a seasoned professional or a newcomer to the field, we hope this book will serve as a valuable resource for advancing your expertise in international program and project management.

This book highlights an impressive international collaboration, featuring contributions from 24 unique authors representing 11 different countries. Scholars and professionals from the United States, Albania, Australia, Brazil, Spain, Canada, Germany, India, Nigeria, Kazakhstan, and Iran have come together to share their expertise. The global nature of this collaboration underscores a comprehensive approach to addressing project management and innovation from various perspectives and cultural contexts. The diversity of contributors showcases the book's commitment to offering a truly international perspective on contemporary challenges and emerging technologies in the field.

The first chapter, titled "Contemporary Project Management: A Macro Environmental Analysis", introduces the evolving trends in project management through a detailed DEPEST (Demographic, Economic, Political-Legal, Ecological, Socio-Cultural, and Technological) analysis. The chapter emphasizes the importance of aligning project management with organizational strategy to enhance value creation, sustainability, and competitive advantage. This chapter highlights how modern project management is no longer confined to traditional boundaries of time, cost, and scope but integrates broader strategic objectives. The authors stress the role of entrepreneurial innovation, sustainability, and cutting-edge technology in ensuring long-term project success. The chapter also examines how external macro-environmental factors are reshaping project management practices across industries. For instance, demographic shifts in the workforce, ecological sustainability goals, and rapid technological advancements are key drivers of change. The authors argue that the successful integration of these factors into project management frameworks is essential for organizations to stay competitive in an increasingly complex and interconnected global business landscape. This analysis provides valuable insights for project managers and organizations to navigate these macro-environmental forces and strategically align their projects for optimal performance and sustainability.

The second chapter, "Brief Comparison of ISO 21500 Series to Predictive and Agile Approaches", explores the challenges and opportunities of integrating ISO 21500 standards with agile project management methodologies. The authors examine the differences between predictive project management approaches, which prioritize detailed upfront planning, and agile methodologies, which emphasize adaptability, continuous improvement, and iterative value delivery. By comparing the principles of ISO 21500 with agile frameworks like Scrum and Kanban, the chapter highlights the need for flexibility and innovation in modern project management practices. The chapter further discusses how ISO 21500 can be adapted to agile environments, providing practical recommendations for integrating these standards. This chapter outlines key steps such as training in ISO and agile principles, iterative planning, and continuous monitoring to ensure both flexibility and compliance. The authors propose hybrid models and phased implementations that balance the structured approach of ISO 21500 with the dynamic nature of agile methodologies, making it a valuable guide for project managers seeking to navigate the complexities of modern project delivery.

The third chapter, "Understanding the Trade-off Between Efficiency and Value in Collaborative R&D Projects—A Complexity Theory Based Conceptual Framework", examines the balance between efficiency and value in managing collaborative R&D projects (CR&DP). The chapter focuses on how these projects, which involve multiple independent organizations, must navigate the inherent trade-offs between meeting short-term goals such as time, cost, and scope, and achieving long-term value, such as innovation and strategic outcomes. Through a complexity theory lens, the authors propose a conceptual framework that addresses the challenges of governance, organizational learning, and equivocality (the presence of multiple interpretations) in CR&DPs. The chapter presents a multiple-case analysis of CR&DPs, examining how project success can be evaluated from both efficiency and value perspectives. This chapter highlights the need for adaptability in governance and leadership to manage the complex, non-linear interactions among project partners. The findings underscore that project success in these dynamic environments cannot be measured solely by traditional efficiency metrics but must also account for the value created through innovation and long-term impact on stakeholders. This approach provides insights for both academics and practitioners into how collaborative projects can better align their management practices to navigate these complexities effectively.

The fourth chapter, "Simplified ISO 21502 Project Management Applied in a Fintech Case Study", presents a practical application of the ISO 21502 framework within the context of the financial technology (Fintech) industry. The chapter provides a comprehensive case study from a New York-based multinational financial firm, demonstrating how the ISO 21502 guidelines can be adapted and simplified to improve project management outcomes. The case study highlights the importance of stakeholder engagement, resource management, and risk mitigation, while addressing the high failure rates in global projects—particularly in IT and software development—through the application of universal project management best practices. The chapter further illustrates the use of key project management tools such as work breakdown structures (WBS), risk analysis (including the PERT technique), and stakeholder registers. By simplifying complex project management guidelines into a more intuitive model, the authors offer a practical approach for emerging project management practitioners. The case study reveals the significant impact of clear project governance, iterative planning, and proactive risk management, emphasizing the value of a tailored approach to ensure project success in an increasingly complex and dynamic business environment.

The fifth chapter, "PMBOK® Challenges and Lessons Learned in the Quebec Public Administration: A Comparative Analysis with ISO 21502 Standard", explores the integration of the Project Management Body of Knowledge (PMBOK®) into Quebec's public sector project management practices. The chapter examines the challenges faced by public managers and project team members, particularly in managing stakeholder relationships and balancing routine activities with project tasks. By employing a qualitative case study of Quebec's public administration, the chapter highlights the difficulties encountered in implementing PMBOK® methodologies in a context where the public sector's primary mission is to provide standardized services and comply with legal frameworks. The analysis reveals that, while PMBOK® is a useful guide for public sector managers, many of its performance domains, such as stakeholder engagement and team management, present significant challenges due to the unique structure and needs of public organizations. The chapter also introduces ISO 21502 as a potential alternative to PMBOK® for public sector projects. This chapter suggests that while PMBOK® addresses technical aspects of project management, ISO 21502 could better support public sector team members by simplifying processes and focusing on practical stakeholder and resource management. Through a detailed comparative analysis, the chapter provides insights into how ISO 21502 could complement PMBOK® in addressing the specific needs of public sector managers and enhancing project success rates. Ultimately, the chapter proposes that a hybrid approach, incorporating the strengths of both frameworks, may offer the most effective solution for public sector project management.

The sixth chapter, "Simplified Stakeholder Management in Project Management: Biopharmaceutical Case Study", introduces a streamlined approach to stakeholder management, using a real-world biopharmaceutical drug discovery project as a case study. The chapter focuses on the challenges of managing stakeholders in complex, high-cost, and high-stakes environments like the biopharma industry, where effective stakeholder engagement can be critical to project success. The author highlights the limitations of traditional stakeholder management techniques, which often borrow from corporate strategy frameworks and are not fully applicable in project environments. Instead, the chapter proposes a new simplified stakeholder management model that emphasizes the identification, analysis, and engagement of those stakeholders directly impacted by project deliverables. Through an empirical study that includes interviews with experienced project managers, the chapter develops two key models: one for stakeholder analysis, focusing on power and impact, and another for stakeholder engagement. These models are tailored to the specific needs of projects, where identifying and managing the right stakeholders at the right time can drastically influence outcomes. The biopharma case study illustrates how these simplified methods can be applied effectively, offering practical insights for project managers navigating stakeholder complexities in high-stakes industries. This approach seeks to bridge the gap between theory and practice in stakeholder management, providing a more focused and actionable framework for project managers.

The seventh chapter titled "Framework for Managing Artificial Intelligence (AI) Projects: Avoiding Harms, Losses, and Damages" introduces a project management framework designed specifically for AI projects, focusing on preventing ethical, social, and legal harms. The chapter underscores that AI projects differ significantly from traditional IT projects, given the complexity of AI systems and their potential for significant societal impacts. The framework integrates principles from the ISO 21502 project management standard with the AI development lifecycle, creating a holistic approach that accounts for the moral responsibilities of AI project stakeholders. By doing so, the authors aim to help project managers anticipate and mitigate the risks associated with AI systems, from privacy concerns to environmental impact and regulatory compliance. The chapter also offers practical guidelines on aligning AI project management with ethical standards, highlighting the importance of transparent processes and stakeholder engagement throughout the AI lifecycle. This chapter presents success factors that project managers can apply, such as ethical frameworks, robust data governance, and continuous risk assessment, to ensure that AI systems benefit society while minimizing harms. The chapter serves as a comprehensive resource for project managers, sponsors, and organizations planning AI projects, emphasizing the need for careful governance to avoid potential legal, financial, and societal damages.

The eighth chapter, "An Exploratory Study of Project Team Agility in Business Intelligence and Analytics Projects Using a Complex Adaptive Systems Lens", investigates the dynamics of project team agility within Business Intelligence and Analytics (BI&A) projects. Using Complex Adaptive Systems (CAS) theory as a framework, the authors aim to address the research gap surrounding the application of agile methods in BI&A projects, which are distinct from traditional IT initiatives due to their exploratory nature and uncertain outcomes. The chapter examines how project teams can effectively manage evolving requirements and navigate the socio-technical complexities inherent in BI&A projects. Through a multi-case study, the authors examine how teams adapt, self-organize, and balance the exploration of new insights with the exploitation of current processes, highlighting the unique demands of analytics-focused initiatives. The findings suggest that project team agility is crucial for the success of BI&A projects, particularly in responding to changing stakeholder needs and technology shifts. The study underscores the importance of iterative feedback loops, effective communication, and adaptive planning in managing the uncertainty that characterizes BI&A projects. By applying CAS theory, the chapter provides valuable insights into the mechanisms that enable agility, offering practical recommendations for project managers looking to optimize team performance in high-uncertainty environments. This chapter is especially relevant for those in analytics and data-driven fields, where agile methodologies must be carefully tailored to meet the distinct challenges of BI&A projects.

The ninth chapter, "Fintech and Financial Inclusion—A Review of Risk Management Strategies", examines the pivotal role that fintech innovations play in enhancing financial inclusion, particularly for underserved populations. The chapter highlights how fintech has transformed the financial services landscape by providing more accessible solutions to traditionally excluded groups, such as low-income households and rural communities. By integrating cutting-edge technology like mobile payments, digital wallets, and blockchain, fintech firms have successfully reduced barriers to financial services. However, with these innovations come significant risks, including cybersecurity threats, data privacy concerns, and regulatory challenges. The authors provide a detailed exploration of these risks and propose comprehensive risk management strategies that ensure the security and sustainability of fintech-driven financial inclusion efforts. This chapter also examines the specific types of risks associated with fintech adoption, such as operational, market, and reputational risks. Through a rigorous analysis of a 10-year dataset, the authors illustrate the need for robust cybersecurity measures, continuous regulatory compliance, and effective risk mitigation to safeguard both consumers and financial markets. Furthermore, the chapter emphasizes the importance of striking a balance between fostering fintech innovation and maintaining market stability, particularly as fintech continues to evolve and expand its reach. By offering a nuanced understanding of the complexities involved in managing risks in fintech, this chapter provides invaluable insights for policymakers, fintech entrepreneurs, and financial institutions working to advance financial inclusion.

The tenth chapter titled "Spotlight on Public Administration and Project Management in Nigeria: Discovery of New Frontier Through Emerging Trends and Technology" explores the transformative impact of new technologies on Nigeria's public administration and project management practices. The authors highlight how bureaucracy and inefficiency have historically hampered project management in the public sector, but the adoption of digital tools, automation, artificial intelligence, and data analytics is revolutionizing governance. These technologies have improved transparency, decision-making, and resource allocation, while fostering citizen engagement through open data platforms and project management software. The chapter emphasizes that embracing these emerging trends opens new frontiers for more effective project execution and governance in Nigeria, promoting collaboration and enhancing public trust in government operations. The chapter also examines the challenges faced by Nigeria's public administration, including mismanagement of funds, project failures, and resistance to change. Despite these hurdles, the authors provide examples of successful technology adoption in public administration, such as AI-powered chatbots and blockchain-based procurement systems, which have increased efficiency and reduced corruption. By integrating these technologies with traditional public service frameworks, Nigeria can overcome systemic challenges and propel its public administration into a new era of governance. This chapter offers valuable insights into how strategic investments in digital transformation can lead to significant advancements in project management and public service delivery in Nigeria.

The eleventh chapter titled "Project Risk Management (PRM) in Situations of High Complexity and Deep Uncertainty" provides a comprehensive analysis of the limitations of traditional project risk management (PRM) methods when applied to complex projects with high levels of uncertainty. The author argues that traditional PRM assumes risks are standalone, non-interacting events, which overlooks the compounding effects of risk interactions and fails to account for the dynamic nature of external project environments. The chapter introduces two fundamental paradigm shifts: the use of nonlinear Monte Carlo modeling to capture risk interactions and the dynamic adaptive methodology (DAM) for managing external uncertainties through ongoing monitoring and adaptation. These approaches aim to address the deep uncertainties and interconnected risks that arise in complex projects, providing project managers with tools to better anticipate and mitigate cascading project risks. The chapter further emphasizes the role of systems dynamics and game theory in shaping these new methodologies. By considering the broader context of risk interactions and external environmental changes, project managers can avoid the pitfalls of "predict-then-act" decision-making and instead adopt a "monitor-and-adapt" approach. Practical examples, including the use of nonlinear Monte Carlo techniques, are provided to illustrate the power of these methodologies in improving project outcomes. This shift from linear to nonlinear risk management is particularly vital in projects with high complexity, where traditional methods often result in project delays, cost overruns, and even project failure.

The twelfth chapter titled "Incorporating Negative Values into the Simple Additive Weighting (SAW) Under Uncertain Conditions: An Applications in Project Manager Selection Problem" addresses the challenges of integrating negative values into the traditional Simple Additive Weighting (SAW) method, particularly in multi-criteria decision-making (MCDM) scenarios with interval data. The chapter explores the necessity of accounting for negative values, which are often overlooked in decision models, despite their significance in areas such as cost and risk factors. By extending the SAW technique using the interval-based methods developed by Salehi and Izadikhah, the author proposes a modified procedure that handles mixed data containing both positive and negative values, ensuring more accurate and realistic outcomes in decision-making. Through a detailed case study on project manager selection, the chapter demonstrates the practical application of the proposed model. This chapter outlines how the integration of interval arithmetic allows decision-makers to assess alternatives with negative data, such as cost-related criteria, and evaluate them alongside positive attributes. The author provides a robust framework that enhances the SAW method's flexibility, enabling better decision quality in uncertain conditions. This approach is particularly useful for scenarios where traditional methods fall short, offering a more comprehensive solution for complex decision-making problems in fields like project management.

The last chapter titled "Human Resource Management in Project Management: Software Engineering Case Study" provides an in-depth analysis of project human resource management (PM HRM) with a focus on a case study from the software engineering industry. The chapter explores best practices in PM HRM, highlighting the importance of recruitment, selection, leadership, and team development within project environments. The author compares traditional organizational human resource management (OHRM) with project-specific HRM, noting that while OHRM focuses on long-term employee management, PM HRM deals with short-term, project-specific needs that often require faster recruitment and performance evaluation cycles. Through the software engineering case study, the chapter illustrates how a project manager's ability to build and lead an effective team can significantly influence project success. The chapter also presents practical tools, such as a staffing pool database, to help project managers keep track of team member skills, performance, and availability. The case study emphasizes the importance of leadership in ensuring team cohesion and high performance, particularly in high-tech, dynamic environments like software engineering. Performance evaluation, a critical component of PM HRM, is discussed in detail, providing insights into how project managers can document and assess team members' contributions throughout the project lifecycle. By demonstrating these best practices in a real-world context, the chapter offers valuable guidance for project managers, especially in industries where rapid adaptability and technical expertise are crucial.

In conclusion, *International Program and Project Management—Best Practices in Selected Industries* offers a comprehensive exploration of project management methodologies tailored to the complexities of diverse industries. Each chapter presents unique insights into how different sectors, such as FinTech, software engineering, public administration, and biopharmaceuticals, navigate the challenges of modern project management. From integrating ISO 21500 and agile methodologies to tackling high-risk AI projects and exploring human resource management in project environments, the chapters emphasize the importance of adaptability, innovation, and strategic governance. The case studies provided serve as practical examples of how theoretical frameworks and standards can be applied in real-world scenarios to overcome both anticipated and unforeseen challenges. Collectively, these chapters underscore the need for project managers to be dynamic, flexible, and well-versed in both traditional and emerging project management practices. Whether it's addressing the ethical concerns of AI projects, managing stakeholder complexities in collaborative R&D, or utilizing tools like the staffing pool database in human resource management, this book highlights the vital role of leadership, risk management, and stakeholder engagement in ensuring project success. With contributions from a range of industries and project environments, this book serves as a critical resource for professionals and academics alike, offering actionable insights and strategies to enhance the effectiveness of international project and program management.

Williamsburg, USA Kenneth David Strang
Blagoevgrad, Bulgaria Narasimha Rao Vajjhala

Contents

Contemporary Project Management: A Macro Environmental Analysis

Md Lutfur Rahman, Desmond Tutu Ayentimi, Mark Wickham, and Kevin Swarts

Abstract This chapter provides a comprehensive overview of future trends in project management through a DEPEST (Demographic, Economic, Political-Legal, Ecological, Socio-Cultural, and Technological) analysis. The DEPEST analysis suggests contemporary project management emphasizes ongoing entrepreneurial innovation, collaboration, and continuous improvement. Recognizing the interconnectedness of sustainability and project success is vital for navigating the evolving landscape of project management. This chapter highlights the importance of aligning projects with organizational objectives, integrating sustainability as a strategic imperative, and leveraging cutting-edge technologies to enhance project execution efficiency. These strategies will help organizations enhance their competitive advantage, ensure sustainable practices, and achieve their strategic objectives. By adopting these insights, project managers and organizations can effectively navigate the evolving landscape of project management, ensuring long-term success in a rapidly changing global project management environment.

Keywords Trends in project management · Contemporary project management · Project management · DEPEST analysis · Project management trends · Entrepreneurial innovation · Sustainability integration · Strategic alignment · Organizational objectives · Technological advancements · Continuous improvement · Competitive advantage · Global project management

---

M. L. Rahman · D. T. Ayentimi (✉) · M. Wickham · K. Swarts
University of Tasmania, Hobart, TAS 7000, Australia
e-mail: desmond.ayentimi@utas.edu.au

M. L. Rahman
e-mail: mdlutfur.rahman@utas.edu.au

M. Wickham
e-mail: mark.wickham@utas.edu.au

K. Swarts
e-mail: kevin.swarts@utas.edu.au

K. D. Strang and N. R. Vajjhala (eds.), *International Program and Project Management—Best Practices in Selected Industries*, Information Systems Engineering and Management 31, https://doi.org/10.1007/978-3-031-80275-1_1

1 Introduction

The integration of project management with strategic objectives is essential for organizations seeking to generate significant value from their project investments [1]. A clear connection between project execution efficiency and the value delivered by the project's products and services is crucial for perceiving project management as strategically valuable [2]. Conversely, when project success is narrowly defined based solely on time, cost, and scope, without considering the link to business value, project management is viewed as providing tactical rather than strategic value. Historically, the Project Management Body of Knowledge (PMBOK) depicted project management primarily as a tactical approach to executing work in alignment with organizational policies [3]. However, recent iterations have acknowledged the importance of linking project management to organizational strategy, recognizing projects as a means of implementing strategic initiatives [4]. Similarly, project management methodologies like PRINCE2 emphasize deriving project goals and scope from an organization's strategic objectives or business opportunities [5].

Recent empirical research has further highlighted the evolving awareness of the strategic importance of project management. Applying the Resource-Based View from strategic management to project management, scholars have identified intangible resources within the discipline, such as tacit knowledge sharing and know-how, as potential sources of competitive advantage [6]. The recognition of the convergence between project management and strategy is increasingly acknowledged as critical for organizational performance and competitiveness in modern business environments [7]. Project management involves the structured planning, execution, and oversight of temporary endeavors aimed at achieving specific objectives, while strategy encompasses the formulation and execution of plans to attain long-term organizational goals.

Central to the relationship between project management and strategy is the alignment of project goals with broader organizational objectives and strategic imperatives. Research suggests that projects appropriately aligned with organizational strategy are more likely to achieve their intended outcomes [7]. This strategic alignment ensures that project management contributes directly to the organization's mission and vision, maximizing their impact and value. Strategic project portfolio management involves the deliberate selection and prioritization of projects that best support organizational strategic objectives [8]. By aligning project portfolios with strategic priorities, organizations can optimize resource allocation, mitigate risks, and enhance their competitive advantage. Moreover, effective project management serves as a crucial enabler for strategy execution, facilitating the translation of strategic plans into actionable initiatives [9]. Project management provides a structured framework for implementing strategic objectives, enabling coordination, communication, and resource allocation across functional areas.

This chapter aims to highlight contemporary trends in project management and elucidate the intricate relationship between project management and strategic management in shaping the future of project management. Furthermore, it delves into the analysis of macro-environmental factors such as demographic shifts, economic conditions, political-legal frameworks, ecological concerns, socio-cultural influences, and technological advancements—and their impact on contemporary project management. The chapter draws from the macro-environmental lens to highlight the future trends of project management. The contribution of the chapter is twofold: firstly, it strengthens the ongoing agenda of managing projects from a strategic perspective to better align with organizational long term objectives; and secondly, it offers insights into contemporary trends in project management, providing valuable foresight for practitioners and scholars alike to improve project benefits.

In summary, the alignment of project management with organizational strategy is fundamental for achieving strategic objectives, creating value, and sustaining competitive advantage in today's dynamic business landscape. By integrating project management practices with strategic planning processes, organizations can enhance their ability to achieve strategic goals more effectively.

2 Strategic Management—Internal and External Assessments

Developing organizational strategies stands out as a paramount challenge in the contemporary competitive landscape. Project managers and corporate organizations are confronted with a plethora of complexities, ranging from the dismantling of trade barriers and regulatory shifts to mounting socio-environmental pressures, all amidst intense rivalry among industry peers [6]. Moreover, the evolving nature of organizations, increasingly driven by technological advancements, adds another layer of intricacy. To confront these challenges effectively, companies need sophisticated organizational structures and adaptable processes capable of harmonizing with the dynamic external environment while bolstering internal capacities.

Furthermore, it is imperative for organizations to ensure alignment between their strategies and their foundational principles, encompassing values, vision, and mission, to realize short-term objectives without losing track of the long-term strategic direction [8]. Primarily, the strategic management process commences with the articulation of the company's mission, followed by the delineation of its future-oriented vision. For example, Tesla's mission statement, "to accelerate the world's transition to sustainable energy," exemplifies how a clear mission can guide a company's strategic direction. This mission drives Tesla's decisions across product development, market expansion, and innovation, ensuring that every project aligns with the overarching goal of sustainability. For instance, the development of the Tesla Model 3 was aimed at making electric vehicles more accessible, thereby advancing the mission of widespread adoption of sustainable energy solutions.

Subsequently, a comprehensive examination of both internal and external factors ensues, informing the crafting of strategies geared towards securing competitive advantage. Notably, the strengths, weaknesses, opportunities and threats (SWOT) analysis framework serves as a cornerstone, facilitating an assessment of internal strengths and weaknesses versus external opportunities and threats. Coca-Cola's SWOT analysis is an ideal example in this regard as shown in Table 1.

By understanding these factors, Coca-Cola can strategize to leverage its strengths and opportunities while mitigating its weaknesses and threats. For instance, the company has expanded its portfolio to include healthier options like bottled water and other beverages to address customer health concerns.

In comprehending the external landscape, particular attention is directed towards assessing industry dynamics and competitive forces. Established methodologies, such as Michael Porter's Five Forces framework, enable a systematic evaluation of the bargaining power of buyers, suppliers, and new entrants, as well as the threat of substitutes and the intensity of industry rivalry [10]. For instance, PepsiCo's analysis provides a clear application of this framework:

- Competitive Rivalry (Strong Force): The Coca-Cola company is one of PepsiCo's biggest competitors. The high aggressiveness of firms in product innovation and marketing, low switching costs for consumers, and a high number of competing firms create a strong competitive force against PepsiCo.
- Bargaining Power of Buyers (Strong Force): Consumers have low switching costs, high access to product information, and many substitutes available, giving them strong bargaining power. PepsiCo must ensure high customer satisfaction to maintain its market position.
- Bargaining Power of Suppliers (Weak Force): The high overall supply of raw materials, low forward integration of suppliers, and moderate size of individual suppliers weaken their bargaining power over PepsiCo.
- Threat of Substitutes (Strong Force): Substitutes such as real fruit juices and brewed coffee are high-performing, easily available, and have low switching costs, posing a strong threat to PepsiCo's products.
- Threat of New Entrants (Moderate Force): Low switching costs for consumers, moderate customer loyalty, and the high cost of brand development moderate the threat of new entrants in the market.

Additionally, competitor analysis, characterized by the systematic gathering of intelligence on industry rivals, aids in forecasting their strategic maneuvers and responses to environmental shifts [11]. Apple's analysis of Samsung's strategies,

Table 1 SWOT analysis of Coca-Cola

Strengths: Strong brand recognition, extensive distribution network	**Weaknesses:** Health concerns related to sugary drinks, reliance on carbonated beverages	Internal factors
Opportunities: Diversification into healthier beverages, expansion in emerging markets	**Threats:** Increasing health consciousness among consumers, intense competition from other beverage companies	External factors

for instance, revealed insights into Samsung's rapid product iteration and aggressive marketing tactics. In response, Apple has focused on differentiating its products through superior design and ecosystem integration, ensuring a loyal customer base despite the competitive pressures.

Integrating these theoretical frameworks with real-world examples not only illuminates their practical application but also underscores the importance of strategic alignment in navigating the complexities of the contemporary business environment. However, while these analytical tools offer valuable insights, a comprehensive examination of project management through a strategic lens necessitates a nuanced understanding of broader strategic imperatives.

3 Strategic Management—A Macro-environmental Assessment

A macro-environmental assessment involves comprehensively understanding external factors that are beyond the control of a company. These factors, categorized into demographic (D), economic (E), political-legal (P), ecological (E), socio-cultural (S), and technological (T) environments, collectively shape the business landscape. Often referred to as DEPEST analysis, this framework provides a well-structured approach for project managers to monitor and respond to changes swiftly and effectively [11]. The DEPEST framework represents the main macro-environmental lens to help project managers comprehensively understand the nature of the wider project management environment. Often, textbooks will refer to fewer factors and label them as PESTEL or other variations of the DEPEST acronym. A brief overview of each factor is given below:

- Demographic Factors: These encompass characteristics such as population size, age distribution, gender, income levels, education levels, and cultural diversity. Changes in demographics can significantly impact consumer demand, labor markets, and market segmentation strategies.
- Economic Factors: Economic factors include indicators such as gross domestic product (GDP) growth, inflation rates, unemployment rates, interest rates, exchange rates, and overall economic stability. Understanding economic trends is crucial for assessing consumer purchasing power, demand for goods and services, and the cost of doing business.
- Political-Legal Factors: This category encompasses government policies, regulations, political stability, trade policies, taxation laws, and legal frameworks. Changes in political and legal environments can affect market entry barriers, trade agreements, taxation policies, and business operations both domestically and internationally.
- Ecological Factors: Ecological factors pertain to environmental sustainability, natural resource availability, energy consumption, waste management, and climate change. Businesses need to consider ecological concerns in their operations to

address sustainability goals, regulatory compliance, and consumer preferences for eco-friendly products and practices.
- Socio-Cultural Factors: Socio-cultural factors include societal values, attitudes, lifestyles, cultural norms, beliefs, and consumer behavior patterns. Understanding socio-cultural trends helps businesses tailor their products, marketing strategies, and corporate social responsibility initiatives to meet the preferences and expectations of diverse consumer segments.
- Technological Factors: Technological factors encompass advancements in technology, research and development (R&D) activities, automation, digitalization, and innovation. Businesses must stay abreast of technological trends to remain competitive, enhance operational efficiency, and capitalize on opportunities for product innovation and market disruption.

These external environmental factors are interconnected, and changes in one domain can potentially influence others with broader implications for project managers and corporate organizations [6]. For instance, technological innovations can disrupt traditional industries and consumer behaviors, while demographic shifts can impact workforce dynamics and market demand.

By conducting a macro-environmental analysis, organizations can anticipate potential risks, identify opportunities for growth, and develop strategic responses to external challenges [12]. This process enables informed decision-making, effective risk management, and the formulation of competitive strategies that align with the evolving external landscape [13].

In exploring what it takes for an organization to stay ahead, key factors include being entrepreneurial, and innovative, and focusing on core capabilities [14]. Competitive advantages can disappear quickly as rivals come up with new and better products or services. So, it is a persistent battle for organizations to build and maintain their competitive edge by leveraging their unique capabilities and resources. Since business success relies on staying competitive, there is growing interest in how project management can help maintain that competitive edge. Contemporary project management principles can reduce risks and add control measures to ensure organizational success and generate greater competitive advantage.

4 Project Management Trends—A Strategic Assessment Using the DEPEST Framework

In the twenty-first century, businesses of all sizes and types have become deeply interconnected within the global business community, making them vulnerable to various social, political, environmental, technological, and economic forces worldwide [15]. The advent of the Fourth Industrial Revolution, globalization, disruptive innovations, external forces and other global factors has further intensified this dynamic and ever-changing business environment. Traditional project management approaches struggle to keep pace with these rapid transformations.

Project management activities are no longer isolated endeavors but occur within multifaceted environments shaped by complex societal structures, values, customs, and systems [16]. As a result, project managers must navigate a diverse array of influences while ensuring project success. Embracing agile methodologies and leveraging technological advancements are crucial strategies for coping with the complexities of contemporary project management. Additionally, fostering adaptability, collaboration, and cross-cultural understanding within project teams is essential for effectively managing projects in today's interconnected and dynamic business landscape. By acknowledging and addressing the interplay between global forces and project management practices, organizations can enhance their ability to thrive and innovate in an increasingly interconnected world.

The evolution of project management reflects a transition from hierarchical and rigid methodologies to adaptive and human-centric approaches [17]. Historically, projects were governed by top-down control and fixed plans, but as businesses faced increasing complexity, adaptive methodologies emerged to emphasize flexibility and collaboration. Human-centric approaches gained prominence, focusing on stakeholder engagement, team dynamics, and leadership skills. Additionally, there's a growing emphasis on incorporating sustainability principles into project management practices. Strategic alignment has become crucial, ensuring project objectives align with organizational strategies. Overall, project management has shifted towards adaptable, holistic, and strategic approaches that prioritize both human factors and sustainability for project success.

The complex interplay between environmental dynamics and industry evolution will continue to shape the field of project management. In order to anticipate the future of project management, it is important to reflect on emerging changes within the macro environment [6]. This section presents a synthesized analysis of these insights, aiming to offer a comprehensive understanding of the future trajectory of project management. In this section, we used the term "project management" in a broader context to include program and portfolio management.

4.1 Demographic Factors and Related Impacts

Demographic factors will have significant impacts on project management practices in the future, influencing team composition, leadership styles, communication methods, and project delivery. One of the influences will be the generational Shifts. The influx of younger generations including generation Z (born between mid-1990s and 2010s) and generation Alpha (born between early 2010s and 2020s) into the workforce and the retirement of older generations will bring about shifts in work culture, communication preferences, and technological proficiency [18]. Project managers will need to adapt their leadership styles and communication methods to accommodate the preferences and expectations of the incoming workforce.

Another critical factor is, as projects become more complex, communication in multiple languages will become common, necessitating multicultural and multilingual capabilities and competencies. Like many other developed countries, Australia's diverse population, shaped by waves of migration from various countries, continues to enrich the nation's multicultural landscape and contribute to its growth [19]. In 2022, Australia's population was approximately 26 million. Projections suggest that by 2071, this figure could reach between 34.3 and 45.9 million people. These projections are based on various scenarios of fertility, life expectancy, and migration rates. Consequently, the workforce will comprise a significant proportion of individuals not born in Australia, leading to a rise in multicultural and multilingual preferences and the need for project managers to build cross-cultural competencies.

In addition, the adaptability of baby boomers, born between 1946 and 1964, to the evolving workforce will significantly impact their employability in the years ahead. By 2030, as baby boomers enter retirement, their wealth of project and functional experience could become increasingly sought after. The growing demand for healthcare and leisure activities among baby boomers will fuel a surge in construction projects, including hospitals, retirement homes, and nursing homes [6]. Additionally, there will be an uptick in projects related to hospitality, travel, second home development, and entertainment.

Overall, demographic shift will continue to shape the composition of project teams, the skills required for project management roles, and the dynamics of project execution. Adapting to these changes will be essential for organizations and project managers to remain competitive and achieve their strategic objectives.

4.2 Economic Factors and Related Impacts

Economic factors exert a significant influence on project management practices, shaping decision-making, resource allocation, risk management, and overall project outcomes. Several countries, including Brazil, Russia, India, China and South Africa (BRICS, an emerging geopolitical cluster), are projected to experience ongoing economic growth and improvements in living standards. India, for instance, is expected to sustain high growth rates, with long-term GDP growth projected at around 6–7% annually until 2035. This growth is supported by a young population, urbanization, and ongoing economic reforms. This global trend toward development will be accompanied by increased globalization, leading to more diverse and potentially multilingual project teams [20]. As a result, collaboration and communication skills will become even more critical in navigating the complexities of project executions.

According to the World Bank, the global GDP growth rate is projected to average around 2.8% annually over the next decade. However, the current global GDP growth rate stands at 3.1% for 2024, with a slight increase to 3.2% projected for 2025. These projections are influenced by greater than expected resilience in several large emerging market and developing economies, along with fiscal support in China. However, these forecasts remain below the historical average of 3.8% from 2000 to 2019 due to elevated central bank policy rates to combat inflation, reduced fiscal support, and low underlying productivity growth.

These economic forecasts imply that project managers must remain adaptable and vigilant, particularly in regions experiencing economic volatility. The ability to manage resources efficiently and mitigate risks will be crucial in maintaining project success amidst fluctuating economic conditions. Furthermore, fluctuations in labor, material, and equipment costs, as well as availability, will affect project budgets and timelines [21]. Project managers must optimize resource allocation, negotiate contracts, and monitor expenses to control costs and minimize budget overruns.

Importantly, as the baby boomer generation retires, many individuals will shift their focus from professional careers to more fulfilling pursuits. A significant number of retirees may choose to engage in volunteer work and travel, particularly in developing nations. This transition could spark a rise in social projects, driven by the values and passions of both retiring baby boomers and the younger internet generation (the Net Generation, born after 1987). These generations share a commitment to social causes, environmental sustainability, and ethical principles, which may fuel an increase in projects aligned with these values [22].

Overall, economic factors will continue to shape project management practices, influencing project financing, resource allocation, risk management, and strategic decision-making. Project managers will need to adapt to dynamic economic conditions, proactively manage risks, and capitalize on emerging opportunities to achieve project success in the evolving global economy.

4.3 Political-Legal Factors and Related Impacts

Political and legal factors can significantly impact project management practices in the future, influencing project planning, execution, and outcomes. Environmental laws, sustainability standards, and carbon emissions regulations impact project design, construction, and operation [23]. Project managers must integrate environmental considerations, conduct environmental impact assessments, and implement sustainable practices to meet regulatory requirements and minimize environmental footprints.

Another crucial aspect is that geopolitical unrest will continue, especially in the Middle East and across some developing countries [24]. Simultaneously, the ongoing wars on terrorism contribute to global instability. These trends will fuel the demand for specialized project managers in specific domains including high technology, military technology, information technology, and security-oriented projects. Given the increasing geopolitical provocations, defense and security projects may increase in size and scope, and it may become harder to address related procurement and logistics needs. Ensuring timely delivery of resources, managing supply chains, and coordinating logistics will be crucial.

In summary, political and legal factors exert a significant influence on project management practices, requiring project managers to stay informed, adapt to regulatory changes, and navigate legal complexities to ensure project success in an increasingly regulated and dynamic business environment.

4.4 Ecological Factors and Related Impacts

Ecological factors play a crucial role in shaping project management practices in the future, as organizations increasingly prioritize sustainability and environmental stewardship. As we look ahead to the future, energy, climate, and environmental considerations will continue to significantly influence project management trends. Project management will be seen as a vehicle for not only economic growth but also societal well-being, necessitating a reevaluation of traditional methodologies to ensure alignment with broader societal goals and values [17]. Stakeholder engagement and collaboration are underscored to foster inclusivity and ownership, while resilience-centric approaches prioritize proactive risk management and flexibility to navigate uncertainties like climate change and global crises.

Sustainability will play a central role in shaping how projects are planned and executed [17]. One notable trend will be the seamless integration of sustainability principles into project management methodologies. Project managers will need to assess the environmental impact of their initiatives and take more proactive steps to minimize carbon emissions, reduce waste, and conserve valuable resources.

Moreover, the growth of renewable energy projects, such as wind farms and solar installations, will be substantial [25]. These initiatives demand specialized expertise from project managers who can navigate the complexities of renewable energy infrastructure. Coordinating with diverse stakeholders—including government agencies, regulatory bodies, suppliers, and local communities—will be essential for successful implementation. Beyond renewable energy, there will be a surge in projects aimed at enhancing energy efficiency across various industries. Project managers will lead efforts to retrofit existing buildings with energy-efficient technologies, optimize industrial processes, and deploy smart grid solutions. These measures contribute not only to sustainability but also to cost savings and improved resource utilization.

Additionally, environmental conservation will take center stage [26]. Project managers will oversee initiatives focused on biodiversity preservation, habitat restoration, and ecosystem health. Engaging stakeholders, conducting thorough environmental impact assessments, and ensuring compliance with regulations will be critical components of these projects.

In summary, ecological factors will increasingly shape project management practices in the future, necessitating a holistic approach that integrates environmental considerations into project planning, implementation, and evaluation. Project managers must prioritize sustainability, resilience, and environmental stewardship to ensure projects align with ecological goals and contribute to a more sustainable and resilient future.

4.5 Socio-cultural Factors and Related Impacts

Socio-cultural factors have a significant impact on project management practices in the future, influencing team dynamics, communication styles, stakeholder engagement, and project outcomes. In the future, diversity and inclusion will become increasingly vital in organizational management. As societies become increasingly diverse, project teams will reflect this diversity in terms of cultural backgrounds, perspectives, and experiences [17]. Cultural intelligence and diversity management skills will be essential for project managers operating in diverse and globalized contexts. Project managers will need to demonstrate sensitivity to cultural differences, foster inclusive environments, and effectively lead multicultural teams, valuing diverse perspectives and leveraging cultural differences as strengths. Cultural competence training and diversity initiatives will be prioritized to enhance collaboration and innovation.

Another important aspect is that the workforce of the future will be characterized by greater flexibility, remote work options, and the gig economy as the young generation including generation Z and beyond dominate the workforce [6]. Project managers will need to cater to their preferences and lifestyles and adapt to managing virtual teams, self-managed teams and coordinating project activities across different time zones and locations. Embracing digital tools and platforms for remote or virtual collaboration will be essential for ensuring project success in this evolving work environment.

In addition, the continued proliferation of collaborative and social networking platforms will transform project management tools and practices [26]. Platforms like Facebook, MySpace (and their successors) have already transformed how we connect and collaborate. As these technologies evolve, they will likely become even more integrated into our work lives. Moreover, online forums, wikis, and other virtual communities provide spaces for individuals to share knowledge, brainstorm ideas, and solve problems collectively. These platforms foster collaboration across geographical boundaries, enabling global teams to work seamlessly. The youth of today, who are digital natives, will carry forward this familiarity with social tools into their professional lives. Their comfort with online collaboration will drive organizational culture and practices.

More importantly, project management skills are increasingly recognized as essential for success in various fields. Beyond traditional project managers, professionals across disciplines benefit from understanding project management principles. Education systems are likely to adapt by incorporating project management education at earlier stages [6]. Imagine early grade school students learning about project planning, teamwork, and resource allocation! This proactive approach aims to equip future professionals with the skills needed to handle complex projects. Whether it is about launching a new product, organizing an event, or managing a team, project management expertise will be invaluable.

In summary, socio-cultural factors profoundly influence project management practices, necessitating cultural competence, empathy, and adaptability on the part of project managers to navigate diverse cultural landscapes, foster inclusive environments, and achieve project success in an increasingly interconnected and multicultural world.

4.6 Technological Factors and Related Impacts

Technological factors will continue to have a profound impact on project management practices in the future, shaping project planning, execution, and delivery. Integrating technology with social values, prioritizing stakeholder engagement, fostering resilience, and upholding ethical principles will be key to achieve sustainable project outcomes [22].

Recent technological advancements, such as artificial intelligence, machine learning, big data analytics and robotic process automation, are automating repetitive tasks, streamlining processes, and enhancing project management efficiency. For instance, AI-powered project management tools can optimize scheduling, resource allocation, and risk management by analyzing vast amounts of data and predicting potential issues before they arise. Project managers must leverage digital tools and automation technologies to optimize project workflows, improve productivity, and reduce manual effort.

As organizations become more global and remote work becomes commonplace, project managers will use more collaborative technology [18]. Advanced video conferencing tools, such as Zoom and Microsoft teams, now seamlessly connect team members across time zones and geographical boundaries, enabling real-time collaboration. Collaborative software platforms like Slack, Trello, and Asana facilitate real-time communication, document sharing, and task tracking, ensuring that all team members are aligned and informed.

Cloud-based solutions are essential for maintaining up-to-date project files, ensuring that everyone works with the latest project files. Features such as version control, access permissions, and secure storage are critical for managing documents in a virtual environment and will be more critical in the future. Project managers must navigate document collaboration, approvals, and archiving in this digital landscape to ensure efficient project execution.

Moreover, big data analytics, predictive analytics, and data visualization tools provide project managers with valuable insights into project performance, risks, and opportunities [26]. Project management software of the future may well allow extensive scenario planning, role playing capabilities, and intuitive verbal data entry to enhance user experience and effectiveness. Real-time dashboards will track project milestones, resource allocation, and risks. Predictive analytics will help project managers anticipate bottlenecks, resource shortages, or potential delays. Early detection of deviations from the project plan will be critical. Analytics can flag potential problems before they escalate. Business intelligence tools will provide insights into project performance, allowing adjustments as needed.

Similarly, Virtual Reality (VR) and Augmented Reality (AR) technologies will revolutionize project visualization, design reviews, and stakeholder engagement by providing immersive and interactive experiences. Project managers must embrace VR and AR applications, leverage virtual mockups and simulations, and engage stakeholders through immersive experiences to improve project understanding and decision-making. With increased reliance on digital technologies, cybersecurity threats and data breaches pose significant risks to project management practices [17]. Project managers must prioritize cybersecurity measures, implement robust data protection protocols, and ensure compliance with data privacy regulations to safeguard project information and mitigate cyber risks.

It is obvious that technological development will drive innovation, efficiency, and collaboration in project management practices, enabling project managers to leverage digital tools, automation, and data-driven insights to optimize project outcomes and deliver value in an increasingly digital and interconnected world.

In our analysis from a DEPEST lens, it is apparent that there are two prevailing themes concerning emerging patterns in project management. Firstly, the enduring significance of human-centric approaches, which is expected to persist; and secondly, a burgeoning emphasis on sustainability within project management practices. These observations are supported by the growing attention devoted to sustainable and ethical leadership studies after a seminal investigation sponsored by the Project Management Institute in 2006 [27] and also in alignment with the more recent survey—Pulse of the Profession—in 2024 by the same institute [28]. This highlights the rising importance of strategic alignment which could evolve into a progressively dominant facet of project management.

5 Conclusion

Drawing upon the above insights along with our extensive expertise in both practical project management and academic inquiry, we provide a comprehensive overview of the evolving trends and future prospects within the realm of project management. When contemplating the horizon stretching beyond 2030, DEPEST analysis illuminates a multitude of global shifts that will continue to shape the landscape of project management. Amidst these changes, project managers will confront persistent external pressures to navigate the triple constraints of time, cost, and scope while adeptly managing stakeholder expectations. To effectively meet these evolving demands, organizational leaders must acknowledge the requisite resources essential for successful project execution and provide unwavering support, recognizing the pivotal role project management plays in steering organizational strategy and ensuring success.

Future project management practices will be profoundly influenced by various demographic, economic, political-legal, ecological, socio-cultural, and technological factors. The influx of younger generations such as Generation Z and Generation Alpha into the workforce, coupled with the retirement of baby boomers, will bring shifts in workplace culture, communication preferences, and technological proficiency. This generational shift will necessitate adaptive leadership and communication styles to align with the expectations and preferences of a diverse workforce.

Economic growth in emerging markets, such as the BRICS countries, will continue to shape project management by emphasizing efficient resource management and risk mitigation. Economic forecasts indicate moderate global GDP growth, suggesting the need for project managers to remain adaptable and vigilant, particularly in regions experiencing economic volatility. Fluctuations in labor, material, and equipment costs will also impact project budgets and timelines.

Political and legal factors, including environmental laws and sustainability standards, will require project managers to integrate environmental considerations into project planning and execution. Geopolitical unrest and global instability will drive the demand for specialized project managers in high-technology, military, and security projects. Ensuring timely delivery of resources and managing supply chains will be crucial in these domains.

Ecological priorities will emphasize sustainability and environmental stewardship in project management. The growing importance of renewable energy projects and energy efficiency initiatives will require project managers to assess the environmental impact of their projects and implement proactive measures to reduce carbon emissions, conserve resources, and enhance energy efficiency.

Socio-cultural shifts will highlight the importance of diversity, inclusion, and remote work. As project teams become more culturally diverse, project managers will need to demonstrate cultural intelligence and foster inclusive environments. The rise of remote work and the gig economy will require the adoption of digital tools and platforms to facilitate virtual collaboration and manage project activities across different time zones.

Technological advancements, including artificial intelligence, machine learning, and automation, will drive innovation, efficiency, and collaboration in project management. Project managers must leverage these technologies to optimize project workflows, enhance productivity, and reduce manual effort. Collaborative tools, cloud-based solutions, and data analytics will provide valuable insights into project performance, enabling real-time decision-making and scenario planning. Additionally, cybersecurity measures will be critical to safeguarding project data and mitigating risks.

In light of the above, we recommend the following set of actions to be undertaken by project managers and organizations to improve project benefits:

1. **Adaptive Leadership and Communication:** project managers will need to develop flexible leadership styles and communication methods that cater to the diverse preferences and expectations of different generations and cultural backgrounds within the workforce. Use inclusive language and foster open dialogue to build strong team cohesion.
2. **Cross-Cultural Training:** Invest in cultural competence training for project managers and team members to enhance their ability to collaborate effectively in

multicultural environments. Encourage the appreciation of diverse perspectives and leverage cultural differences as strengths in project execution.

3. **Resource Management:** Implement robust resource management practices that optimize the allocation of labor, materials, and equipment. Develop contingency plans to address potential economic fluctuations and ensure that project budgets and timelines are adhered to.
4. **Regulatory Compliance:** Stay informed about political and legal changes, particularly those related to environmental regulations, health and safety, particularly psychosocial safety and sustainability standards. Ensure that projects meet all regulatory requirements and minimize their environmental footprint through sustainable practices.
5. **Sustainability Integration:** Integrate sustainability principles into project methodologies by conducting thorough environmental impact assessments and adopting proactive measures to reduce carbon emissions, waste, and resource consumption. Promote renewable energy projects and energy efficiency initiatives to align with broader societal goals.
6. **Leveraging Technology:** Utilize advanced technologies such as AI, automation, and collaborative tools to streamline project workflows, enhance productivity, and improve project outcomes. Embrace digital platforms for remote collaboration and data-driven decision-making to stay competitive in an increasingly digital and interconnected world.
7. **Cybersecurity:** Prioritize cybersecurity measures to protect project data and ensure compliance with data privacy regulations. Implement robust data protection protocols and continuously monitor for potential cyber threats to safeguard project information.

Envisioning an exciting trajectory, we anticipate the integration of cutting-edge technologies such as AI and business intelligence into project selection and scoping processes, complemented by the adoption of technology-driven productivity tools tailored to the needs of professionals operating within dispersed work environments. Central to achieving success will be ensuring seamless alignment between project objectives, market dynamics, and overarching organizational strategic imperatives. As the adoption of project, program, and portfolio management methodologies continues to gain momentum, a heightened appreciation for the strategic significance of project management is anticipated among executives. While some may swiftly grasp this value, others may require additional time due to the absence of robust measures to assess the discipline's contribution in tangible business terms. Nonetheless, we foresee project management evolving into an indispensable skill set for both practitioners and executives, facilitating the mitigation of inadvertent project management pitfalls through targeted investments in training, resources, and staff support.

Already acknowledged as a prerequisite skill set across diverse sectors, project management is poised to undergo a transition towards formal professionalization in the ensuing decades. We anticipate a burgeoning trend towards the integration of project management education at early grade school level, reflecting the discipline's burgeoning significance across varied fields. The demand for comprehensive training and education in project management is projected to endure, with a sustained emphasis on vital topics such as stakeholder management, sustainable leadership, governance, and program and portfolio management. Particularly, the prominence of human resources management is expected to escalate, driven by the collaborative nature of contemporary work environments and the increasing reliance on knowledge workers within matrix-based organizational structures.

Looking ahead, our analysis underscores a burgeoning diversity in project types, underscoring the necessity for project management to remain an experiential discipline, best cultivated through immersive practices such as hands-on experience, mentorship, and job shadowing. Emphasizing the interpersonal dimension of project management, we advocate for the integration of communities of practice into project management frameworks, fostering collaborative knowledge sharing and enhancing collective expertise. Recognizing that a substantial portion of workplace learning occurs informally, prioritizing the creation, acquisition, capture, sharing, and effective utilization of project knowledge emerges as an imperative [29].

In conclusion, guided by our rigorous analysis and research, the future of project management manifests as a promising landscape characterized by ongoing innovation, collaboration, and an unwavering commitment to continuous improvement. Interestingly, central to the nexus between project management and strategy is sustainability given the increasing attention to the triple bottom line of people, prosperity and planet. The integration of sustainability into project management has transcended from mere ethical considerations to become a strategic imperative [30]. The environmental, social, and economic dimensions of sustainability are intricately woven into the core objectives of project management, necessitating a holistic approach. Environmental factors, such as resource efficiency and pollution reduction, are indispensable for ensuring the long-term viability of projects. Social and ethical considerations, such as stakeholder engagement and community impact, are vital for upholding the social license to operate. Moreover, economic sustainability is paramount to ensuring that projects remain financially viable and contribute positively to overall economic development. As organizations increasingly recognize the interconnectedness of sustainability and project success, embracing these insights will be crucial for navigating the evolving landscape of project management in the years to come. Sustainability will continue to take center stage in the nexus between project management and strategy, serving as a guiding principle for achieving both organizational and societal goals in a rapidly changing world.

While this study provides important insights into contemporary trends in project management and their impacts on future project management practices, several avenues for future research could further advance our understanding of this subject. This chapter relied on secondary data, which may not capture the latest practices and trends in the profession. Future research could benefit from combining primary and secondary data sources to gain a more comprehensive understanding of the topic. Second, this chapter focused on contemporary trends in project management, emphasizing the strategic importance of managing projects in an increasingly dynamic business environment. Future research could investigate how projects can be better aligned with organizational objectives, exploring comprehensive strategies and frameworks for strategic and entrepreneurial project management. Third, the chapter highlighted the central role of sustainability in the nexus between project management and strategy. Future research could explore the integration of sustainability into project management as a strategic imperative, going beyond ethical considerations to examine the interconnectedness of environmental, social, and economic sustainability with project success. These areas of research will contribute to a deeper understanding of the evolving field of project management and provide valuable insights for practitioners and researchers aiming to advance the discipline.

References

1. Crawford, L., Pollack, J., England, D.: Uncovering the trends in project management: journal emphases over the last 10 years. Int. J. Project Manage. **24**(2), 175–184 (2006)
2. Papke-Shields, K.E., Boyer-Wright, K.M.: Strategic planning characteristics applied to project management. Int. J. Project Manage. **35**(2), 169–179 (2017)
3. Project Management Institute. A Guide to the Project Management Body of Knowledge (PMBOK Guide). Project Management Institute, Newtown Square, Pennsylvania (2000)
4. Project Management Institute. A Guide to the Project Management Body of Knowledge (PMBOK Guide). Project Management Institute, Newtown Square, Pennsylvania (2021)
5. Office of the Government Commerce Homepage. https://www.ogc.gov.uk/methods_prince_2.asp. Accessed 18 March 2024
6. Jugdev, K., Müller, R., Hutchinson, M.: Future trends in project management: a macro-environmental analysis. In: James, S.P. (eds.) Project Management Circa, pp. 229–240 (2024)
7. Paulmakesh, D.A., April, Y.M.Y.: Implementation of project management for strategy realisation. J. Univ. Shanghai Sci. Technol. **23**(10), 887–892 (2021)
8. Jamieson, A., Morris, P.W.: Moving from corporate strategy to project strategy. In: The Wiley Guide to Project, Program, and Portfolio Management, vol. 3, pp. 34–62. Wiley & Sons, Hoboken, NJ (2007)
9. Martinsuo, M., Geraldi, J., Gustavsson, T.K., Lampel, J.: Actors, practices, and strategy connections in multi-project management. Int. J. Project Manage. **38**(7), 389–393 (2020)
10. Dathe, T., Helmold, M., Dathe, R., Dathe, I.: Corporate strategy. In: Implementing Environmental, Social and Governance (ESG) Principles for Sustainable Businesses: A Practical Guide in Sustainability Management, pp. 57–67. Springer International Publishing, Cham (2024)
11. Grant, R.M.: Contemporary Strategy Analysis: Concepts, Techniques, Applications, 6th edn. Backwell Publishers, Malden, Massachusetts (2007)
12. Bargorett, E., Williams, B.: Macro-environmental Analysis for the Alberta Midstream Oil and Gas Sector. Athabasca University, Alberta, Canada (2014)
13. Ibeh, C.V., Asuzu, O.F., Olorunsogo, T., Elufioye, O.A., Nduubuisi, N.L., Daraojimba, A.I.: Business analytics and decision science: a review of techniques in strategic business decision making. World J. Adv. Res. Rev. **21**(02), 1761–1769 (2024)
14. Mbiru, J.E., Wickham, M., Ayentimi, D.T.: Entrepreneurial project initiation processes for social enterprises in a developing economy context. J. Soc. Entrepreneurship **12**(2), 1–27 (2021)
15. Omenazu, S.: Strategic management, decision making and organizational performance: case study of construction industry Malaysia. J. Positive School Psychol. **6**(3), 6100–6113 (2022)

16. Rahman, L.: Participatory monitoring and evaluation in development projects of Bangladesh. Int. J. Proj. Organ. Manage. **11**(2), 93–109 (2019)
17. Tornjanski, V., Knežević, S., Vulević, B.: Perspectives of project management sustainability in the society 5.0 context: moving forward towards human-centricity. Eur. Project Manage. J. **13**(1), 61–73 (2023)
18. Hall, N.G.: Project management: recent developments and research opportunities. J. Syst. Sci. Syst. Eng. **21**, 129–143 (2012)
19. Australian Bureau of Statistics Homepage. https://www.abs.gov.au/statistics/people/population/national-state-and-territory-population/latest-release. Accessed 27 April 2024
20. Suleh, E.O., Osuji, G.E.: Multilingualism and its Contribution to Sustainable Socio-economic Development. The Catholic University of Eastern Africa Press, Nairobi, Kenya (2017)
21. Omar, A.H., Omar, N.H.: Factors affecting construction projects' cost estimating. Fayoum Univ. J. Eng. **5**(2), 26–41 (2022)
22. Pollack, J., Adler, D.: Emergent trends and passing fads in project management research: a scientometric analysis of changes in the field. Int. J. Project Manage. **33**(1), 236–248 (2015)
23. Zhang, J., Ouyang, Y., Ballesteros-Pérez, P., Li, H., Philbin, S.P., Li, Z., Skitmore, M.: Understanding the impact of environmental regulations on green technology innovation efficiency in the construction industry. Sustain. Cities Soc. **65**(1), 102647 (2021)
24. Yakubu, M.A., Oyigebe, P.L.: The current Israel-Palestine conflict and its impacts on Israel's economy and diplomatic ties. Kashere J. Polit. Int. Relat. **2**(1), 213–221 (2024)
25. Hassan, Q., Viktor, P., Al-Musawi, T.J., Ali, B.M., Algburi, S., Alzoubi, H. M., Jaszczur, M., et al.: The renewable energy role in the global energy transformations. Renew. Energy Focus **48**(1), 100545 (2024)
26. Paredes, C., Ribeiro, P.: Future trends in project management. In: Paredes, C., Ribeiro, P. (eds.) International Conference on Intelligent Systems 2018, pp. 637–644. Institute of Electrical and Electronics Engineers (IEEE) (2018)
27. Müller, R., Turner, R.: Choosing Appropriate Project Managers: Matching Their Leadership Style to the Type of Project. Project Management Institute, Newtown Square, Pennsylvania, United States (2006)
28. Project Management Institute. Pulse of the Profession. The Future of Project Work: Moving Past Office-Centric Models. Project Management Institute, Newtown Square, Pennsylvania (2024)
29. Livingstone, D.: Mapping the Iceberg. NALL Working Paper. Ontario Institute for Studies in Education, Toronto, Canada (2002)
30. Orieno, O.H., Ndubuisi, N.L., Eyo-Udo, N.L., Ilojianya, V.I., Biu, P.W.: Sustainability in project management: a comprehensive review. World J. Adv. Res. Rev. **21**(1), 656–677 (2024)

Brief Comparison of ISO 21500 Series to Predictive and Agile Approaches

Nelson José Rosamilha and Luciano Ferreira da Silva

Abstract In contemporary society, marked by rapid technological progress, cultural diversity, and various environmental and political issues, organizations must continuously adapt and strategize to remain competitive and meet societal demands. In this context, public and private organizations must develop projects and adopt project management methodologies to navigate complex changes effectively. Thus, project managers and other professionals involved in project activities must continually adjust their strategies to align with prevailing circumstances. This chapter examines the integration challenges of agile and predictive methodologies with established project management standards, ISO 21500 series. It emphasizes the importance of monitoring project progress and quality beyond traditional documentation, advocating for using Agile project management tools to enhance speed, efficiency, and overall quality. Furthermore, it explores the convergence of the ISO 21500 series, Agile Project Management, and the difficulties encountered in implementing these standards alongside agile approaches such as Scrum and Kanban. The chapter proposes practical solutions for implementing ISO 21500 series standards within agile frameworks.

Keywords ISO 21500 · Agile project management · Agile · Predictive project management · Project management · Project management integration · Scrum · Kanban · Project progress monitoring · Efficiency enhancement · Quality management · Implementation challenges

1 Introduction

Contemporary society is experiencing constant change due to technological advances, cultural mixes, and environmental and political crises. Public and private organizations need to adapt and design short, medium, and long-term actions to

N. J. Rosamilha (✉) · L. F. da Silva
Nove de Julho University, São Paulo, Brazil
e-mail: rosamilha@uni9.edu.br

K. D. Strang and N. R. Vajjhala (eds.), *International Program and Project Management—Best Practices in Selected Industries*, Information Systems Engineering and Management 31, https://doi.org/10.1007/978-3-031-80275-1_2

respond to increased competition, climate change, and political crises such as wars, among other demands of contemporary society. This need for adaptation to evolutionary advances and adversities leads us towards a projectification of society [1].

In this context, it makes sense to understand the future estimate by PMI (Project Management Institute), which until 2027, activities related to project management will generate USD 20 trillion, as well as 88 million job positions [2].

A recent example of using projects to respond to society's demands was the fight against the COVID-19 pandemic. Public and private organizations had to change their structures and processes, as well as make necessary adaptations in people's lives, to deal with the social distancing caused by this situation.

Another recent example is the changes in the technological field. Since 2022, we have witnessed the spread of a technology that caused disturbance inside and outside organizations: Generative Artificial Intelligence, such as ChatGPT and BARD.

Witte [3] explored the impacts on people's lives with the adoption of this technology, which brings challenges related to copyright, work processes (writers, designers, voice actors, attendants, etc.), products, and educational processes.

Therefore, in a context where technological advances and health adversities test the capabilities and structures of public and private organizations, project development and management become natural activities to meet current demands and prepare for future challenges. Thus, project management is a relevant activity in the allocation and management of resources to generate time-bound results and products. Furthermore, several discussions emerged regarding project management approaches.

The traditional approach is guided by planed-based orientation, where resources are applied linearly based on a life cycle that seeks predictability and control [4]. On the other hand, adaptive approaches follow iterative delivery cycles that evolve over. This approach consists of more flexible processes than the first, built by breaking down small parts of a whole that are designed and evolve as innovative solutions.

In this context, the ISO 21500:2021 standard and its series emerge as a valuable tool, offering guidance and guidelines for project management. The ISO 21500 series offers guidelines for managing projects, programs and portfolios, which mainly follow a planning orientation, but are also adapted to the iteration cycles of adaptive project management approaches. This standard is the result of an update to the ISO 21500 standard, first published in 2012 by the International Organisation for Standardization (ISO). Therefore, the ISO 21500:2021 standard became the matrix of the ISO series that covers the "3P" of project, program, and portfolio management.

The ISO 21500:2021 standard and its series allow the adoption of appropriate tools and activities to face the challenges of contemporary and projectised society. Although the value of the ISO 21500 series is recognized, it is necessary to understand how it fits into adaptive project management approaches. This controversial point is what stimulated the development of this chapter. In this sense, we highlight that projects of different types require guidance regarding their development and management.

This chapter is structured to provide a comprehensive analysis of the integration between ISO 21500 series standards and agile project management methodologies. Initially, it explores the foundational principles of ISO 21500 and its relevance in contemporary project management. Following this, the chapter delves into a detailed comparison between predictive and adaptive approaches, highlighting the benefits and challenges of each. Subsequent sections discuss the practical implementation of ISO 21500 within agile frameworks, offering specific recommendations and step-by-step guidelines. The chapter concludes with a discussion on the potential for hybrid approaches and phased implementations, aiming to harmonise the rigour of ISO standards with the flexibility of agile practices.

2 Theoretical Background

The objective of the section is to provide a foundational understanding of effective project management within the context of the ISO 21500:2021 series. This section elucidates how the ISO 21500 standards, encompassing guidelines for project, program, and portfolio management, establish a comprehensive framework essential for organisational success in a dynamic global environment. By detailing the purpose and scope of each standard within the series—such as ISO 21502's guidance on project management and ISO 21505's focus on governance—this section aims to highlight the systematic approach and uniform principles introduced by ISO 21500.

It further underscores the importance of these standards in mitigating uncertainties, enhancing operational efficiency, and ensuring project success. Additionally, the section addresses the applicability of ISO 21500 in various project contexts, including adaptive and agile methodologies, thereby setting the stage for more profound discussions on integrating these standards into modern project management practices.

2.1 ISO 21500:2021 and Its Series

Effective project management is a fundamental part of the success of organizations in a highly dynamic global scenario. The ISO 21500:2021 series, entitled "Project, program and portfolio management—Context and Concepts", is an international standard developed to provide a comprehensive set of guidelines that assist organizations in effectively managing projects of any nature and scope.

This standard was developed by a technical committee and provides a comprehensive understanding of essential project management concepts and processes essential to project performance. To understand these processes and practices, refer to the guidelines presented in ISO 21502, ISO 21503, ISO 21504, and ISO 21505, as described in Table 1.

Table 1 ISO 21500 series structure

Standard name	Purpose
ISO 21500:2021—Context and concepts	Specifies the organisational context and underlying concepts for managing projects, programs, and portfolios. It also provides guidance for organizations to adopt or improve project, program, and portfolio management using the standards prepared by ISO/TC 258
ISO 21502:2020—Guidance on project management	Provides guidelines for project management regardless of approach
ISO 21503:2017—Guidance on program management	Guides program management and high-level descriptions of relevant terms, definitions, concepts, prerequisites, and practices, including roles and responsibilities that constitute good practice in program management
ISO 21504:2022—Guidance on portfolio management	Provides guidance on project portfolio management principles
ISO 21505:2017—Guidance on governance	Describes the context in which project, program, and portfolio governance is conducted and provides guidance for project, program, and portfolio governance
ISO/TR 21506:2018—Vocabulary	Establishes a set of terms and definitions related to project, program and portfolio management
ISO 21508:2018—Added value	Guides and establishes principles for the application of Added Value as a performance management technique in projects and programs
ISO 21511:2018—Package analytical structure	Guides and establishes principles for the development and use of Package Breakdown Structures in project and program management contexts

This table describes breaks down the various standards within the ISO 21500 series, detailing each standard's name, year of publication, and purpose. The ISO 21500:2021 standard and its series represent a significant advancement in the field of project, program, and portfolio management by establishing a uniform set of principles and processes.

Thus, as a standardizing instrument, ISO 21500:2021 and its series guide professionals and companies, mitigating uncertainties and misconduct in project development and management. Furthermore, its comprehensive approach, logical structure, and emphasis on standardization make it a valuable tool for improving operational efficiency, mitigating risks, and boosting project success [5]. The content of the ISO 21500 standard series serves as guidance for practitioners, as well as a useful tool in certifying and auditing projects.

Thus, as evidence of the increasingly pressing need for projects, an instrument of this type contributes to beginners and senior professionals in such a complex and diverse field. In this way, the guidance obtained with this content means that professionals and companies can orient themselves with guidelines that go beyond the development of professional practices for conducting projects. The systemic view obtained with this content allows an understanding of specific aspects such as work package management, as well as discussing high-level aspects of project governance.

However, although the ISO 21500:2021 standard and its series promote relevant advances in the field of project management, discussions are still needed about its use in the context of adaptive, incremental, or iterative projects, such as those using the agile approach.

2.2 Different Agile Project Approaches and the ISO 21500 Series

Projects can follow life cycles that are predictive, adaptive, or hybrid. These cycles cover different phases related to developing products, services, or results. Regarding the development of each type of cycle, these models manifest themselves as predictive, adaptive, iterative, incremental or hybrid, and predictive models are commonly referred to as cascade life cycles, or gates [4, 6].

Adaptive lifecycles, also known as agile or change-driven lifecycles, can adopt iterative or incremental delivery approaches. In the iterative delivery approach, project scope is generally determined early in the project life cycle, but time and cost estimates are routinely modified as the project team's understanding of the product increases [6]. Iterations consist of repeated cycles for product development, or the construction of a solution by a partial deliverable known as a minimum viable product (MVP). In the incremental approach, deliverables are produced through successive iterations, like sprints, until the final product meets the client's requirements. The final iteration results in a complete deliverable, containing the necessary and sufficient capacity to meet the demands of a project client [6].

The hybrid delivery approach combines predictive and adaptive elements. This mix of practices and processes can be adjusted based on the project's needs. For example, when developing new features for a product with some uncertainty in resource allocation and the need for innovation, an incremental adaptive delivery approach may be the most suitable.

On the other hand, in cases where the creativity in innovative processes is not necessary, but the scalability of processes with a high degree of certainty of resource allocation and adherence to the schedule. The predictive delivery approach turns out to be the most appropriate. Thus, aspects of the project characterized by well-known or fixed requirements follow a predictive development life cycle, while those still evolving adopt an adaptive, incremental or iterative development life cycle [6].

In the general context, implementing the ISO 21500 series with agile delivery approaches can be an effective strategy for organizations that want to combine the flexibility and adaptability of agile practices with the structured principles of the ISO 21500 series standard. Therefore, an adaptive delivery approach is necessary, considering the flexibility inherent to agile methodologies based on the prescriptions of the ISO 21500 series.

2.3 *ISO 21500 Series in an Agile Environment*

Table 2 outlines in each line detailed explanations the suggested steps for implementing the ISO 21500 series in an agile environment. It elaborates on what needs to be done, why it is important, and how it contributes to the integration of ISO 21500 standards with agile practices.

The recommendations outlined in Table 2 aim to harmonize the ISO 21500 series framework with agile principles and practices, providing an integrated approach to project management that combines flexibility and adaptability with a standards-driven framework. It's important to remember that agile practices and processes focus on increasing the value by evolving the project product throughout its development.

Therefore, the key to successfully implementing the ISO 21500 series in an agile environment is flexibility and the ability to adapt the standard's principles and practices in accordance with agile values. This ensures effective project management that meets the dynamic needs of agile projects. One of the main aspects of this integration is understanding the delivery of benefits from a project. According to the ISO 21500:2021 standard, a benefit is an advantage, value, or other positive effect resulting from the project. This concept should guide the adaptation of practices and processes, aligning them with agile schedules and the need for predictability in resource allocation.

There are several agile methods and frameworks available for adopting project management delivery approaches where it is possible, depending on the context, to make the necessary adjustments to adapt to the ISO 21500 series. Among these frameworks, we can mention Scrum, which is an agile framework designed to assist individuals, teams, and organizations in creating value through the implementation of flexible solutions to complex issues. Scrum uses an incremental and iterative methodology to improve predictability and manage risk effectively [7].

2.4 *ISO 21502 and Scrum Framework*

The ISO 21502:2020 standard offers guidelines for project management, while Scrum is an agile framework focused on iterative and incremental development. Integrating ISO 21502 into Scrum projects requires combining Scrum's agile principles with ISO 21502's comprehensive guidelines. For example, ISO 21502 provides detailed guidance on planning a project.

However, in the context of Scrum, this means creating an agile Project Plan that aligns with Scrum's philosophy of responding to change rather than following a rigid plan. Documents such as the Product Backlog and Sprint Backlog can be enhanced to

Table 2 Adoption of ISO 21500 series and agile suggested steps

Step name	Description
Understanding agile principles	Make sure the team and leadership understand the principles and values of the Agile Manifesto. This is essential to align the agile delivery approach with the implementation of ISO 21500
Training in ISO 21500 and agile methodologies	Provide training on ISO 21500 concepts and requirements across project, portfolio, and program dimensions. Make sure the team is familiar with the chosen agile methodologies (e.g., Scrum, Kanban etc.), and understand how these practices can be integrated into the standard
Assessment of necessary adaptation	Guide an assessment of existing agile practices in relation to ISO 21500 requirements (e.g., assess non-adoption of timelines and work packages developed in linear steps). Identify gaps and adaptations needed to effectively integrate the standard's principles into an agile environment
Definition of roles and responsibilities	Adapt roles and responsibilities to reflect the agile delivery approach. Consider the self-organizing and collaborative nature of agile teams when assigning responsibilities as needed. Do not forget that the project manager's roles and responsibilities are distributed among other professionals in the agile approach
Iterative planning	Use agile planning practices, such as sprint meetings and release planning, to integrate ISO 21500 iterative planning. Ensure plans align with project and customer objectives in iterative or incremental cycles
Continuous monitoring and performance assessment	Adopt agile monitoring practices, such as daily scrums, sprint reviews, and retrospectives, to continually evaluate project performance. Integrate agile metrics, such as team velocity, lead time, work in progress, and cycle time, among other possible metrics to measure the progress of a project
Agile change management	Implement agile change management practices, enabling quick adaptations to project requirements. Additionally, professionals must maintain a mindset of continuous learning and improvement
Light and visual documentation	Adapt ISO 21500 documentation requirements to be lighter and more visual, using agile tools such as Kanban boards, Scrum boards or collaborative wikis
Agile risk management	Integrate agile risk management practices, such as using burndown charts to monitor progress and continually identify impediments and risks
Continuous improvement	Encourage a culture of continuous improvement through agile practices, such as regular retrospectives. Use learnings to adjust practices and processes as needed
Adaptation to customer needs	Professional must adopt value continuous adaptation to meet customer needs. Use regular feedback and agile practices, such as sprint reviews, to ensure the project is aligned with customer expectations. The roles of Product Owner and User Experience are essential to integrating the project process and customer satisfaction

(continued)

Table 2 (continued)

Step name	Description
Agile reviews and audits	Perform reviews and audits in an agile manner, integrating them into regular project lifecycle practices. Ensure that inspections and adaptations occur efficiently and in line with the agile approach
Leadership commitment	Ensure leadership commitment to the agile delivery approach and implementation of ISO 21500. This is critical to creating a culture of collaboration, learning and continuous improvement

include more formal planning elements proposed by ISO 21502. For example, when managing the Product Backlog from the perspective of ISO 21502, some adjustments to the standard's guidelines may be necessary. Table 3 presents some valuable recommendations in this regard.

Table 3 provides a detailed explanation of the specific sections of ISO 21502 and how they can be applied to Product Backlog practices in an agile environment.

Another agile delivery approach used is Kanban, an approach to improving value stream efficiency through a method that employs visual facilitation and constraint of work in progress in a pull system.

2.5 *ISO 21502 and Kanban*

This approach works as a management method to improve existing work processes or methodologies [8]. Therefore, using the ISO 21502 guidelines, Kanban projects can benefit from a more formal approach to planning, especially in the creation of clear and comprehensive documentation. This may include creating specific documents to define objectives, scope, and acceptance criteria, aligning with Kanban transparency principles. For example, section 7 of ISO 21502 standard addresses project management practices that can be integrated with Kanban. Table 4 describes some recommendations for this integration.

Table 4 offers specific recommendations for integrating the practices described in each ISO 21502 section with Kanban practices. These recommendations suggest how to align the structured guidelines of ISO 21502 with the visual and iterative nature of Kanban methodology, ensuring effective project management throughout the project lifecycle.

In addition to the information previously presented, it is worth highlighting that ISO 21502 emphasizes the importance of stakeholder involvement and the competencies needed to conduct projects. In the context of Kanban, this means ensuring that stakeholders are informed about progress and also that the team has the necessary competencies to fulfil their responsibilities. This can be integrated with the Kanban method to improve team communication and effectiveness.

Table 3 ISO 21502 × product backlog practices

ISO 21502 section	Description	Recommendation
4.4—Project life cycle	Provides guidelines on how projects should be managed throughout their life cycle, from conception to closure	Align Product Backlog management with the project life cycle defined by ISO 21502, considering the different phases of the project and their implications on the backlog. Understand each iterative or incremental cycle as a project delivery cycle
4.5.10—Interested in the project	Provides guidance on identifying, engaging and managing stakeholders throughout the project lifecycle	Ensure adequate stakeholder involvement in creating and maintaining the Product Backlog. Use ISO 21502 to identify and engage stakeholders effectively, ensuring a common understanding of project needs. It is worth remembering that new roles are needed in the agile approach, such as Product Owner and User Experience
5.3—Continuous improvement	Provides guidance on how organizations can establish and maintain an environment conducive to continuous improvement in project management	Apply continuous improvement practices to the Product Backlog, reviewing and adjusting priorities based on feedback received during iterations. Daily and retrospective meetings should be important sources of improvement
6.55—Initial planning	Provides guidance on how to carry out the initial planning of a project, laying the foundations for the success of the project throughout its life cycle	Integrate the planning elements of ISO 21502 when managing the Product Backlog, considering goal setting and developing an agile plan. A project can have a continuous cycle of open scope, such as a digital platform in development. Thus, some backlog items will still be future benefits of a project that depend on some conditions
7.16—Information and documentation management	The purpose of information and documentation management is to enable relevant and reliable information (physical and digital) to be available to those who are performing work and making decisions	Ensure the Product Backlog is documented, with items prioritized in a transparent for all stakeholders. Use documentation to provide a comprehensive overview of the requirements and functionalities planned for the project, without necessarily leaving the project management tied up in documentation
7.8—Risk management	The purpose of risk management is to increase the likelihood of achieving project objectives	Consider risk management aspects when maintaining and updating the Product Backlog, and evaluating possible risks associated with priority items. However, often taking risks to test a project's hypotheses in one iteration is an essential part of the agile approach

Table 4 ISO 21502 × Kanban practices

ISO 21502 section	Description	Recommendation
7.10—Change control	Control changes to the project and deliverables, and formalize the acceptance or rejection of these changes	Adapt the Kanban system to include specific change management practices. Consider how Kanban can provide an agile approach to dealing with changes in project requirements and priorities
7.18—Lessons learned	The purpose of lesson learning is to leverage experience, avoid repeating mistakes, and disseminate improved practices to benefit current and future project teams	Include regular Kanban retrospective practices to identify opportunities for continuous improvement in the project process. ISO 21502 can guide the establishment of formal improvement processes. Taking advantage of visual and collaborative management is an advantage of using Kanban as a management methodology
7.2—Planning	The purpose of planning is to define the requirements, deliverables, results, developments and constraints, and determine how the project objectives should be achieved	Think about how work-in-progress (WIP) limits in Kanban can influence project planning. Consider how setting limits helps manage workflow and task prioritization
7.2.2—Executing the plan	Planning should be a collaborative activity, whenever possible, involving team members providing advice on planning their work	Consider how Kanban uses a pull system for project implementation, where tasks are pulled according to the team's capacity, rather than being pushed from one step to the next
7.2.3—Monitoring the plan	Progress against the plan baseline should be regularly monitored and analysed	Integrate Kanban board principles to schedule and visualize project progress. Use columns to represent project stages and allow easy identification of task status
7.3.3—Monitoring benefits	Monitor progress throughout the project lifecycle towards achieving results and their impact on realizing intended benefits	Adopt Kanban metrics, such as lead time, cycle time and throughput, to evaluate project performance. These metrics provide insights into workflow efficiency and can be integrated into the overall project assessment. Another relevant aspect of this process is the concept of "done"

2.6 *ISO 21502 and Lean Startup*

Another agile delivery approach used is Lean Startup, where, through a continuous cycle, a simpler version of the product is delivered that meets the needs of interested parties. Thus, through hypothesis testing, rapid experimentation and feedback, adjustments are made to the new version to be improved or redone [9]. As an example of the integration of ISO 21502 and this approach, Table 5 presents some recommendations to be adopted when developing hypotheses for an MVP.

Table 5 offers specific recommendations for integrating the practices described in each ISO 21502 section with Lean Startup methodologies. Based on this, it is evident that maintaining flexibility, agility, and the ability to respond to changes are crucial guiding principles.

This allows the project team to continually validate and adjust their hypotheses throughout the project lifecycle. The project product is adjusted to increase the capacity to deliver value to stakeholders, which is aligned with what the ISO 21500:2021 standard describes as project benefits.

Table 5 ISO 21502 × lean startup practices

ISO 21502 section	Description	Recommendation
4.2—Context	The context of a project can impact project performance and the likelihood of success. The project team must consider factors both inside and outside the organization	Understanding the organizational context, which is relevant when formulating hypotheses about the market and customers, is necessary. In this way, it helps to define the conditions under which the hypotheses will be tested, as well as establish premises and restrictions
6.5.5—Initial planning	An initial plan for the project should be developed with milestones or decision points based on the project life cycle, combined with a detailed plan for at least the immediate phase of the project	Four phases are required for initial planning: (1) Include a specific step for developing hypotheses during initial project planning. Identify key assumptions about the market, customers and proposed solutions; (2) Define metrics and key performance indicators (KPIs) from the beginning of the project. This is in line with the idea of measuring progress and validating hypotheses throughout the project lifecycle; (3) Consider creating an MVP as part of initial planning. Determine essential features that will be developed and released quickly to gather early feedback; (4) Include specific activities to test hypotheses identified during initial planning. This may involve experiments, interviews with potential customers, or other agile validation techniques

2.7 ISO 21502 and Scrum@Scale

The use of agile frameworks for project management is evident in public and private organizations. However, although organizations adopt agile methods and frameworks to generate project deliveries iteratively or incrementally, the search for scalable forms of agile for the entire organisation is a challenge. For example, to apply the Scrum framework on a broader scale, specific practices and principles are required that organizations should follow in different contexts [10].

For this, the Scrum@Scale Guide was created [10]. This guide describes the minimum set of elements that create a team environment that drives innovation, customer satisfaction, performance, and satisfaction. To achieve this, the guide recommends radical transparency and a series of formal events to provide opportunities to inspect and adapt a team and project deliverables [10].

Therefore, when we examine this guide from the perspective of ISO21502:2020, adaptations required by the standard are necessary. For example, one of the topics covered regarding scalability concerns the role of the Chief Product Owner (CPO). The CPO coordinates priorities with the Product Owner Team. Together, they align backlog priorities with stakeholder and customer needs. The CPO may be a Product Owner from a specific team who also performs this role, or it may be a person specifically assigned to this role. In this context, Table 6 presents some analyses of the relevant sections of ISO 21502:2020 and respective recommendations for integration with the Scrum@Scale framework.

These recommendations suggest how to align the structured guidelines of ISO 21502 with the scalable and collaborative nature of Scrum@Scale, ensuring effective project management throughout the project lifecycle.

The practices presented in Table 6 seek to effectively integrate the responsibilities of the CPO with the Project Manager guidelines in ISO 21502, promoting a collaborative and effective approach to project management. In this sense, it must be verified that roles and processes prescribed in ISO 21500 series are more adherent to predictive approaches. They must be understood mainly in terms of their objectives and benefits to be integrated more flexibly into agile delivery approaches. Other frameworks define different activities, but in essence, a project has basic characteristics and defined functions, which can be integrated with new actors such as the Product Owner.

Likewise, the content set out in the ISO 21500 series standard must be studied to facilitate project management, and not be a constraint to the project management process that adopts the agile delivery approach. In this sense, when thinking about program and portfolio management, it is also necessary to understand the limitations and benefits of scaled agile. The standards in this series should be thought of as guidelines to incorporate good project management practices, with the activity of tailoring becoming the solution for the ISO 21500 series to increase maturity in project management in public and private organizations.

Table 6 ISO 21502 × Scrum@Scale practices

ISO 21502 section	Description	Recommendation
4.5.6—Project manager	The project manager is accountable to the project sponsor or project board for completing the project	Two activities are necessary to fulfil the role of the project manager: (1) Implement practices that promote effective collaboration between the CPO and the Project Manager, to ensure synergy between responsibilities; (2) Establish a clear definition of the Project Manager's responsibilities and integrate them with the CPO's activities to avoid overlaps and gaps
7.2.3—Monitoring the plan	Progress against the plan baseline should be regularly monitored and analysed	Coordinate backlog priorities between the CPO and Project Manager, ensuring an integrated approach when monitoring project progress
4.5.10—Interested in the project	People, groups, or organizations that have an interest in affecting, being affected by, or perceive themselves to be affected by any aspect of the project	Align CPO activities in stakeholder management with ISO 21502 practices, ensuring a cohesive approach to meeting stakeholder needs
7.14—Managing organizational and social changes	The objective of managing organizational and social change is to enable the delivery of desired project results	Integrate CPO change management practices with ISO 21502 guidelines to ensure changes to product requirements are aligned with organizational changes
7.16—Information and documentation management	Enable relevant and reliable information (physical and digital) to be available to those who are carrying out work and making decisions	Ensure that the documentation produced by the CPO is aligned with the documentation requirements defined in ISO 21502, to maintain a clear view of the project status

2.8 *ISO 21505 Governance and Agile Approaches*

Another aspect to be considered refers to the ISO 21505 standard, which deals with the governance of projects, programs, and portfolios. Specifically, about the project governance context, which must be supported by processes, procedures, and standards to meet governance requirements [11]. Thus, while ISO 21505:2017 provides guidelines for project governance, agile delivery approaches such as Scrum and Kanban are specific project management methodologies that focus on iterative and incremental delivery of value. Although it may seem that these two approaches are divergent, it is possible to integrate them to obtain the benefits of both. Table 7 presents the practices for this adaptation.

Table 6 provides a brief explanation of what each ISO 21505 section covers. It describes the key focus or purpose of the section, detailing the principles and practices recommended by ISO 21505 for effective project governance.

Therefore, while ISO 21505 provides guidelines for project governance in general, agile methods prioritize delivering value in small, frequent increments, allowing for adaptations throughout the process. Additionally, ISO 21505 provides a clear framework for project management, facilitating the consistent and organized application of agile practices across the organization. In this sense, public and private organizations should consider project management offices. Thus, while agile approaches in general are more oriented towards delivering value from the project product, organizational structures that include governance aspects must be addressed to support the adoption of predictive, adaptive, incremental, iterative, or even hybrid delivery approaches.

Consequently, aligning agile practices with a recognized international standard, such as ISO 21505, enables integration with global project management practices, facilitating communication and collaboration with international partners. Thus, while agile delivery approaches value individuals and interactions more than processes and extensive documentation, it is important to maintain an adequate amount of

Table 7 ISO 21505 practices × agile approaches

ISO 21505 section	Description	Recommendation
6.2—Project governance body	Project governance involves defining a governing body responsible for making strategic decisions and ensuring that the project is aligned with organizational objectives	In the agile context, this translates into establishing a Product Owner, which is responsible for defining and prioritizing the product backlog, and a Scrum Master, who acts as a facilitator of the Scrum process
6.3—Guidelines for project governance	This section establishes guidelines for project governance, including the definition of policies, processes, and procedures	In the agile context, guidelines can include practices such as holding daily Scrum meetings, sprint reviews, and retrospectives to ensure transparency, constant inspection, and adaptation
6.4—Framework	This sub item talks about the need for a framework to guide project governance	In the case of agile, the Scrum framework offers a clear structure with defined roles, artefacts, and ceremonies that help with project governance. In addition to Scrum, other agile methodologies such as Kanban or Lean can be considered, depending on the needs of the project

documentation and traceability to ensure transparency and accountability. In this sense, ISO 21505 can help define requirements, without burdening the team with unnecessary bureaucracy. Accordingly, organizations can gain a solid framework and guidance to implement agile practices effectively, ensuring consistency, continuous improvement, and compliance with global project management standards.

3 Discussion

This section addresses the integration challenges of combining these methodologies into a balanced project management framework.

3.1 Predictive Versus Agile Methodologies

The ISO 21500 series emphasizes a planning-oriented project management approach, where defining scope, schedule, and budget at the project's outset is crucial. This traditional method aims for predictability and control, essential in many organizational contexts, especially those involving regulated environments or large-scale projects. However, this approach can be rigid and may not accommodate the dynamic nature of many modern projects.

Agile methodologies, on the other hand, prioritize flexibility and continuous adaptation, allowing project teams to respond swiftly to changes in requirements and priorities. Agile approaches encourage self-organizing behaviour, collaboration, and iterative value delivery. While this can lead to higher responsiveness and customer satisfaction, it may pose challenges in environments where strict adherence to predefined plans and detailed documentation is necessary.

3.2 Hierarchical Structures Versus Self-organizing Teams

ISO 21500 promotes a hierarchical structure with defined processes and responsibilities, which can be beneficial for maintaining organizational clarity and ensuring accountability. This structured approach helps in complex projects where roles and workflows need to be clearly delineated.

In complement, agile methodologies such as Scrum and Kanban promote self-organizing teams and collective decision-making. Roles like Product Owner, Scrum Master, and UX professionals are integral to agile teams but can conflict with the traditional project manager role outlined in ISO 21500. This divergence necessitates a careful balancing act to integrate these roles without causing confusion or overlap in responsibilities.

3.3 Documentation Versus Direct Collaboration

ISO 21500 places a strong emphasis on comprehensive documentation, including detailed project plans, status reports, and change logs. This is advantageous for maintaining control and transparency in projects, especially those requiring rigorous oversight. However, extensive documentation can become cumbersome and counterproductive in fast-paced environments.

In contrast, agile methodologies reduce the emphasis on formal documentation, focusing instead on direct collaboration and continuous delivery of functional product increments, which approach facilitates quicker adjustments and real-time problem-solving but may lack the thoroughness required for complex audits or regulatory compliance.

3.4 Integration Challenges

Integrating ISO 21500 with agile practices involves significant challenges due to their differing underlying philosophies. The structured and control-focused nature of ISO 21500 can seem at odds with the adaptive and value-driven approach of agile methodologies. However, successful integration requires tailoring project management practices to harmonize these differences, fostering a culture that values both compliance and flexibility.

One alternative viewpoint suggests that a hybrid approach, combining elements of both predictive and agile methodologies, can offer the best of both worlds. By leveraging the planning and control strengths of ISO 21500 while incorporating agile's adaptability and customer focus, organizations can create a more balanced project management framework.

Another perspective advocates for a phased implementation of agile practices within the ISO 21500 framework. This gradual integration allows for a smoother transition, giving teams time to adapt to new roles and responsibilities while maintaining some level of predictability and control.

4 Final Considerations

This chapter has explored the integration challenges between the ISO 21500 series and agile methodologies such as Scrum and Kanban. The focus has been on understanding how traditional, planning-oriented project management practices can be harmonized with agile principles that prioritize flexibility and continuous adaptation.

The primary objective was to examine how the ISO 21500 standards, known for their structured and hierarchical approach, can be adapted to fit within agile frameworks. This involves reconciling the differences in documentation practices, roles, and processes to create a cohesive project management strategy that leverages the strengths of both approaches.

In this sense, we point out that the ISO 21500 series has a bias towards planning-oriented project management practices and processes, emphasizing the importance of clearly defining the scope, schedule, and budget from the beginning of the project. This approach seeks predictability and control, fundamental aspects in many organizational contexts. On the other hand, agile methodologies value flexibility and continuous adaptation. This delivery approach allows teams to respond to changes in requirements and priorities in an agile way, encouraging self-organizing behaviour in project teams. It also prioritizes the delivery of value to the customer in smaller and iterative increments.

The ISO 21500 series standards provide a hierarchical structure of processes and responsibilities, outlining specific roles and formal workflows. This can be beneficial in environments where organizational clarity is essential. In contrast, agile approaches promote intensive collaboration among team members and value autonomy and self-organization. Furthermore, agile teams are encouraged to make collective decisions and adapt their processes according to the needs of the project. Additionally, roles such as Product Owner, UX, CPO, and Scrum master are covered in the agile approach and can compete with the project manager role and responsibilities prescriptions of the standards in this series. Therefore, it is necessary to understand the adaptations in roles and responsibilities aligned with the practice of tailoring.

ISO 21500 tends toward a significant amount of documentation, including detailed project plans, status reports, and change logs. This is an advantage in regulated environments or complex projects that require a high level of control. On the other hand, agile methodologies prioritize delivering value to the customer through direct collaboration and continuous delivery of functional product increments. This often means less emphasis on formal documentation and more focus on human interaction and responding quickly to customer needs.

Thus, while ISO 21500:2021 and ISO 21502:2020 provide valuable guidance for managing projects in a wide range of organizational contexts, it is important to recognize that their prescriptions may not be fully compatible with agile methodologies. Therefore, adaptations will be required when adopting agile approaches, with their emphasis on flexibility, collaboration and continuous value delivery, which represent a significant paradigm shift concerning more traditional project management models.

Therefore, organizations that wish to adopt agile delivery practices must be prepared to face challenges of integration and reconciliation between agile principles and the requirements of the ISO 21500 series standards. This may involve adaptations in project management practices, a more organizational culture oriented towards collaboration and flexibility, and a more holistic approach to compliance with standards and regulations.

Finally, success in implementing agile practices as per the ISO 21500 series standards will depend on organizations' capacity to strike a balance between structure and flexibility, documentation and value delivery, and hierarchy and autonomy. Therefore, further debate is needed on how to integrate agile principles with the ISO 21500 series standards, ensuring that organizations leverage the best of both worlds.

References

1. Wagner, R., Huemann, M., Radujković, M.: An institutional theory perspective on the role of project management associations for projectification of society: the case of Germany. Int. J. Manag. Proj. Bus. **15**(8), 111–134 (2022)
2. The new project-based economy. International Standardization Organisation—ISO. https://www.iso.org/contents/news/2022/06/newsproject-based-management.html. Accessed 25 Jan 2024
3. Witte, M.: How will ChatGPT change the way we think and work? Stanford scholar examines. Stanford scholar examines (2023)
4. PMI. A Guide to the Project Management Body of Knowledge. 6th Project Management Institute, Pennsylvania (2017)
5. TSO, Axelos: PRINCE2 Agile®, 3rd edn. TSO, Norwich (2018)
6. PMI.: Project Management Institute: Process Groups: A Practice Guide, 1st edn. Project Management Institute, Pennsylvania (2022)
7. Scrum Guides. https://scrumguides.org/index.html. Accessed 17 Jan 2024
8. The Official Guide to The Kanban Method. https://kanban.university/wp-content/uploads/2023/04/The-Official-Kanban-Guide_A4.pdf. Accessed 17 Jan 2024
9. Ries, E.: The Lean Startup: How Today's Entrepreneurs Use Continuous Innovation to Create Radically Successful Businesses. Crown Currency, New York (2011)
10. Scrum@Scale Framework. https://www.scrumatscale.com/scrum-at-scale-guide/. Accessed 18 Jan 2024
11. ISO 21505:2017—Project, Programme and Portfolio Management—Guidance on Governance, Second Edition, Reference number ISO 21505:2017 (E)

Understanding the Trade-off Between Efficiency and Value in Collaborative R&D Projects—A Complexity Theory Based Conceptual Framework

Francesc Miralles, Juan Augusto Nihoul, Laurentiu Neamtu, and Juan Antonio Castro-Fernández

Abstract The concept of project success is a heavily researched topic within the field of project management, typically viewed through two contrasting lenses: efficiency and value. This chapter aims to illuminate the functioning of these perspectives within collaborative R&D projects, where outcomes are inherently unpredictable and hinge on complex partner interactions. This chapter seeks to unravel the interplay between equivocality and governance, including the moderating effects of participants' learning capabilities and their impact on project success, measured by either efficiency or value. Consequently, a conceptual framework grounded in the complex adaptive system model is outlined. Through a multiple-case analysis of collaborative R&D projects, an inductive grounded theory approach is used to gain a deeper understanding of the four cases. Empirical data was gathered through open-ended interviews and archival data. The proposed framework delineates the findings of this chapter, enhancing comprehension of collaborative R&D projects by modeling the interactions of governance, equivocality, and organizational learning as a complex adaptive system. From an academic perspective, this framework offers additional insights into the planning and control functions within collaborative R&D projects. For practitioners, it provides a deeper understanding of management practices and their implications for success, emphasizing the nonlinear interactions among the three constructs.

F. Miralles (✉) · J. A. Nihoul · L. Neamtu · J. A. Castro-Fernández
La Salle, Ramon Llull University, Sant Joan Baptista de La Salle, 42, 08022 Barcelona, Spain
e-mail: francesc.miralles@salle.url.edu

J. A. Nihoul
e-mail: juanaugusto.nihoul@salle.url.edu

L. Neamtu
e-mail: laurentiu.neamtu@salle.url.edu

J. A. Castro-Fernández
e-mail: juanantonio.castro@salle.url.edu

K. D. Strang and N. R. Vajjhala (eds.), *International Program and Project Management—Best Practices in Selected Industries*, Information Systems Engineering and Management 31, https://doi.org/10.1007/978-3-031-80275-1_3

Keywords Collaborative R&D projects · Project success · Complexity theory · Complex adaptive system · Equivocality · Governance · Organizational learning · Multiple-case analysis · Grounded theory approach · Nonlinear interactions

1 Understanding Success in Collaborative R&D Projects

Collaborative efforts are increasingly recognized for their role in fostering innovation and competitiveness in both public agencies and industry [1]. This chapter refers to these initiatives as Collaborative R&D Projects (CR&DP), encompassing R&D projects within the industry or regional innovation systems that involve multiple legally independent organizations committed to the success of the collaborative endeavor [2].

Success is one of the main challenges that CR&DPs face. This chapter assumes that CR&DP must overcome the trade-off between efficiency and value. One the one hand, CR&DP should comply with the triple restriction, scope, cost, time, that funding agencies specify. On the other hand, partners interaction can be a source of value due to creative ideas and innovation.

Project management guidelines and methodologies based on standard good practices help to harmonize governance and facilitate monitoring and supervision through an efficiency perspective. However, deterministic approaches to project management have difficulties to promote innovation for value. Systems theory rooted in non-deterministic approaches try establishing new insights to a broader perspective for value. A multiple case analysis of CR&DPs, using an exploratory inductive approach, analyzed four cases, allowing data to be gathered from open-ended interviews and archival data exploration. This section outlines the challenges faced by CR&DPs, the reference structure for CR&DPs in this research, the key characteristics of the empirical effort used to test the conceptual proposal and concludes with a brief overview of the chapter's structure.

1.1 Delimitating Collaborative R&D Projects and Elucidating Their Success

Collaborative projects are essential for achieving innovation within the involved organizations [3]. These projects are driven by the necessity to develop complex and unique outputs that cannot be produced by individual organizations working alone [4]. Often, the goals of such projects are unclear, ambiguous, and subject to change throughout the project lifecycle. One of the cases of the empirical effort, LOCUTIOS, is an example for that. Building a new voice simulator (see Table 1) that could speak in any language has to incorporate basic principles that require the participation of cutting-edge scientific groups.

Table 1 List of collaborative R&D projects included in the multiple-case analysis

Project	Description	Type of Project
LOCUTIOS	Building a new voice simulator based on physical basic principles to an unprecedented degree. The goal is to build a voice simulation engine, that could speak in any language, or sing in any style	EU project (7th framework)
GAMMA	Providing new and innovative on-site quality assurance tools which help to achieve the efficiency targets in a virtual construction management platform	EU project (H2020)
EPSILON	Creation of a new platform to provide access to widely dispersed energy-related data. The platform integrates data from multiple domains using semantic technologies	EU project (7th framework)
BMOI	Building blockchain frameworks and tools that can be used across different industries. Partners work together to drive innovation and accelerate the adoption of blockchain solutions	Open innovation project

Collaborative R&D projects can be funded by either private or public organizations [5]. In both scenarios, a consortium of partners collaborates to tackle a common challenge. The dynamics within each consortium can vary based on the sponsoring organization. Private organizations, driven by specific market needs, often partner with suppliers, universities, and other entities to meet those needs [6]. Conversely, public organizations, such as the European Commission (EC), fund collaborative projects through well-known programs like the Seventh Framework or the Horizon 2020 program (H2020) [7]. The diversity of partners in a project makes a difficult balance in terms of goals, following the triple restriction, and innovation for value. GAMMA (see Table 1) combined a constellation of partners from different origins that challenged to provide an innovative on-site quality assurance virtual platform for partners in construction management.

The concept of project success has been extensively studied within the field of project management, resulting in two coexisting paradigms: success from the perspective of efficiency and success from the perspective of value [8]. The efficiency paradigm focuses on short-term goals, assessing how effectively the project meets initial objectives regarding cost, time, and scope/quality [9]. It emphasizes the project's contribution to the final output during the implementation phase [3, 10, 11]. The value paradigm, on the other hand, takes a long-term view, considering the project's contribution to the organization's overall outcomes [11]. This approach emphasizes strategic value and long-term benefits for stakeholders and their environment [8, 12]. The LOCUTIOS project is an example for this kind of situation. In this case, the value of the project was relevant due to the number of scientific publications that were generated by the research activity.

The uncertain nature of collaborative projects, combined with their innovation objectives, ambiguous and ambitious goals, and the limited predictability capacity of managers, makes it incomplete to assess success based solely on the efficiency paradigm. Evaluating the success of collaborative projects only in terms of inputs versus outputs, with a focus on their efficient development, poses challenges because producing outputs does not necessarily equate to delivering value [8]. Innovation is inherently linked to value rather than efficiency [6, 12]. Therefore, measuring project success as a dynamic and subjective process of value realization and capture among stakeholders during and after the project is more suitable for collaborative innovations, complementing the efficiency perspective [13]. The EPSILON project is an example for a capture of value from a dynamic perspective among stakeholders (see Table 1). EPSILON can be considered as one of the most successful projects according to the evolution of the scientific productivity. The project's outcomes have been used in further research in the field.

Instead of favoring one perspective over the other, this research emphasizes the importance of both paradigms in measuring project success. The challenge lies in the absence of a single method to evaluate success, as each partner has its own approach, aligning more closely with either the efficiency or value paradigm. As the project progresses, inherent conflicts and misunderstandings are likely to arise, necessitating governance adaptations. Each decision is influenced by the differing (and sometimes opposing) interpretations of "success" among project partners and leaders, which increases the complexity of project coordination and cooperation [13].

1.2 Structure and Schemata for CR&DP

There is a board debate on the role of standard practices for project management. Most of them try to provide structures and schemata that allow for the delimitation of management practices. ISO 21502:2020 refers to process groups. PMBOK (Project Management Book of Knowledge) seventh version (2021) [14], by Project Management Institute (PMI), made a conceptual switch, moving from the description of a five process groups-based standard[1] to the broader concept of a value delivery system that includes twelve foundational principles of project management.[2] These principles are expected to "*provide guidance for the behavior of people involved in projects as they influence and shape the performance domains to produce the intended outcomes*" (PMBOK v7, [14]).

Most project management practices follow a deterministic approach, relying on *phase-project-planning* methodologies[3] and a success framework based on the efficiency perspective [11]. However, this deterministic approach has been challenged by non-deterministic approaches rooted in systems theory, which adapt

[1] Previous versions of PMBOK propose the following groups of processes for the practices of project management: *Initiating process group*, *Planning process group*, *Executing process group*, *Monitoring and Controlling process group*, and *Closing process group*.

[2] PMBOK seventh version proposes the following foundational principles of project management practices: *Stewardship*, *Team*, *Stakeholders*, *Value*, *Systems Thinking*, *Leadership*, *Tailoring*, *Quality*, *Complexity*, *Risk*, *Adaptability*, and *Change*.

[3] The deterministic paradigm has been named as "*operations research paradigm*", PMI's execution-based model and cybernetic model of planning and control.

complex models to project management practices [8]. Non-deterministic approaches are still in the process of establishing their place in the field of project management. Non-deterministic approaches have been illustrated in the case of megaprojects as described in the Hinkley Point C nuclear power plant case [12]. A non-deterministic design allows for causal analyses among crucial project events and considers self-organizing patterns in the implementation phases.

Fundamental principles of Collaborative R&D Projects (CR&DP) include fostering a collaborative team environment, effectively engaging with stakeholders, recognizing, evaluating, and responding to system interactions, embracing adaptability and resilience, tailoring approaches based on context, and facilitating change to achieve the envisioned future state. Additionally, CR&DP must manage the collaborative efforts among partners and address how governance can be influenced by the diversity of these partners [2, 3, 6, 9]. BMOI is an exemplary case of a collaborative innovation project (see Table 1). Innovation in the project was rooted in the effort of all partners' joint work towards the project's common goals.

Project governance involves utilizing systems, agreements, authority structures, leadership, actions, decisions, and processes to allocate resources, coordinate, and control activities to define and achieve project goals [2]. In a collaborative context, governance not only aims to achieve project goals but also seeks to enhance collaborative capacity among partners [6].

When firms collaborate, they bring together individuals with diverse experiences, skills, resources, capabilities, and perspectives, leading to conflicting interpretations of tasks, routines, goals, and information [10]. As a result, misunderstandings can arise due to partners' differing interests, terminologies, orientations, and cultures [9]. Unclear and evolving project objectives, coupled with high levels of uncertainty, amplify the diversity of interpretations and increase equivocality [4]. Equivocality refers to having multiple and conflicting interpretations of goals, situations, or tasks [2].

Therefore, governance serves as the mechanism to align partners' expectations, interests, and backgrounds, thereby reducing conflicts and misunderstandings (equivocality) to ensure the project meets its objectives. Project managers focus on optimizing coordination efforts or increasing interpersonal exchanges to enhance the consortium's collaborative capacity.

When it comes to project success, the challenge lies in the absence of a single method to evaluate it, as each partner has a unique approach to measurement, leaning towards either the efficiency or value paradigm [13]. Instead of favoring one perspective over the other, this research highlights the importance of considering both paradigms in assessing project success. As the project progresses, inherent conflicts and misunderstandings are likely to arise, necessitating governance adjustments. Each decision is influenced by the different, and sometimes opposing, interpretations of "success" among project partners, which increases the complexity of project coordination and cooperation.

1.3 Multiple Case Analysis

This study focuses on a multiple-case analysis of the different perceptions and implications of success in four collaborative R&D projects in Open Innovation and European projects (such as the Horizon 2020 program) (Table 1). These projects span various research fields, including urbanism (architecture), acoustics (engineering), modeling methods (engineering/mathematics), and smart cities/ consultancy (management). All projects exhibit high level of scientific production (more than 150 scientific outputs) and technology transfer products (more than 20 prototypes). None of the projects was affected by the COVID-19 pandemic. All of them had finished when the outbreak started. The empirical effort centered around the study of the project evolution and the leaders' experience in managing specific projects. The project itself serves as the unit of analysis. The objective was to narrow the analysis and maintain a focused approach, ensuring that the examination remained closely aligned with the specific unit of study.

The method path started by building a theoretical framework to perform the multiple-case study. A mixed approach was conducted to collect the references for the theoretical framework. The primary source of article resources was WOS (Web of Science). Searches were refined to find the most relevant references for each topic. Based on the initial searches, deeper searches were performed in the top journals on the topic. Like, International Journal of Project Management, Project Management Journal, Technovation, Project Leadership and Society, some of the journals in Academy of Management, and IEEE, Journal of Knowledge Management, Journal of Innovation and Knowledge, R&D Management, and so on. More than 150 articles were used, and a subset of them was selected for this chapter.

In addition to a deep analysis of the archival data of each case (see Table 1 for a list of the projects analyzed), project leaders and scientific leaders for each project were interviewed to understand how different success perspectives coexist in a collaborative project. Interviews lasted, on average, 45 min and were conducted from July 2021 to July 2022. A protocol was used as the starting point for the interview, and the discussion turned to different interests following interviewee's perceptions. Although their experience levels vary, all interviewees have over 15 years of experience working on R&D projects, qualifying them as senior project participants.

For exploratory studies, open-ended interviews are suitable for gaining a deeper understanding. They allow the researcher to delve into emergent insights rather than focusing on specific themes [15]. The data analysis in this chapter followed the methodology outlined by Corley and Gioia [16], which enables empirical comparison of models based on empirical data.

The findings from the case analysis are categorized into themes based on their alignment with existing literature. Through inductive analysis [16], the data facilitate a comprehensive exploration of both the pre-established framework and the new insights that emerged organically during the analysis process.

1.4 Layout of This Chapter

Following the introduction, Sect. 2 examines the nuances of project success and the evolving nature of Collaborative R&D Projects (CR&DP). Section 3 describes the dynamic character of CR&DP evolution. Section 4 discusses the elements necessary to achieve equilibrium and a stable outcome in CR&DP. Section 5 focuses on the design of the framework to understand CR&DP. Finally, Sect. 6 presents the conclusions and suggestions for further research.

2 Success Perception in the Analyzed Case Studies

Project leaders shared both perspectives on success perception in the case studies. According to the interviews, evaluating project success, whether based on efficiency or value perspectives, is a relevant issue in most projects. This evaluation is closely linked to the interests of the project partners. Some partners may prioritize delivering valuable outcomes, while others may focus on developing outputs in the most cost-efficient manner, using minimal resources [13]. Strategic decisions depend on how partners measure project results, which can vary based on their expectations and impact on the project. Given the diverse range of interests and expectations in collaborative projects, there is a corresponding variety in what success means for each partner, whether it pertains to the efficient development of outputs or the valuable delivery of outcomes.

For instance, in projects like GAMMA, the partners allocated resources upfront to deliver the initial scope (output) on time. From an efficiency perspective, this might be considered a success. However, there is no evidence that the outcomes of the project are being perceived by stakeholders (such as the European Commission) in a long-term process of value realization [17]. Based on the value paradigm (value added by the outcomes), this project cannot be considered a success, according to the data collected.

> Considering the GAMMA, there were more than 25 partners, and none knew what the other ones do. Each one of them were focused on delivering what it was supposed to deliver. I don't have any idea if someone is perceiving the benefits of the outcomes produced to the project. But, I think all the partners dedicated the time and resources agreed on the contract to deliver "something"; What we delivered is providing a differential value to the society? I have no idea; and I don't think so (GAMMA project manager).

On the other hand, in projects such as LOCUTIOS and EPSILON, key partners had to invest more time and resources than initially allocated according to the efficiency perspective. These additional resources were not compensated by the sponsor (European Commission). Despite this, they could deliver the expected results on time by investing more resources. From the value perspective, both projects can be considered successful, as several organizations are perceiving and capitalizing on their outcomes. For example, they have achieved a high number of citations and received recognition and awards from the scientific community [3, 6].

> If you ask to me, I have to say we have spent much more resources than we expected to and also, more than what we were allowed to. But at the end, we were able to finish the project on time, and accomplishing 98% of the agreed scope. If we consider the value that we added to the scientific community, we have to admit that the LOCUTIOS was an example of success. (LOCUTIOS project manager).

> If we follow the number of citations that the EPSILON project has, it can be considered as one of the most successful projects developed for us, according to the scientific recognition. I know there are also several projects that continue the line that we had created, even though it is not associated with citations. But, I (as a scientific coordinator) and the main partners, have dedicated more resources (extra hours) that had not been economically compensated (EPSILON project manager).

In summary, the challenges in balancing value versus efficiency underscore the significant implications of this issue. The diversity of success perspective interests and involvement exists not only in evaluating project results, which is the primary unit of analysis [18], but also in how partners perceive success based on their interests and involvement in the project [13]. This adds a managerial challenge, as decisions are made collectively by the partners. Each partner perceives the implications of these decisions on the project results differently, with project success holding varying meanings for each of them. This variation makes aligning partners—whether with the efficiency or value perspective—more complicated.

3 Dynamic Evolution of Collaborative R&D Projects

This research highlights that leadership in Collaborative R&D Projects (CR&DP) requires a specific management perspective often overlooked by standard project management practices. CR&DPs face unique challenges compared to traditional projects, such as fostering collaboration among stakeholders, ensuring a shared vision of project outcomes, balancing value and efficiency, and moving beyond a deterministic approach centered on a single sponsor. In summary, this research assumes that the co-creation of project outputs and outcomes follows a dynamic evolution influenced by these challenges [1–3, 6, 7, 11–13, 18, 19].

Several components influence the management practices that should govern this dynamic evolution. This section describes three main components to understand the dynamics in the evolution of a CR&DP from a project management perspective. Firstly, leadership and governance must include actions towards enhancing collaborative capabilities among partners. Secondly, partners' diverse experiences and cultural backgrounds create divergent interests and orientations, leading to higher levels of uncertainty and increased equivocality. Finally, the interplay between leadership and diversity must adapt to the changing dynamics of project evolution to achieve the project outcome.

Each of these components and their interactions are delineated in the rest of this section.

3.1 Governance and Leadership

Governance has been defined in various ways in the literature. In this study, we adopt the perspective of Gulati and Nickerson [20], which views governance implications through the lens of relational and contractual mechanisms' complementarity or substitution effects [21]. Evidence supports that relational and contractual governance can function as complements (enablers/compensators), leading to improved outcomes when combined [6, 9, 22]. Conversely, better results might be achieved when one mechanism dominates the other, functioning as either/or [21].

Recent research works [23, 24] reported that in co-exploitation projects, where contextual uncertainty and output novelty are lower, formal governance leads to better outcomes compared to relational governance. Similarly, additional outcomes show that during the exploitation phase of R&D collaborative projects, contractual governance yields superior results over relational governance, and the lack of contractual agreements negatively impacts project outcomes [23, 24]. In exploration phases, where uncertainty and novelty are higher, relational governance facilitates better adaptation to unpredictable issues, resulting in improved outcomes [24]. During these phases, the combination of relational norms and contractual agreements can sometimes create additional ambiguity, potentially leading to partner confusion, conflicts, or mismatched expectations, thereby negatively affecting project results [23, 24]. In such cases, substitution effects have been linked to better results in project success measures.

However, additional research efforts [22] identified positive effects from the complementary role of governance approaches, serving as enablers/compensators in public–private partnerships. They argue that while contractual agreements are essential for regulating partner behaviors, they are insufficient on their own to ensure partner involvement. Relational approaches are necessary to translate contractual provisions into actual partner contributions. This relationship, characterized by mutual causality, is linked to improved project outcomes when present [22].

Leadership is one of the most extensively studied topics in organizational sciences, particularly in the context of collaborative projects where exploration and exploitation activities alternate throughout the project [23]. The primary demands for innovation involve both exploration and exploitation, as well as the flexibility to transition smoothly between these two opposing activities [25]. In this regard, while leaders possess competencies for a bureaucratic paradigm that emphasizes efficiency in project management, new competencies are necessary to navigate innovative and collaborative environments [25]. Leadership focuses on the evolution of interactions and informal relationships among participants [26] and enables a project to adapt to the varying (and competing) knowledge demands. Adaptations in leadership are linked to better project outcomes in collaborative contexts [27].

Contractual governance and its implications are closely tied to coordination, a close leadership style, efficiency, and, therefore, exploitation activities. In contrast, relational governance and its implications are strongly linked to cooperation, an open leadership style, value addition, and, consequently, exploration activities [23].

3.2 *Diversity and Equivocality in Collaborative Contexts for Innovation*

The diversity among partners in CR&DP, arising from cross-cultural backgrounds, multidisciplinary experiences, varied expertise, and comprehensive competence training at multiple levels, leads to situations of equivocality. Managing equivocality is a central challenge in CR&DP [28], particularly in the context of innovation [1, 13].

In recent years, equivocality has been conceptualized as a double-edged sword in collaborative projects [28]. On the one hand, it fosters divergent and competing interpretations, which can enhance the recombination of knowledge and facilitate the exploration of different possibilities [13]. This also encourages inter-organizational teams to engage in discussions that challenge existing understandings of information and tasks [28]. On the other hand, high levels of equivocality can negatively impact project results [1, 13]. It can hinder the establishment of a common understanding among project participants regarding development needs, project goals, and subsequent procedures [1, 3, 10]. This lack of alignment and clarity, arising from conflicting interpretations, can persist throughout the project lifecycle, potentially leading to undesirable outcomes [28].

Predictive Project Management approaches, which are *phase-project-planning* oriented, follow the paradigm of regulation and control [11]. Situations of equivocality might affect triple constraint objectives, and methodologies focused on planning steer the project back to its initial plan and stability [11]. In this sense, predictive approaches, with a high emphasis on planning, regulation, and control, help project managers reduce equivocality situations, keeping the project closer to the initial plan. When faced with equivocality, teams tend to exploit existing knowledge and understanding of reality to return to system stability [11, 28]. However, this can be detrimental to the emergence of novel and unexpected outcomes, as exploration is the main vehicle for managing equivocality.

3.3 *The Interplay Between Governance and Equivocality in CR&DP*

Despite the independent significance attributed to both constructs (equivocality and governance) and the extensive body of knowledge surrounding them, they are intricately interdependent in CR&DP. Most research has focused on examining either equivocality or governance and their respective impacts on project outcomes. However, we assert that each of these constructs not only influences project outcomes but is also influenced by the behaviors of the other [19].

On the one hand, contractual agreements play a crucial role in clarifying roles and responsibilities, thereby reducing the range of potential interpretations and minimizing equivocality [29]. Relational approaches facilitate knowledge exchange and deeper communication, which are necessary to align different perspectives and provide a common understanding, thus reducing equivocality [30]. Relational exchanges, such as face-to-face meetings, significantly foster trust by promoting a deeper understanding and creating an environment conducive to collaboration and consensus [21]. In this sense, the implications of equivocality on project outcomes are positively moderated by the governance approaches employed to manage it. Positive implications are encouraged by more relational approaches, while negative implications are mitigated by either relational or contractual governance.

On the other hand, relying on contractual agreements might limit the emergence of novel or innovative solutions, reducing the flexibility and creativity needed to solve unknown issues [13]. Specific contractual agreements can constrain the emergence of different perspectives, thereby diminishing the positive implications of equivocality on innovativeness. In uncertain situations, contractual agreements might also become a source of misunderstandings, as partners may interpret the unknown situation and its contractual implications in very different ways [3]. Thus, contractual agreements serve not only as a means to mitigate equivocality but also as a source that amplifies the diversity of conflicting interpretations, increasing tensions between partners.

Furthermore, increasing communication through relational approaches might also be detrimental, as tensions between partners could be exacerbated, leading to a negative loop of arguments and counterarguments. This increases the time and resources invested in communication while reducing the likelihood of reaching common agreements [6]. Both contractual and relational approaches might therefore exacerbate equivocality issues and their negative implications.

In summary, equivocality and governance approaches are closely interconnected, each influencing and being influenced by the behaviors of the other. This relationship is mutual and dynamic rather than static or linear, producing varying outcomes based on the evolution of their interaction [19].

3.4 A Value Perspective to Challenge the Dynamic Evolution of CR&DP

In collaborative projects, understanding the implications of the trade-off between equivocality and governance approaches on project results relevant for several reasons.

Firstly, the primary objective of collaborative projects is to foster innovation [1]. Evaluating the impact of equivocality and governance solely on output or outcome success measures is insufficient. Thus, comprehending the evolution of this relationship and its implications for project results, assessed from various success perspectives, remains an underexplored area deserving further investigation.

Secondly, most knowledge in project management is framed within the operational perspective (efficient development of outputs) rather than the strategic perspective (value delivered by outcomes) [11]. Many innovative projects fail to meet stakeholder expectations [21], partly due to the heavy reliance on plan-oriented techniques that prioritize reducing equivocality, often at the expense of novel outcomes. This affects stakeholders' value assessments. Further research is needed to explore the potential positive implications of equivocality on project results and how governance can support this [28].

To navigate towards a value perspective in the dynamic evolution of CR&DP, teams and project managers must develop new competencies. This involves inter-organizational teams learning to collaborate effectively [29]. These learning capabilities must be developed throughout the project lifecycle, both at the strategic (top-level) and operational levels within each interconnection (coupling), requiring distinct learning processes for different levels [31].

Additionally, we hypothesize that learning behaviors can help inter-organizational teams overcome equivocality situations through governance implementations [32]. To fully grasp the complexity of the issue, it is essential to consider the interplay between equivocality situations, governance, and the learning behavior of the involved organizations.

3.5 Organizational Learning

Given that partners possess diverse backgrounds and varying interests in the project, aligning their expectations becomes a challenge. This necessitates that partners and coordinators develop mutual understanding and foster relationships to successfully create joint outputs. Learning capabilities enable partners to transfer individual knowledge into a collective knowledge output and build trust among themselves [31]. Different learning behaviors and actions occur depending on the coupling, whether at the strategic level, inter-organizational, or intra-organizational level.

> The key success factor is to gain the respect and authority of the partners (as a coordinator), since the first minute until the end of the project, and that creates trust among all the consortium of partners (EPSILON Project Manager)

Bridging knowledge gaps among partners without sharing previous experiences also involves learning. As the project progresses, it is expected that the learning capacity will enable teams to bridge these gaps and improve collaboration.

> it was a matter of learning. It was like the intensity of the collaboration gradually increased toward the end of the project and say, when we saw that the deadline was approaching, it was like an incentive (LOCUTIOS Work Package leader).

Organizational learning is defined as the process through which organizations change or modify their mental models, rules, processes, or knowledge, thereby maintaining or improving their performance [33]. This capability is crucial in all types of projects, but it is even more important in collaborative contexts where multiple partners may lack prior experience working together [34]. This research assumes that organizational learning moderates the relationship between equivocality and governance. Different learning behaviors influence how the interplay between equivocality issues and governance approaches is managed during change situations [35].

The literature extensively discusses the relationship between equivocality, governance, and inter-organizational learning in a linear causal manner, following simple and deterministic cause-effect analysis. For instance, studies have shown that equivocality acts as a driver for slower learning processes [32] or that governance approaches serve as a vehicle to reduce equivocality [36]. However, in more unstable environments, these relationships become more complex, non-linear, and characterized by feedback loops [19].

We assert that the learning capacities of partner organizations in a CR&DP influence their reactions and responses to change situations, whether due to the self-organization of partner relationships or the emergence of new opportunities [12]. To navigate these change situations, organizations must adapt their governance approaches and manage the positive or negative implications of equivocality. We argue that the effectiveness of both processes is moderated by how organizations learn to overcome change situations.

According to Argyris and Schön [33] and their Organizational Learning approach, an organization might follow two types of learning processes to react to change situations: single-loop and double-loop learning. Single-loop learning involves adapting current methodologies and practices to meet the requirements of the change situation. In contrast, double-loop learning involves rethinking and redefining current methodologies and practices to address the underlying requirements of the problem [31].

Single-loop learning occurs when organizations adjust or fine-tune existing knowledge to address change situations without critically reviewing the underlying foundations of that knowledge [33]. In contrast, double-loop learning involves challenging and reconsidering the underlying assumptions about a concept, problem, or procedure, representing a deeper and higher-level learning process [31]. In both learning approaches, team learning moderates the interplay of governance and equivocality in CR&DP.

4 CR&DP Success. A Dynamic Equilibrium Through Convenient Couplings, Overcoming Uncertainty, and Recombination for System Evolution

The model described in the previous sections cannot be studied or understood through deterministic or Newtonian schemes. The multiple causality of the constructs that govern a CR&DP requires a systemic perspective. Therefore, in this section, we propose a systems theory perspective. Furthermore, the model exhibits dynamic evolution resulting from the interaction of all its components, which is essential for understanding CR&DP. A systemic vision allows us to study the CR&DP as a whole rather than being limited by the behavior of its individual components.

This work follows the theory of complex models to understand the systemic behavior proposed for a CR&DP, utilizing the complex adaptive system (CAS) model to conduct this study. This section introduces complexity theory to explore the study of CR&DP, outlines the components of CAS, and suggests how CR&DP can be viewed as a CAS. Doing so aims to enhance the understanding of CR&DP behavior and contribute to a better grasp of management practices that can influence a CR&DP.

4.1 Complexity Theory and Complex Models

Complexity theory has been proposed to study the behavior of phenomena requiring a systems perspective [37]. Within this domain, various theories and models exist, including chaos theory, wholeness theory, and complex systems [12]. Our approach is to analyze the dynamic evolution of a CR&DP through systems theory and use a complex system model to understand CR&DP behavior better. The concept of complex systems is rooted in physical, biological, and social systems [38]. Complex systems differ from merely complicated systems [37, 38]. If a system can be described by its individual constituents and their linear relationships (even if there are many constituents), it is a merely complicated system [38]. However, if the interactions among constituents are such that the system cannot be fully understood by analyzing its components, it is a complex system [39]. In simpler terms, a complex system consists of heterogeneous parts that interact in non-linear relationships, and the entire system evolves in ways that cannot be understood as merely the sum of its parts [37, 38]. The evolution of each part is influenced by non-linear interactions with other parts and environmental effects. Consequently, the entire system results from the dynamic and non-linear relationships among all parts of the complex system [12].

4.2 Complex Models and Project Management

Systems theory and complex models have been proposed as a perspective to address organizational science phenomena [37, 38] that are better understood as mutual non-linear dependencies among the parts of the whole organizational system. We suggest that using a systems perspective with a complex model should provide new insights into the behavior of a CR&DP. In recent years, a new wave of project management scholars has begun applying non-deterministic approaches to understand the complex nature of projects [8, 12]. This research aligns with this trend by considering collaborative projects as a complex phenomenon [39] and by using complexity theory to holistically capture the dynamic evolution of the three components: equivocality, governance, and organizational learning in a CR&DP.

The interplay between equivocality and governance, along with the moderating effect of organizational learning in collaborative R&D projects, is critical for project success [13]. Existing literature suggests revisiting conventional principles and considering the dynamic and emergent interactions of CR&DP components. By focusing on self-organization and self-transcendence, this approach aims to analyze such complex phenomena, moving beyond the deterministic approaches of traditional project management practices [12]. Applying non-deterministic approaches should provide a more comprehensive understanding of the non-linear, non-deterministic interrelationships between CR&DP constructs during the project's dynamic evolution.

This research suggests adopting a complex model approach to examine CR&DP as complex adaptive systems (CAS) to understand their dynamics. This approach follows four characteristics applicable to organizational phenomena [12, 13, 40].

1. Participant partners follow explicit schemata: Partners, acting as agents within the project system, adhere to cognitive structures that guide their behavior and activities. These structures consist of guidelines, rules, and practices that evolve over time and are influenced by the environment. Additionally, partners have their own interests and expectations regarding the project, which can be explicit or hidden [2]. Project managers lead the project by managing the interactions, interests, and expectations among partners.
2. Self-organizing networks of partners: Partners operate through interconnected networks that create feedback loops, shaping their collective behavior. In collaborative projects and their partner networks, this is particularly significant because decision-making is shared collectively among partners [1]. Unlike contractual relationships governed by customer-supplier agreements, no single partner has the formal authority to impose deadlines, objectives, or specific requirements [3]. Therefore, project members must find ways to organize themselves to address the various project needs.
3. Coevolution to the edge of chaos: The collective behavior and feedback loops among partners lead to their coevolution towards a dynamic equilibrium. This evolution demonstrates that the whole is greater than the sum of its parts, indicating the presence of emergence within the system. In collaborative projects,

change situations such as shocks or stressors [41] can push the system components to the edge of chaos or temporary instability, necessitating adaptation and learning to restore the system's stability [13].

4. Recombination, adaptation, and system evolution: Partners adapt their behavior to successful states. The system evolves over time through the addition, removal, or transformation of its components [12]. The system's components or agents are formed by recombining previously successful elements. This recombination allows the system to evolve and continue, adapting or transcending its initial state [12].

The CAS model has been applied in various areas, including testing megaprojects [12], managing portfolios [40], handling disruptions in R&D projects [13], and exploring project leadership [25]. This study uses the CAS model to analyze CR&DP, focusing on understanding the dynamics behind their success. The goal is to examine self-organization, emergence, and adaptation patterns that contribute to the evolutionary success of collaborative R&D projects.

4.3 Towards a Dynamic Equilibrium in CR&DP

Collaborative projects function in environments of high uncertainty, where managers often lack sufficient information for accurate estimations and predictions [1]. As a result, they frequently encounter pivotal decisions that affect both the project's ongoing viability and the interests of the involved partners, often without a clear understanding of the consequences. R&D projects funded by the European Commission are expected to be self-organizing, with partners adapting their actions according to the project leader's direction and the guidelines outlined in the Grant Agreement and the Description of Work from the funding agency. It is presumed that these documents provide all necessary guidelines for the project's evolution and that partner behaviors will align with a predefined pattern, where each partner collaborates with others to achieve the project's objectives [7].

In the context of CR&DP, project managers struggle to predict outcomes based on input interactions due to the inherent unpredictability [12]. The project plan often serves merely as a hypothesis for the final product, lacking clear guidelines on the methods and processes required to achieve it [42]. Ambiguous and shifting goals [4] combined with limited forecasting ability [42] result in CR&DP outcomes emerging from the dynamic and evolving interactions among partners, rather than being entirely predefined from the outset [12, 13].

In situations of equivocality, tensions between partners can be intensified [43], potentially leading to either positive or negative impacts on project outcomes, depending on the governance strategies employed to address these tensions [1, 13, 28]. However, shifts in governance approaches can also influence the level of equivocality, sometimes exacerbating or alleviating the situation, creating a counteracting effect [28]. Moreover, we argue that this reciprocal, dynamic, and evolving relationship is moderated by the team's learning capacity to manage change [38]. The learning behavior of the team directly impacts how the relationship between equivocality and governance evolves [32] and affects the outcomes of navigating change. Thus, project results (both outputs and outcomes) are shaped by the dynamic interplay between governance and equivocality, as well as the moderating influence of learning, rather than by each factor in isolation [19].

In such a challenging environment, positive leadership behaviors [25] play a crucial role. They foster trust between partners and project managers, particularly during conflict-laden decision-making processes [20]. This trust helps minimize misunderstandings that arise from varying interpretations. Positive leaders are adept at aligning diverse partner perspectives toward a common and unified goal [25–27]. Thus, effective project leadership also serves to synchronize partners and reduce misunderstandings.

> The role of the coordinator is relevant to align all the parties. You need to gain the respect of the partners and that would enable to gain their trust. (EPSILON Project Manager)
>
> It is the role of the coordinator to align all the expectations and understandings. During the project, you need to gain the partner's respect, and afterwards, trust will emerge. (GAMMA Project Manager).

However, not all ~~the~~ leadership styles might be suitable to deal with collaborative projects.

> "The leadership doesn't work based on authority, because everything is subjective and subject to interpretation. In this sense, if leadership would be ruled by authority, the whole project will explode, mainly due to EGO, different implications, etc." (…) "Our role (as coordinators) were to harmonize the workflows between the different groups" (LOCUTIOS Work Package leader)

Trust is not only built through positive relationships among partners but is also closely linked to the authority held by project coordinators or managers. Additionally, sharing common scientific interests and mutual commitment to the project contribute to reducing misunderstandings and fostering trust between partners.

> If both parties are willing to cooperate, they will overcome these misunderstandings easily and that is really nurturing. Science is the common language (LOCUTIOS Project Manager)
>
> When you respond according to the expectations you start gaining trust with the other partners. But, everything starts when you show results, when you show you are committed with the project. (EPSILON Project Manager).

All the cases studied show the need to adapt the leadership style applied in the project. This adaptation arises from the existence of balancing the tensions between the project partners and the occasions of equivocality and governance and depends on the characteristics of the scientific field and the level of interaction between the partners.

4.4 Self-organization in CR&DP Through Looking for Convenient Couplings

In this subsection, we aim to highlight how collaboration in CR&DP results in varying levels of interaction among partners. Partners seek to identify the most suitable collaborators by forming strategic connections with other participants in the project. Collaborative projects involve a network of interconnections among partners at multiple levels [5]. At the highest level, managerial decisions are made by the leaders of the principal partners. At the operational level, collaboration deepens among different organizations, addressing not only managerial aspects but also technical and methodological challenges [5].

It is important to consider this aspect because the nature of interactions and collaboration varies depending on the level of analysis [31]. Given that this research focuses on collaborative projects, the analysis is confined to two levels: collaboration at the strategic level (the top level) and the interconnections among different organizations at the operational level (couplings).

The term "coupling" refers to the joint activities and exchange of knowledge between independent organizational actors, characterized by both the intensity and strength of the connection [1]. The nature and intensity of collaboration and interactions are dependent on the level of these couplings.

> You have three levels: the macro level, where you see the overall vision among partners; then you have the micro level, either among partners internally or between work packages. Then there's the micro-micro level, which is internally within the team in your own organization (LOCUTIOS Work package leader)

The nature of cooperation varies across different levels. At the strategic level, team leaders focus on high-level management issues such as project objectives, resource allocation, and communication with sponsors. Interactions at this level are infrequent and less intense. In contrast, at the operational level, which involves inter-organizational relationships within or between work packages, interactions are more frequent and intense as partners address both managerial and technical issues. Here, partners must align their expectations, working methodologies, and backgrounds, and work to bridge knowledge gaps. These operational interactions are characterized by greater frequency and intensity compared to strategic ones. At the micro-level, team members within each organization engage in daily collaboration and coordination of activities, which is typically explained by traditional team management models.

> You have these different levels, and within each space, you have leadership. So, there was a structure with a different hierarchy depending on the partners. (GAMMA Project Manager)

This represents one of the most significant challenges in collaborative projects. Managerial issues are handled at the executive level, where achieving partner alignment through a shared objective is crucial. However, the intensity of collaboration increases at the coupling level, where diverse partners and their varied interests intersect not only in managerial but also in operational and technical aspects. This heightened intensity amplifies the complexity of interactions [5].

Examining the different levels of analysis in inter-organizational relationships among partners is essential for understanding a project's collaborative capacity and its success outcomes. Distinguishing between the project level and the coupling level is important for research, as situations of equivocality may impact each level differently.

Self-organization is characteristic of CR&DP projects, as it arises from opportunities for interaction and collaboration between partners. However, the way in which this interaction occurs requires different perspectives depending on the level at which the interaction occurs. For example, it can happen at a high or low level and can even arise from interactions between levels of the hierarchy. Projects with a clear hierarchical structure such as GAMMA, EPSILON or BMOI can develop an interaction adapted to the hierarchy. Whereas a project such as LOCUTIOS, where the project structure requires less hierarchy, may show a tendency to self-organize in a hierarchy different from the project primary.

4.5 *Uncertainty, Coevolution, and Innovation Drive to Emergence in CR&DP*

Understanding the evolving relationship between equivocality and governance is a promising research area with clear implications for project outcomes. This exploration becomes even more relevant during moments when unexpected issues arise. In projects marked by high uncertainty, the team's ability to navigate and adapt to unforeseen challenges is crucial, as these challenges are inherent to the project's nature [29]. When faced with uncertain and unforeseen situations, equivocality can be intensified. Each partner's unique background and interests lead to varied perceptions and interpretations of the uncertainty based on their individual perspectives. This can result in misunderstandings, both in assessing the causes of the unexpected issue and in deciding on the appropriate actions to address it [43].

> The important thing is to strive to reach the initial objective, but the key is to understand that many unexpected things can happen, forcing you to adapt. Change management becomes crucial. It's about finding ways to confront the unknown and still seek solutions to try to achieve the initial goal (EPSILON Project Manager)

Thus, investigating the implications of the interplay between equivocality and governance during change situations arising from uncertainty is highly relevant. To advance research in this field, it would be beneficial to shift the focus from the entire project lifecycle to specifically examine scenarios stemming from unexpected issues [43]. This focus is justified by the fact that CR&DP are especially susceptible to unforeseen issues, which increase equivocality and necessitate adjustments in governance approaches [41].

Managing equivocality emerged as a key challenge in the interviews, along with addressing unexpected issues arising from the high uncertainty inherent in these projects. This challenge was especially pronounced in collaborative projects aimed at advancing scientific knowledge or conducting basic research (Lower Technology Readiness Level - TRL).

> "In the world of ideas, everything works. But reality is revealed when problems arise, and no one knows exactly what or how to do." (...) "And in those cases, we fared well because we had solid knowledge foundations that enabled us to face issues derived from uncertainty" (GAMMA Project Manager)

> The important thing is to strive to reach the initial objective, but the key is to understand that many unexpected things can happen, forcing you to adapt. Change management becomes crucial. It's about finding ways to confront the unknown and still seek solutions to try to achieve the initial goal (EPSILON Project Manager).

> The risk in these projects is extremely high, and failure is the most common outcome. When you have Low Technology Readiness Levels (TRL), proposals are often very ambitious, resulting in a high failure rate (LOCUTIOS Work Package leader).

> You sort of are expected to deliver something that is in some sense predictable, which kind of runs against the grain of what it's usually considered to be basic research (EPSILON Project Manager).

In exploratory scientific projects (like LOCUTIOS, GAMMA, and EPSILON), the implications of uncertainty are even more pronounced compared to projects closer to market readiness or focused on knowledge exploitation [10]. As a result, these projects are particularly vulnerable to unexpected challenges arising from their intrinsic uncertainty. The success of such projects is closely tied to their ability to effectively manage unforeseen issues while continuing to generate outputs and deliver value [42]. In these contexts, the collective behavior of partners, positive leadership, and feedback loops foster coevolution among partners, guiding the system toward a dynamic equilibrium. This evolutionary process demonstrates that the whole is greater than the sum of its parts, and an emergent quality is evident in CR&DP [41].

4.6 *Recombination, Adaptation and System Evolution in CR&DP*

Drawing from the interdependencies and interactions highlighted in existing literature, we propose that mutual or circular causality could elucidate the relationship between equivocality, governance, and organizational learning during the dynamic evolution of CR&DP. Our research treats these three concepts as components of a complex system, characterized by heterogeneous elements that interrelate with each other and with their environment.

The relationships between partners operate under the principle of self-organization, with decisions made collectively, requiring partners to self-organize to address unforeseen issues [1]. As equivocality increases, adjustments in governance are required, initiating a feedback loop where mutual reactions either mitigate or exacerbate equivocality [19]. The system may approach the *edge of chaos*—a state of limited instability caused by the evolution of components or external factors/decisions—where the project must adapt or change to regain stability [37]. The edge of chaos represents a phase of transformation, where a complex system has the capacity to adapt or evolve, leading to system recombination and modification of its properties to meet environmental demands [38]. All the cases studied are exemplary of a transformation behavior to evolve towards environmental demands.

In summary, to grasp the evolution of the relationship between equivocality, governance, and learning during the development of a CR&DP, adopting a complex systems perspective is crucial. This approach allows researchers to comprehensively understand the intricate and non-linear interactions among components, where the outcomes are shaped by the interactions themselves rather than being a mere sum of the individual components [37, 38].

5 Project Management Practices for CR&DP, Leveraging the CAS Perspective

This section introduces a framework developed to describe CR&DP through the lens of complex theory, aiming to offer new insights into management practices for these projects. The proposed framework utilizes the Complex Adaptive Systems (CAS) model, with adaptations tailored to CR&DP. This adaptation considers the specific aspects of CR&DP, such as the trade-off between efficiency and value, and incorporates elements like dynamic evolution, governance, equivocality, and organizational learning. It also aligns mechanisms for achieving dynamic equilibrium, self-organization, emergence, and recombination and adaptation within the CAS model for CR&DP. The section concludes by highlighting its contributions to project management practices for CR&DP.

5.1 CR&DP Under the Lens of Complex Adaptive Systems

Building on the analysis from the previous section, CR&DP can be defined using the components of a CAS model. This research posits that framing CR&DP within the CAS model offers theoretical and methodological contributions to understanding project management practices for these projects. By applying complexity theory and complex models, this research advances the metaphor of complexity theory and provides practical applications for project management in CR&DP.

The following traits can define the contribution of the CAS model to understanding CR&DP practices:

- The perception of success in CR&DP cannot be confined to the efficiency paradigm alone. Partners, acting as agents, adhere to the schemata outlined in the Grant Agreement and the Description of Work provided by the funding agency. They work towards achieving success by navigating a dynamic equilibrium shaped by governance and equivocality, with project learning playing a moderating role.

While initial planning provides a foundation, maintaining dynamic equilibrium towards the success goal necessitates ongoing adaptation of the project plan, which in turn requires active involvement and positive leadership.

- CR&DP self-organization is driven by the interconnected networks of partners, which create feedback loops and shape collective behavior. This self-organization depends on factors such as the perception of success, couplings, uncertainty, and leadership, all of which are influenced by formal and relational governance at various levels for each partner. Additionally, the opportunities for coupling and the evolution of leadership impact the systematic revision of the project's governance structure and decision-making processes. Self-organization occurs at multiple levels—*macro, meso,* and *micro*—each corresponding to different decision-making layers within the project. At each level, effective self-organization requires that the control function accounts for the collaborative involvement of partners affected by the self-organization dynamics.
- CR&DP coevolution arises from situations of uncertainty and innovation opportunities, which are shaped by the interplay between governance and equivocality and moderated by partners' learning. These uncertain situations create emergent opportunities within the project, enabling partners to explore and capitalize on transformative possibilities. To effectively manage these emergent opportunities, partners need to collaboratively share planning decisions, ensuring that all participants benefit from the project's overall success.
- CR&DP recombination represents the final mechanism that maintains dynamic equilibrium and balances the trade-off between efficiency and value paradigms. Recombination involves managing the interaction between governance and equivocality, with partners' learning as a moderating factor. This process requires partners to adapt continuously to the evolving project landscape, guided by positive leadership. Effective leadership facilitates shared control over project management functions, enabling the project to progress towards success.

Table 2 shows a schema of the proposed framework.

5.2 Project Management Recommended Good Practices for CR&DP Based on a CAS Model

From a project management practices perspective, this research focuses on project's management planning and control functions [11]. It aims to explore how CR&DP management practices can benefit from the CAS model analysis conducted in the previous sections. The CAS perspective on CR&DP challenges the conventional project management paradigm, particularly questioning the traditional cybernetic-control model concerning the two main groups of functions.

The conditions of stability and equilibrium in a CR&DP diverge from the predictive *Plan-Do-Check-Act* model of project management. In CR&DP, partners are

Table 2 Main components of the framework for CR&DP through the lens of the CAS models

CAS model characteristic	Relevant behaviors in CR&DP	Action to develop and manage in CR&DP	Actionable project management practice
Dynamic equilibrium	Varying interpretation of the goals of the project Specific interests of partners Misunderstandings	Synchronize partners Foster trust among partners	Positive leadership
Self-organization	Network of interactions and collaborations among partners Varying nature of interactions and collaborations Macro, meso, micro level for interactions and collaborations	Joint activities and exchange of knowledge among partners Specific attention to "couplings"	Sustain a hierarchy view of interactions among partners
Emergence	Uncertain and unforeseen situations High uncertainty Partner's backgrounds lead to varied perceptions	Investigate implications of the interplay between governance and equivocality Coevolutionary process among partners	Design scenarios of potential evolution
Recombination	Unforeseen issues Difficulties in maintaining the balance between governance and equivocality Mutual or circular causality between governance, equivocality and organizational learning	Understand non-linear interactions Transform: adapt or evolve Tensions between mitigate or exacerbate circular causalities	Comprehensive understanding of the project system dynamics

actively engaged and committed to achieving a shared success goal. Positive leadership is paramount, and the management process cannot be solely governed by a bureaucratic project manager. These dynamics necessitate a reevaluation of the planning and control functions within CR&DP, emphasizing the importance of adaptive and collaborative approaches over rigid, conventional methods.

The planning functions

Given the interrelationships among partners in CR&DP, the planning function requires a particular focus on at least two key activities:

1. First, as CR&DP must adapt the project plan to achieve the success goal, a collaborative effort is essential to align all partners and maintain dynamic equilibrium. This requires balancing the trade-off between efficiency and value to reach the project's success goal.
2. Second, the planning function will face challenges when uncertainty and innovation opportunities lead to emergent outputs that could significantly impact the entire project. Therefore, the planning function should be a collective responsibility among partners, ensuring that all are involved in planning and addressing these emergent opportunities.

The control functions

According to the CAS model, the control function faces significant challenges under two main scenarios:

1. Self-Organization Opportunities: these opportunities arise at various decision-making levels within the project—macro, meso, and micro. At each level, effective control requires the collaborative participation of all involved partners. Without this collaboration, conflicting interests might undermine self-organization efforts and hinder the project's ability to adapt dynamically.
2. Recombination for System Evolution: as CR&DPs evolve, recombination activities—driven by adaptation needs—demand a shared approach to control functions. If control responsibilities are not collectively managed, divergent control practices may emerge, potentially jeopardizing the effectiveness of adaptation and evolution processes.

In both cases, ensuring collaborative control is crucial for facilitating self-organization and successful recombination, thereby supporting the project's ongoing evolution.

5.3 Collaborative Layer. Actionable Practices for Project Leaders in Real-World CR&DP

The CAS model approach for CR&DP assumes that project managers can develop actionable practices to overcome the *Plan-Do-Check-Act* approach by adding a specific collaborative layer that pays attention to the interplay of governance, equivocality, and organizational learning among partners.

The collaborative layer has to involve all partners and should promote a dynamic equilibrium perspective of the trade-off between efficiency and value. It should contemplate mechanisms of partner self-organization in the complete hierarchy of the project participants. It should expect emergence situations due to high uncertainty and prepare the project for recombination due to unforeseen issues.

For all that, this research proposes to consider a Collaborative layer in leading CR&DP that should allow project leaders to include project management practices (refer to Table 2, third and fourth columns).

- Positive leadership approaches to foster trust, synchronize partners interests, and harmonize common goals
- Make interactions and collaborations among partners to flourish and be shared by the project as an asset of the project. Interactions and collaborations can appear at all levels of the hierarchy.
- Contemplate different scenarios for the evolution of the interplay among partners due to uncertain and unforeseen situations. Scenarios should be driven by the varied perceptions due to partners' backgrounds.
- Build a systems perspective of comprehensive understanding of the non-linear interactions among partners. Promote feedback loops that facilitate transformation when unforseen issues are detected.

In the previous paragraphs of this section, a set of practices is proposed derived from a CAS model for CR&DP. The term Collaborative layer has been coined to identify them. These practices are derived from the reflection on the CAS model for CR&DP and are proposed as actionable practices that have practical value for a CR&DP. Although CR&DP project leaders should feel free to consider their application in their projects, most CR&DP results from funding from organizations such as the European Commission (EC) or can be based on the application of project management standards such as the PMI. At this point, this research pay attention to the promoters or policymakers of CR&DP projects. In this sense, funding agencies and institutions promoting standards in project management should be able to consider how to adapt their frameworks to help project leaders to follow a project management approach that could adapt better to the peculiarity of CR&DP.

In this sense, the EC promotes frameworks for funding projects that include aspects such as social impact, risk analysis of the project, and responsible research and innovation. All of them are in addition to the basic project management practices of the project. Also, organizations such as the PMI evolved the perspective of their good practices from the processes that make up project management to the fundamental principles for this management. With this, aspects such as the team, the system vision, value, and leadership are introduced.

The vision derived from the CAS model could be incorporated into the management frameworks promoted by the EC or the standards derived from PMI's good practices. In this way, the proposed Collaborative layer could be included in the formal description to accommodate a vision derived from the CAS model.

6 Conclusion, Limitations and Future Research

Collaborative R&D projects face numerous challenges in defining project success. Tensions arising from governance and equivocality—whether due to imprecise requirements or cultural diversity among partners—prompt self-organizing actions that align individual partners' aspirations with the overall project success.

To gain a deeper understanding of how partners interact towards a common goal, this chapter employs a complex theory approach to propose a framework grounded in the complex adaptive systems (CAS) model. This framework offers a novel perspective on the relationships among partners. Using a multiple case study analysis, the chapter identifies specific characteristics related to CR&DP through the lens of CAS models.

The primary contribution of this chapter is developing a novel framework for CR&DP based on the key components of a complex adaptive systems (CAS) model: self-organization, emergence, and recombination. By examining various subsystems within CR&DP—such as governance, equivocality, and organizational learning—the chapter proposes alternative patterns for project management practices, particularly concerning the planning and control functions that diverge from traditional approaches. The term Collaborative layer has been coined to identify this framework.

Additionally, it reframes the trade-off between efficiency and value in CR&DP not as a limitation but as a crucial aspect that can be better understood through CAS models. The chapter also offers new insights into the theory of emergence in project management, enhancing understanding of the nuances in CR&DP management practices.

From a practitioner's perspective, the Collaborative layer offers project leaders valuable insights into adapting classical planning and control paradigms to suit CR&DP. Specifically, they should focus on fostering collaboration among all partners and ensuring that planning and control functions are shared collectively. Furthermore, positive leadership should be recognized as a crucial transformative element in CR&DP, and project management should avoid relying on bureaucratic approaches. Moreover, hierarchical interactions among partners should contemplate self-organization opportunities, scenarios of coevolution among partners should help face highly uncertain situations, and a systemic perspective should help understand tensions between mitigating or exacerbating circular causalities. Furthermore, the Collaborative layer should help policymakers and promoters of standard practices for project management to complete their frameworks to include the subtleties of CR&DP.

The limitations of this chapter stem from the specific cases examined, all of which involve research projects funded by the European Commission. Since various entities can sponsor collaborative R&D projects, future research should include case studies from different sponsors to gain a more comprehensive understanding of CR&DP behavior through the lens of CAS models. Also, the analysis performed in this chapter has yet to distinguish the specific characteristics of each project. Some of the characteristics of the project suggest a different behavior within the Collaborative layer. These limitations provide opportunities to improve the Collaborative layer further and adapt to the facets of the distinct characteristics of real-world projects. In this vein, the TRL level of the project and a specific perspective for the different levels of the partner (top-level, scientific level, and operational level) should be considered in further research efforts.

References

1. Klessova, S., Engell, S., Thomas, C.: Dynamics of couplings and their implications in inter-organizational multi-actor research and innovation projects. Int. J. Project Manage. **40**, 547–565 (2022)
2. Malherbe, M.: Cooperating in interorganizational innovation projects: toward a better understanding of coupling with the permanent ecosystem. Int. J. Project Manage. **40**(8), 871–885 (2022)
3. Barbosa, A.P.F.P.L., Salerno, M.S., de Souza Nascimento, P.T., Albala, A., Maranzato, F.P., Tamoschus, D.: Configurations of project management practices to enhance the performance of open innovation R&D projects. Int. J. Project Manage. **39**(2), 128–138 (2021)
4. Gama, F., Sjödin, D.R., Frishammar, J.: Managing interorganizational technology development: project management practices for market- and science-based partnerships. Creativity Innov. Manage. **26**(2), 115–127 (2017)
5. Klessova, S., Thomas, C., Engell, S.: Structuring inter-organizational R&D projects: towards a better understanding of the project architecture as an interplay between activity coordination and knowledge integration. Int. J. Project Manage. **38**(5), 291–306 (2020)
6. Nisula, A.M., Blomqvist, K., Bergman, J.P., Yrjölä, S.: Organizing for knowledge creation in a strategic interorganizational innovation project. Int. J. Project Manage. **40**(4), 398–410 (2022)
7. González-Piñero, M., Páez-Avilés, C., Juanola-Feliu, E., Samitier, J.: Cross-fertilization of knowledge and technologies in collaborative research projects. J. Knowl. Manag. **25**(11), 34–59 (2021)
8. Laursen, M., Svejvig, P.: Taking stock of project value creation: a structured literature review with future directions for research and practice. Int. J. Project Manage. **34**(4), 736–747 (2016)
9. Wu, A., Wang, Z., Chen, S.: Impact of specific investments, governance mechanisms and behaviors on the performance of cooperative innovation projects. Int. J. Project Manage. **35**(3), 504–515 (2017)
10. vom Brocke, J., Lippe, S.: Managing collaborative research projects: a synthesis of project management literature and directives for future research. Int. J. Project Manage. **33**(5), 1022–1039 (2015)
11. Daniel, P., Daniel, C.: Complexity, uncertainty and mental models: from a paradigm of regulation to a paradigm of emergence in project management. Int. J. Project Manage. **36**(1), 184–197 (2018)
12. Daniel, E., Daniel, P.: Megaprojects as complex adaptive systems: the Hinkley point C case. Int. J. Project Manage. **37**(8), 1017–1033 (2019)
13. Nihoul, J.A., Miralles, F., Neamtu, L.: Understanding response to perturbations in collaborative projects from a complex generative and adaptive systems perspective. Proj. Leadersh. Soc. **4**, 100106 (2023)
14. PMBOK Guide. Seventh version (2021)
15. Corbin, J.M., Strauss, A.: Grounded theory research: procedures, canons, and evaluative criteria. Qual. Sociol. **13**(1), 3–21 (1990)
16. Corley, K.G., Gioia, D.A.: Identity ambiguity and change in the wake of a corporate spin-off. Adm. Sci. Q. **49**(2), 173–208 (2004)
17. Smyth, H., Lecoeuvre, L., Vaesken, P.: Co-creation of value and the project context: towards application on the case of Hinkley Point C Nuclear Power Station. Int. J. Project Manage. **36**(1), 170–183 (2017)
18. Ika, L.A., Pinto, J.K.: The "re-meaning" of project success: updating and recalibrating for a modern project management. Int. J. Project Manage. **40**(7), 835–848 (2022)
19. Majchrzak, A., Jarvenpaa, S.L., Bagherzadeh, M.: A review of interorganizational collaboration dynamics. J. Manag. **41**(5), 1338–1360 (2015)
20. Gulati, R., Nickerson, J.A.: Interorganizational trust, governance choice, and exchange performance. Organ. Sci. **19**(5), 688–708 (2008)
21. Cao, Z., Lumineau, F.: Revisiting the interplay between contractual and relational governance: a qualitative and meta-analytic investigation. J. Oper. Manag. **33–34**, 15–42 (2015)

22. Benítez-Ávila, C., Hartmann, A., Dewulf, G., Henseler, J.: Interplay of relational and contractual governance in public-private partnerships: the mediating role of relational norms, trust and partners' contribution. Int. J. Project Manage. **36**(3), 429–443 (2018)
23. Solís-molina, M., Hernández-espallardo, M., Rodríguez-orejuela, A.: Governance and performance in co-exploitation and co-exploration projects. J. Bus. Indus. Mark. **5**, 875–894 (2020)
24. Henttonen, K., Hurmelinna-Laukkanen, P., Ritala, P.: Managing the appropriability of R&D collaboration. R&D Manage **46**(S1), 145–158 (2016)
25. Swart, M.: Complexity Leadership Theory: An Exploratory Study in the South African Context. Stellenbosch University, South Africa (2023)
26. Uhl-Bien, M., Arena, M.: Leadership for organizational adaptability: a theoretical synthesis and integrative framework. Leadersh. Quart. **29**(1), 89–104 (2018)
27. Arena, M.: Adaptive space. Manage. Bus. Rev. **01**(01), 273 (2021)
28. Eriksson, P.E., Patel, P.C., Sjödin, D.R., Frishammar, J., Parida, V.: Managing interorganizational innovation projects: mitigating the negative effects of equivocality through knowledge search strategies. Long Range Plan. **49**(6), 691–705 (2016)
29. Yang, X., Wang, L., Zhu, F., Müller, R.: Prior and governed stakeholder relationships: the key to resilience of inter-organizational projects. Int. J. Project Manage. **40**(1), 64–75 (2022)
30. Sakka, O., Barki, H., Côté, L.: Relationship between the interactive use of control systems and the project performance: the moderating effect of uncertainty and equivocality. Int. J. Project Manage. **34**(3), 508–522 (2016)
31. McClory, S., Read, M., Labib, A.: Conceptualising the lessons-learned process in project management: towards a triple-loop learning framework. Int. J. Project Manage. **35**(7), 1322–1335 (2017)
32. Marcandella, E., Guèye, K.: Tensions in collaborative innovation projects and higher-level learning. Learn. Organ. **25**(4), 248–259 (2018)
33. Argyris, C., Schön, D.: Organizational learning: a theory of action perspective (77), 345–348 (1997)
34. Bäck, I., Kohtamäki, M.: Joint learning in innovative R&D collaboration. Ind. Innov. **23**(1), 62–86 (2016)
35. Mu, T., Yang, J., Zhang, F., Lyu, C., Deng, C.: The role of task conflict in cooperative innovation projects: an organizational learning theory perspective. Int. J. Project Manage. **39**(3), 236–248 (2021)
36. Rizova, P.S., Gupta, S., Maltz, E.N., Walker, R.W.: Overcoming equivocality on projects in the fuzzy front end: bringing social networks back in. Technovation **78**(March 2017), 40–55 (2018)
37. Anderson, P.: Complexity theory and organization science. Organ. Sci. **10**(3), 216–232 (1999)
38. Desai, D.A.: Co creating learning: insights from complexity theory. Learn. Organ. **17**(5), 388–403 (2010)
39. Arranz, N., de Arroyabe, J.C.F.: Joint R&D projects as complex systems: a framework of analysis. IEEE Trans. Eng. Manage. **55**(4), 552–565 (2008)
40. Aritua, B., Smith, N.J., Bower, D.: Construction client multi-projects—a complex adaptive systems perspective. Int. J. Project Manage. **27**(1), 72–79 (2009)
41. Naderpajouh, N., Matinheikki, J., Keeys, L.A., Aldrich, D.P., Linkov, I.: Resilience and projects: an interdisciplinary crossroad. Project Leadersh. Soc. **1**(April), 100001 (2020)
42. Wied, M., Koch-Ørvad, N., Welo, T., Oehmen, J.: Managing exploratory projects: a repertoire of approaches and their shared underpinnings. Int. J. Project Manage. **38**(2), 75–84 (2020)
43. Burström, T., Wilson, T.L.: The texture of tension: complexity, uncertainty and equivocality. Int. J. Manag. Proj. Bus. **11**(2), 458–485 (2018)

Simplified ISO 21502 Project Management Applied in a Fintech Case Study

Kenneth David Strang

Abstract This chapter provides best practices by illustrating how a simplified ISO 21502 project management (PM) framework was applied in a New York, USA-based financial-technology (Fintech) industry case study. According to the literature, the rationale was based on the established metric that over half of global projects have failed. The research goal was to improve the body of knowledge to assist stakeholders around the world, particularly emerging PM practitioners, by sharing simplified PM best practices. The research question was how and why a simplified ISO PM framework was applied in a Fintech case study using several critical PM tools, techniques, artifacts, theories, and models. The ISO 21502 standard was leveraged to describe the best practices because this standard is international and accessible to all practitioners without any potential bias towards a specific PM association or country.

Keywords ISO 21502 · Project management framework · Fintech industry · Best practices · Case study · Project failure rates · Emerging PM practitioners · PM tools and techniques · Stakeholder engagement · Global project management

1 Introduction

ISO 21502 is based on the Project Management Institute (PMI) Project Management Body of Knowledge (PMBOK) fifth edition, which focused on how a project manager (PM) can apply tools, processes, and techniques across the ten project management knowledge areas [1]. PMI [2, 3] significantly revised the PMBOK in the seventh edition to focus on how a PM applies 12 project management principles across eight project performance domains using models, methods, and artifacts. There are

K. D. Strang (✉)
W3-Research, New York, NY 11041, US
e-mail: professor@kennethstrang.com

University of the Cumberlands, Williamsburg, KY 40769, US

RMIT University, Melbourne, NSW 3001, Australia

K. D. Strang and N. R. Vajjhala (eds.), *International Program and Project Management—Best Practices in Selected Industries*, Information Systems Engineering and Management 31, https://doi.org/10.1007/978-3-031-80275-1_4

a dozen other project management frameworks or approaches, including Agile, and these books may cover any of them [1, 2, 4–11].

1.1 Rationale for Stakeholders: High Global Project Failure Rate

With over five decades of evidence, the global project failure rate is consistently high. The PM body of knowledge has failed stakeholders. Statistically, a sponsor could use a coin toss with a 50/50 probability to predict the success or failure of a proposed project. Academic scholars Borbath et al. [4]; Eckerd and Snider [5]; Pace [6]; Shoar et al. [7] argued there was longitudinal evidence pointing to a consistently high project failure rate of approximately 50%. In some industries, the failure rate was even higher than 50%—Masticola [8] claimed the IT software project failure rate was 71%.

If that is not enough evidence of the problem, there is more. The best-known and highly cited global source of information technology (IT) industry project performance is the U.S.-based Standish Group. The longevity of their data and sample size makes these project performance statistics reliable, credible, and valid to discuss. As described by Kurek et al. [9], Standish Group International has developed a 'Chaos database' to track IT project success and failure since 1984. The Chaos database has over 120,000 registered IT projects from over 1000 organizations [9]. Standish claimed that 53% of IT projects ended significantly over budget, and many missed their scheduled deadline, while 18% outright failed to meet the sponsor's cost, time, scope, and quality requirements [8, 9]. In a separate empirical study, Pace [6] found that, on average, most organizations had at least one project failure in the previous year. More so, Masticola [8] asserted that IT project failures have caused the underlying organization to experience financial constraints and lose shareholder confidence. The IT project failure rate was 41% in the American defense industry based on a retrospective analysis of 2692 projects [10].

Despite all the empirical research, scholars have yet to identify precisely what is causing projects to fail beyond identifying a few predictors with small effect sizes ranging from 2 to 6% [11]. One interesting study in 2022 identified professional certification as a predictor of project success [10]—the three-factor model composed of project manager experience, project manager certification, and contractor ISO quality certification yielded a significant 12% effect size. This suggests that the project management body of knowledge learned through professional certification could help project managers improve the likelihood of project success. Nevertheless, a 12% impact to improve the chance of success is far from what decision-makers would expect when evaluating multi-million-dollar budgeted projects. The authors assert better project management guidance is needed.

Additionally, the authors argue the entire PM profession is motivated to improve the body of knowledge and increase the discipline's reputation. That is why we wrote these books. The underlying research question was to develop a simplified adaptation of the universal ISO project, project, and portfolio management guidelines, with examples, to help practitioners and decrease the global project failure rate.

1.2 Rationale for Practitioners: Guidance Complexity

This author asserts that the various project management guidelines need to be simplified, theoretical, and practical examples, especially for specific industries. There is almost a dozen recognized project management frameworks published in the literature, including the UK's Association of Program Management (APM) and the well-known Project Management Body of Knowledge (PMBOK) controlled by the USA-based Project Management Institute (PMI). The PMBOK complexity is apparent. At the time of writing, the PMBOK had over 420 pages, containing 47 processes, written like scientific assembly instructions, and the document is a moving target because a new version is published every four years [3].

The International Standards Organization (ISO) is the go-to source for global, formal, and international standards, including project, program, and portfolio management. Nevertheless, the ISO guidelines are also complicated, written like shorter scientific assembly instructions, and the documentation is currently spread over nine volumes! ISO has extended the 21500:2012 version three times to 21502 with the eight supplemental guideline documents as of the current writing [12, a–i].

More than one scholar has reiterated that the PMBOK, ISO, and alternative standards are overly complicated, too theoretical, and lacking practical examples [4–10]. Čabarkapa [1] recently compared the PMBOK 6th edition (2019) with the ISO 21500 (2021), which is an exciting paper demonstrating the similarity and complexity of PMBOK and ISO on a component-by-component basis. Neither guideline provides industry or professional practical examples [1, 2]. The author of this chapter is a licensed project management professional (PMP) from PMI (initially certified in 1996 and currently recertified) with many years of applied project experience since 1983. The author claims ISO 21502 and the PMBOK [3] are complicated for emerging PM practitioners, and the PMBOK lacks examples.

The above shortcoming of overly complex PM guidance further substantiates the primary research question driving this book. The research question was to develop a simplified interpretation of the ISO project management guidelines, using a Fintech case study to help practitioners and decrease the global project failure rates.

2 Simplified Universal Project Management

First, the universal project management (UPM) model is briefly explained. The UPM model consists of a diagram and brief explanations of the model. The authors have leveraged their practical experience (as discussed above) to develop the UPM model.

ISO 21502 is leveraged and cited as the basis, where applicable, to facilitate the dissemination, understanding, and adoption of these concepts. Since ISO adapted its 21502 guidelines from the PMBOK and other sources, there is compatibility across the common bodies of knowledge.

Second, the UPM model is adapted, recast, and transformed into a workflow and applied to an industry case study, cited from the literature. This transformed workflow is presented as a table, showing how and why the UPM activities were undertaken by the project manager. This practical approach is known as experiential learning rather than a theoretical lecture. A simple project is used in this paper—later chapters will explore other essential PM concepts applied in selected industries or professions.

2.1 ISO 21502-Grounded Project Management Phases, Functions, Activities

The significant UPM difference is in control and risk management. Also, scientific, legal type numbering was eliminated in the UPM (compared to the PMBOK and ISO)—the UPM model is designed to be intuitive and not an assembly schematic.

The UPM 41 PM activities contrast the 47 or so in the PMI PMBOK and are like other industry-specific PM bodies of knowledge. It could also be mentioned in these books that the term phase corresponds to 'process group' in ISO or 'cycle' in PMI. Similarly, in these books, the term PM function or activity function group is identical to the word 'subject' in ISO 21500, like the 'knowledge area' label from the PMI. Another difference between UPM ISO and PMBOK is that the UPM model has four phases (initiate, plan, execute, close), whereas the ISO and PMBOK have five process groups or cycles. The author asserts that only four phases are practically needed because in ISO, the controlling process group, and in PMI's PMBOK, the monitoring process cycle is not sufficient because they are disjointed from the ongoing PM functions during the execution phase. The reason for there being only four phases in the UPM model is that once a project is initiated and planned, then the PM must focus on the implementation of the project, which requires both applying the identified tasks, managing the team and monitoring or periodically checking for specific risks—risk management is not a by-stander activity waiting for an event to happen or an occasional task—it is ongoing and regular. The author asserts that the lack of integration of risk into the PM daily task execution could be a factor impacting the high project failure rates.

In Fig. 1, the PM activities are grouped into the PM functional groups, starting with the most important at the top left: Scope, time, cost, quality, risk, stakeholder, integration, communication, resource, and procurement. Each PM activity falls into one of the four project phases: Initiate, plan, execute, and close. Intuitively, the activities in the initiate phase are done at the start of the project, the plan comes before execution, and in a similar fashion, the activities in the close phase are done at the end.

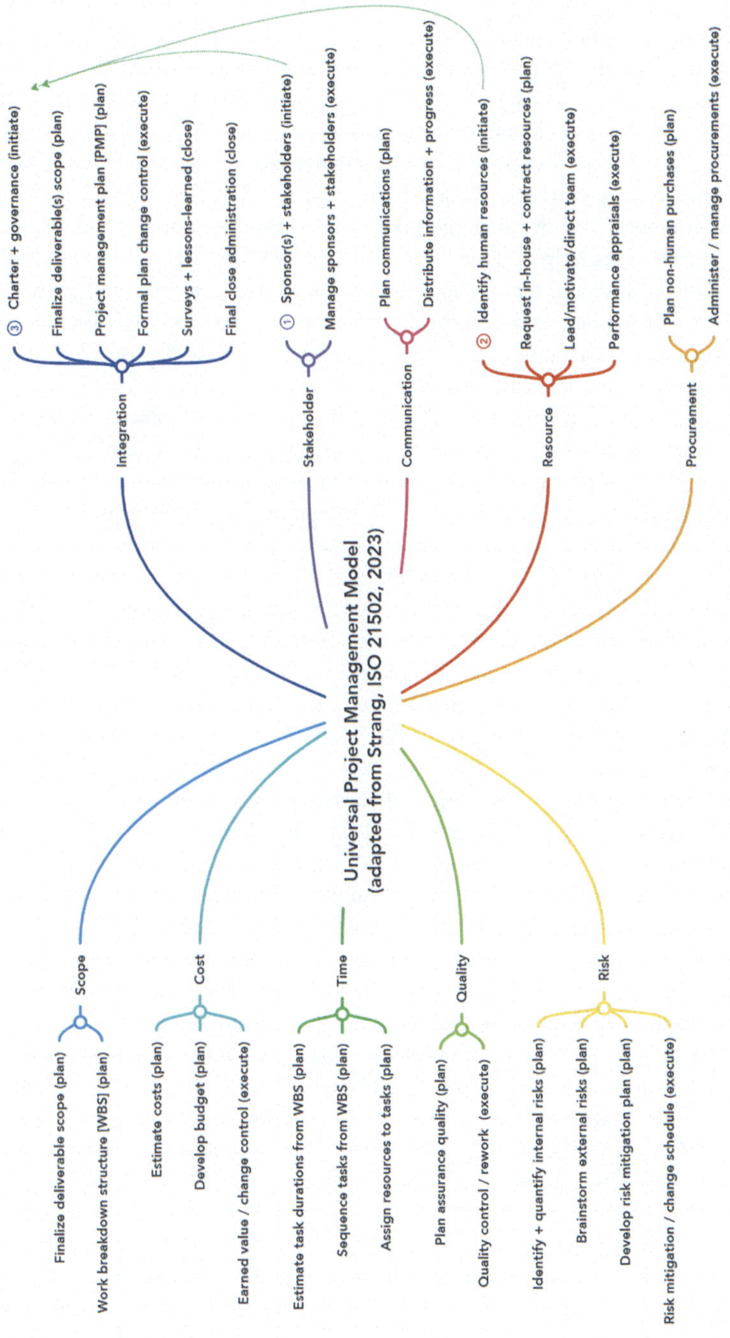

Fig. 1 Universal project management model, 10 functions in 4 phases (adapted from [10, 12])

The author specifies the four phases (initiate, plan, execute, close) in this PM profession because they intuitively suggest the sequence for the PM to follow. The 10 PM functions (integrate, stakeholder, scope, time, cost, quality, risk, communication, resource, procurement) represent the source of the interdisciplinary knowledge—for example, resource derives the theories, concepts, and models from the human resource management and organizational behavior disciplines. In similar nomenclature, cost borrows concepts from financial and economics management, while time leverages operations research, scope tapes into strategic management, and integration closely follows managerial business administration. The PM functions and activity phases help organizations collect performance data to measure the effectiveness of the PM and team. Here is a reason to use functions and phases in performance measurement: If sponsors take a long time to authorize charters in the initiating process, the PM cannot mitigate that, so it will likely delay the project, but the data will prove where the fault lies.

Figure 1 is a multi-dimensional UPM model consisting of phases, functions, workflows, data transmission, hidden relationships, process interactions, and other facets. For simplicity, only three types of characteristics are shown in Fig. 1. The most essential 41 pm functions are listed with the phases in parenthesis, plus there are two crucial workflow relationship arrows superimposed for the initiate phase. As can be seen from Fig. 1, the activities within the initiate phase fall into three different types of PM functions: Integration, stakeholder, and resource. Table 1 is a two-dimensional version of the same information from Fig. 1 (the author will explain this later). Table 1 (Fig. 2) shows unlisted managerial functions of the PM—they are not to be listed in a plan or shared with team members (unless the PM wishes to). In that sense, Table 1 (Fig. 1) is essentially the PM's job role description, the disciplinary mental model.

Initiate is the first phase of any project. The two arrows on the top left of Fig. 1 show essential data for the initiate phase entering the charter-governance activity within the integration function. Subsequently, the first arrow from the 'sponsor(s) + stakeholders' activity, drawn to the charter, is numbered 1. This refers to stakeholder identification (see chapter in this book on stakeholder management). The second arrow from activity 'in-house + contracted team' is numbered 2, and it is connected to the charter. This is intended to show that the PM must determine who the initial team members will be to finalize the charter (even if this is not the final team—a task in the planning phase will be to request internal and external team members.

The term 'deliverable' is used in this discipline to describe the outputs produced in a project, such as products and or services. Some deliverables are internal and used only by the PM, such as the charter. The project charter is a project authorization document, somewhat like an informal contract between the PM and the sponsor(s)/steering committee. Development of the project charter is the first and most important function for the PM—it cannot be overlooked and must be signed or acknowledged by the sponsor or the steering committee. In some organizational cultures, the term project champion is substituted for project sponsor. The charter is not cast in stone because it can be revised with approval from the sponsor(s) during the execution phase 3 through the formal change control activity.

Table 1 Universal project management matrix, functions across phases

PM functions	Project phases			
	Initiate	Plan	Execute	Close
Stakeholder	Stakeholder identification	Stakeholder analysis	Stakeholder engagement	
Resource	Identify human resources	Request in-house + contract resources	Lead/motivate/ direct team	
Resource			Appraise performance	
Integration	Charter + governance	Finalize deliverable scope		Surveys + lessons-learned
Integration		Project management plan	Formal change control	Final close administration
Scope		Work breakdown structure [WBS]		
Scope		Define WBS tasks for deliverable(s)		
Time		Estimate task durations from WBS		
Time		Sequence tasks from WBS		
Time		Assign resources to tasks		
Risk		Identify + quantify internal risks		
Risk		Brainstorm external risks		
Risk		Develop risk mitigation plan	Risk mitigation/ change control	
Quality		Plan quality assurance	Quality control/ rework/signoff	
Procurement		Plan non-human purchase governance	Administer/ manage procurements	
Cost		Estimate costs/ budget	Formal change control/earned value	
Communication		Plan communications	Distribute information + progress	

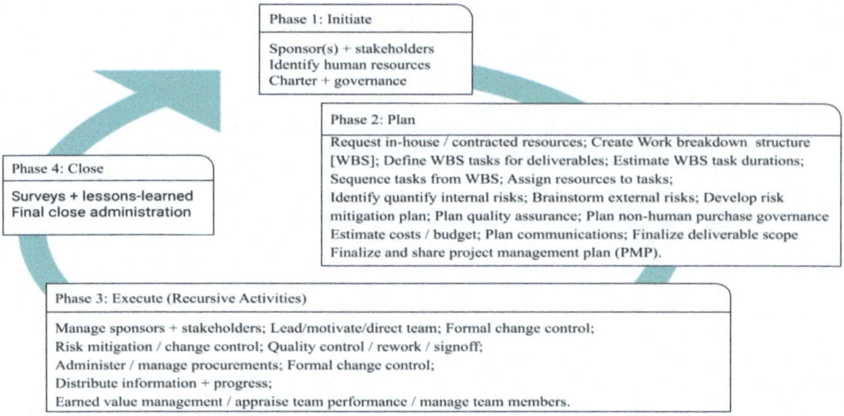

Fig. 2 UPM matrix applied to Fintech case study (adapted from [2])

An important point to share with practitioners and readers is that the project management plan (PMP) is the second critical internal deliverable after the project charter. The PMP is created by the PM as an ongoing activity during the phase 2 plan to contain everything from phase 2, such as deliverable scope, budget, schedule, resources, quality plan, risk management, communications, and procurement (if needed). As a best practice, the PMP should embed the charter (with governance policies and stakeholders identified using the stakeholder register) or reference it from phase 1 initiation.

In phase 2, the stakeholders are analyzed to document their power and impact. Power indicates a stakeholder's ability to provide resources to the PM or the formal authority in the organization or government, which could be used to assist or at worst, to stop the project. Impact refers to how the stakeholder may be affected by either WBS work package activities (e.g., if the stakeholder is giving up their resources to help the project or their processes are changing to accommodate the project), as well as if the deliverables (when completed) will cause the stakeholder's functional area to change their processes. The PM will use the PMP to manage and share the project with the team.

Once the initiate phase 1 and plan phase 2 are complete, the PM kicks off the project in the execute phase 3. Phase 3 is unique because it is not top-down (waterfall) compared to phases 1–2. Tasks in the execute phase are recursive for the PM. It should also be clarified that the PMP and schedule contain what is required for the project regarding how to produce the external deliverables. Consequently, the PMP will not usually list tasks from Table 1 (or Fig. 1) unless a PM wishes to do that. Phase 3 execution includes activities to manage the team and risk, usually called change control or monitoring actions.

The PM engages the stakeholders throughout this phase, which is recursive because if new (phantom) stakeholders are identified, or some leave the project (e.g., retirement), the PM must update the stakeholder register with new stakeholder analysis. Stakeholder engagement refers to communicating with key stakeholders, sometimes asking them for help, resources, or simply providing status update. Once the execution phase is completed, the PM will ask the sponsor to approve the deliverables and then complete phase 4's close activities. The PM will objectively perform performance evaluations on all resources (ensuring each person has input to their evaluation) and forward the final evaluations to the organization's HRM department, as well as to each resource's functional manager. The PM will usually prepare a lesson learned report, identifying the weaknesses and positives from the project, including any insights from the performance evaluations. This activity is known as the lessons learned activity in the closing phase.

Finally, the reader or practitioner should be aware that the UPM model from Table 1 or Fig. 1 is not mandatory—it is recommended. Many UPM model alternatives are available in the literature, or the PM could construct their own disciplinary mental model for managing their project. Table 1 and Fig. 1 are offered to help emerging PMs as well as to help stakeholders better understand this discipline.

3 UPM Applied to Financial Industry Case

A financial technology (Fintech) industry case study [2] is utilized to illustrate an application of the UPM. Figure 2 depicts how the UPM was applied, in sequence, for the Fintech project case study, by modifying the matrix from Table 1. A PM practitioner does not necessarily need to complete all the tasks in Table 1, nor in the sequence depicted in Fig. 2—customization of the UPM model is allowed and very common. The task sequence can be changed within reason (discussed later), some steps can be eliminated, and additional steps added. This discussion is merely an applied example.

It is common in scholarly research to discuss the study sample and method. The population under consideration was the approximately 7000 large USA-based asset management companies regulated by the Financial and Insurance Regulatory Authority (FINRA) as well as by the U.S. Securities and Exchange Commission (SEC). FINRA as well as SEC were involved due to the multiple state products and size of the assets managed. This was a single case study sample arising out of convenience. The method was an action research project, where the current chapter author was a participant, a researcher, and the PM. The company had revenues of $2.3 billion USD, it was a multinational corporation employing 4900 people full-time in offices around the world, while the head office and case study site was in New York City. The remaining demographic characteristics were withheld by client request.

Case Study Project Phase 1: Initiate

The first goal for the PM in the case study was to complete phase 1—initiate, particularly the charter, as summarized in Fig. 2. The charter defined the goals, which included the deliverables (outputs) known at that time. Most projects have a ballpark maximum budget to achieve the deliverables, milestone dates, a starting team, and of course a sponsor or group of sponsors in a committee. All these details can be revised later in the plan phase—they are specified in the charter to provide a starting scope. The duration of the initiate phase was 3 days which included 2 h on site for the meeting, plus travel, and elapsed time for writing the charter and obtaining approval.

The PM was requested to manage the case study project. The PM met with a small group of executives from the case study company at their head office, where one senior director was identified as the sponsor/project authority. If the sponsor had not been identified, the PM would have requested the company to identify a person or committee—it is critical to have the sponsor formally identified.

During the 'initiate' phase 1 meeting, the PM asked the stakeholders which types of human resources—or named staff—were going to be provided to the project—it was important to do that upfront. The author asked permission to record the meeting to avoid looking distracted when taking detailed notes. Meeting notes should be taken. The PM then uses the meeting notes to develop the stakeholder register and staffing pool database. As part of stakeholder identification, the PM asked key stakeholders such as the sponsor, and any existing stakeholders to identify more stakeholders (using the snowball technique).

During the initiate phase meeting, the sponsor named the project case study as the 'Asset Manager Index' (AMI). The deliverables involved examining asset managers from the New York Stock Exchange, in terms of how they performed their duties, and then building an AMI competency model for internal company use. A draft budget was set at $650,000 USD. The corporation provided all internal resources (researchers, programmers, database administrators, network security specialists), and equipment, so procurement was unnecessary. The deliverables were:

1. AMI model—attributes to benchmark portfolio manager capability;
2. AMI prototype—software application and database to hold and serve AMI.

The recording was later reviewed to create a draft of the project charter, with governance (the standard operation procedures requested by the sponsor or PM for the project), which was later presented to the sponsor for revision, constructive feedback, and subsequent approval. The sponsor approved the charter through an email reply.

The most important information for the author to share with practitioners and readers about phase 1 in this case study, as well as in any project, was that the PM must convince the sponsor(s) or steering committee to take ownership of the overall responsibility for initiating the project. In some difficult project situations, or in nonprofit situations, the sponsor may be assigned by a more senior level supervisor, such as in government departments or large companies, and therefore the sponsor may hesitate to actively accept any ownership accountability. In those cases a PM should note when the charter was delivered, and who received it, even if it lacked signatures. That follows the brute force principle of legal service, but it works well in practice.

Case Study Project Phase 2: Plan

The next project phase 2 (plan) was started by the PM immediately after the charter was approved (when phase 1—initiate—was completed—as shown in Fig. 2). One week was allocated and five days were needed. The plan phase involved all team members, in a brainstorming session, including the sponsor for certain tasks. The sponsor participated only a few hours at the start to identify the deliverable specifications (the asset manager construct), the equipment, the names of the team participants allocated to the project full time, the progress communication requirements, quality level, and milestone target dates. The stakeholder registers and staffing pool database was updated to reflect any changes. The PM performed stakeholder analysis by adding power and impact attributes for each stakeholder to the stakeholder register.

The author can share that it may seem like a catch-22 situation for a PM to 'request in-house + contract resources' because often the PM is assigned resources without any options to request more. In some cases, there may be no resources offered, so the PM must use procurement to contract the team. Therefore, the first task for the PM in the case study was to meet with the sponsor to clarify which resources could be available, and on what basis (full or part time). In the case study, seven resources were being provided full-time and the PM was told no more resources could be procured.

Since the sponsor had already identified the resources, the PM extended that into the project management plan—this was the 'assign resources to tasks' step in Fig. 2, although the exact task level allocation was done later in this plan phase, during schedule development. The PM verified the two deliverables with the sponsor and confirmed the project scope was to develop the AMI structure, as well as develop an internal software product (second deliverable) to record the index data and display it. The important limitation was that a 'prototype' was requested, rather than a complex software program as in a mobile smart phone application. The PM drew a conceptual picture of what the prototype might look like, to have the sponsor approve it.

The next step was to document the work breakdown structure (WBS) for each deliverable (the AMI structure and the software prototype). It can be noted that the WBS is an important requirement for 'each' deliverable (unless the deliverables are intertwined to make it more practical to develop only one overall WBS). The WBS is a top-down hierarchy of every facet needed to produce the deliverable(s). In the AMI project case study, the model was copyrighted so it cannot be displayed in this chapter.

The first WBS resembled the top portion of an investment fund manager prospectus, which is because the AMI was an extension of that. Each proposed new attribute was listed with a likely source or formula to build it. The WBS for the AMI prototype looked like a typical software design and development blueprint, showing the underlying database, the user menus, security, and maintenance functions. Requiring a WBS makes the next steps easier to define, sequence, assign work, and operationalize the tasks.

The next step was to define tasks for WBS and develop a milestone schedule from the approved chart. The PM then met with the team to estimate task effort required for each WBS item and sequence the tasks from the WBS. The PM worked iteratively, during that first week with the team, to finalize the tasks, sequences, and then develop a proposed schedule, by working backward from the deliverable milestone dates. In the case study, the process worked, no extra resources were thought to be needed. However, if a PM is unable to create a schedule to achieve the deliverable milestone dates by using the identified resources, then an immediate change request ought to be raised with the sponsor—otherwise this would be a fatally flawed project schedule.

At this point, the draft project schedule was finished, but the project management plan (PMP) was not yet finalized. The PMP was completed by adding the schedule, attaching the charter, and including any other important documentation.

Risk management must be completed by the PM to finalize the PMP. First, the PM identified internal risks, by determining if the team members could realistically achieve the goals, as outlined in the schedule. For example, the PM would look at the resumes of the team members, to determine if they were capable in the past of the work needed on the current project, and if not, meet with the member to discuss exactly how the work could be done, and where necessary, how the requisite experience could be incubated—in some situations a mentor can be assigned for inexperienced team members.

An interesting discovery revealed in the case study was that the PM quantified the internal risks using the Program Evaluation Review Technique (PERT), a theory originally developed by the U.S. Department of Defense, and by adapting a technique from the literature [13]. Internal risks in a project can be estimated by the amount of variance in times for the tasks, by comparing the estimates from several different subject matter experts, calculating a standard deviation (SD), and then using the Sd to identify tasks with higher risks of going longer. As a best-practice, tasks on the critical path, with no slack time available between them (e.g., no idle calendar time), should be buffered as a hedge against risk, by adding 10% extra time. Alternatively, the PM could add a 10% contingency buffer to the critical path (often that is easier for less-experienced PMs wishing to avoid statistical analysis, because it does not require SD calculations). To help other practitioners, it can be noted that if PERT cannot be calculated, such as when there is only one team member available (e.g., small projects), the generic 10% contingency buffer ought to be added to the critical path in the schedule. PERT and other valuable techniques will be discussed in later chapters.

Another interesting aspect of the case study was the external risk identification process. External risks are defined as potential negative events which may occur outside of the project context (positive events are not usually listed since mitigation is not needed—if things go better than planned, a PM does not change the plan!). Brainstorming with the team (and sponsor) is commonly used to identify the external risks. The lessons learned from previous projects and the literature may also be reviewed. In the case study, the PM and sponsor discovered from a literature review that cybercrimes were on the increase, so they agreed that a secure firewall needed to be established for the AMI prototype. This was added as a task in the PMP. Additionally, the PM was required to use a private virtual network when connecting to the project site remotely, to avoid causing vulnerabilities to the internal site.

In a project, the quality plan refers to how the sponsor will know the delivery is sufficiently completed within scope. The best practice for a quality plan is to develop a quality assurance set of test cases, typical transactions or behaviors, along with exceptional circumstances to stress-test the deliverables. This strategy works well for products and services, it just takes a bit of imagination to document the typical use-cases for quality assurance testing. Functional experts in the company (or externally) are usually consulted for quality assurance, but team members can also participate as long as a team member is not creating their own use-case tests. In the case study, the steering committee provided the basic quality assurance test scenarios, and the team members brainstormed the specific use-case tests with the PM overseeing. In summary, the quality plan required the AMI software to work.

There were no non-human purchases planned, since the company had already stipulated, they would provide all equipment, software, hardware, networks, and security. As noted above, the security was very helpful to mitigate against the external risks. The author can share that many government projects will require procurement. If procurement is required, this starts with the governance for procurement in plan phase 2, which describes how items and human resources must be contracted. It could be mentioned that the procurement is administered in the execute phase 3, where there is typically a request for quotation (RFQ) released by the PM, contracting companies submit bids, and the PM selects the appropriate proposals. The proposals can be for a subsection of the project (e.g., outsourced project components) or as a call-up for specific goods or services. The author can share that procurement is a lot of work if needed. Thus, the author can suggest a best practice to help practitioners. If the sponsor is a government department or company, the PM could request an in-house procurement administrator be assigned to assist with the procurement functions.

Going back to the case study, at this point, the PM had completed everything in the plan phase 2. The PM presented the PMP to the team, making sure to review the schedule, and accept constructive feedback from every team member. The author can point out that if a team member is adamant that a change is needed to the PMP, then the PM ought to accommodate the change if possible—team resources are considered experts on the detailed tasks needed. Once the PMP was again finalized, it was shared electronically with the sponsor and to the team members—as a courtesy.

Case Study Project Phase 3: Execute

Execute phase 3 involves ongoing, recursive PM activities, following the PMP, managing the team, including monitoring and controlling for potential risks. The PM steps in this phase do not necessarily need to be performed in a specific sequence, but the first time through the tasks, the PM is likely to follow what is suggested in Fig. 2.

Overall, the PM engaged the stakeholders by answering emails and meeting with them as requested, including any planned milestone updates (usually every week on Mondays). For example, at the end of the previous phase 2 (plan), the PM would have communicated (shared) the PMP to the stakeholders. The PM also performs generic managerial tasks such as leading, motivating, monitoring, and directing/controlling the team members.

Although it was not needed in the case study, the PM may have to reassign resources if conflicts arise or if work is not completed as requested. This leads to the performance appraisal process that a PM must consider; even if done informally, it is highly advised that a PM keep notes about each team member (by monitoring work and measuring team member performance against the allocated schedule in the PMP). This technique is known as earned value [2]; it is complex, so another chapter in this book is dedicated to explaining how that methodology functions. The PM not only directed the team but also offered advice to the company's network staff to pass it on to the internet security provider (under the procurement contract). Note there was no procurement in the case study, but many projects, particularly government programs, rely heavily on contracted resources as well as equipment.

In the UPM, control, and risk management are simplified into two activities, respectively, but it can be shared with the reader that in the alternative bodies of knowledge, those two items often require dozens of steps and pages of guidelines. In the case study, one recursive critical activity in phase 3 execution, called change control, was carried out by the PM to manage scope, time, and cost. This was done by monitoring the schedule for deviations beyond the planned time but before the 10% contingency buffer was exhausted for critical path tasks. There were no significant deviations in the case study, but hypothetically speaking, if time deviated, the PM must act by meeting with the resource(s) to attempt to resolve the constraint, or if the delay cannot be determined, the PM must document a change request and present it to the sponsor for approval. Change requests are not a best practice in any project, so the PM will try to avoid that by proactively monitoring the team, motivating them, and, where necessary, substituting resources (even replacing people if absolutely required).

Risk management is carried out by observing any external non-team event that negatively impacts the schedule, cost, scope, resource, or procurement. Quality is taken care of separately, as discussed below. If the PM discovers a risk event, for example, a technology or software component malfunctioning, a new legal constraint emerging, etc., then the PM must contact the sponsor. Often, the mitigation will be to document a change control request and submit it to the sponsor for approval. If the change is approved, and if it requires a shift in the schedule, scope, or cost, the PMP may be re-baselined by the PMP (and reissued), which is a tactic to avoid a project being completed as required but looking like it was late, over budget or beyond scope.

In the case study, the PM controlled scope by ensuring team members stayed within what was required. The PM referred to this as when team members added extra 'bells and whistles' features to the software program—the PM politely reiterated that everyone was being paid to produce what the sponsor had dictated in the two deliverables—if that was achieved, then no problem, the team members could make the screens look fancy and attractive. In other words, the PM acted only when the scheduled tasks took longer than planned in the PMP. This activity also leads to the PM calculating schedule variances by asking team members to record their work against the planned schedule at the end of every day in a project management software database. Sometimes, the team members forgot to do that, especially on Fridays, so the PM reminded them the next business day.

The author can add, for the benefit of emerging PMs, that scope creep can become a risk in any project. In the case study, the sponsor invited executives to the presentations, and these visitors often suggested new features. The PM carefully controlled the scope by referring to the charter and PMP. This is a tactic that the PM can leverage to deflect adding new features by directing such questions to the sponsor and then offering to recalculate the new cost, time, risk, and quality needed for the revised scope. Usually, sponsors will clue into this subtle maneuver and offer to consider new features in a subsequent project, thus supporting the PM's strategy.

Quality assurance measurement was done by regularly testing the AMI software within the team—usually, this was done early at the start of the week, on Mondays, when everyone was returning refreshed from the weekend, but before anyone had started their tasks. Not surprisingly, the PM had included a quality assurance task for specific senior team members in the schedule within the PMP. Quality control was completed by presenting the AMI software prototype to the sponsor and inviting executives. This was done before the deliverables were due to ensure nothing was missing. In the PM discipline, quality control may be considered the user acceptance test of specific deliverables for business functionality.

The PM also communicated progress periodically to the stakeholders, as well as to the team members. In fact, during the case study, the PM met with the team (with stakeholders invited) on a weekly basis. At the end of the meeting, the PM would also informally appraise team performance based on monitoring and earned value calculations and then suggest changes to the team members if needed. The PM scheduled monthly progress meetings with the sponsor and encouraged the sponsor to invite key stakeholders such as the steering committee executives. As a tip to practitioners, it is better to receive critical constructive feedback, especially from external arm-chair philosophers, during the project when there is time to make appropriate adjustments or to prepare strategies designed to increase future acceptance of the deliverables and of the project.

Case Study Project Phase 4: Close

In the case study, phase 4 closed very quickly. The sponsor approved both deliverables, and the PM was paid. Nevertheless, a few more details are worth mentioning, particularly concerning risk management.

The PM performed performance evaluations for each resource, meeting with each person individually and privately. The PM conducted performance evaluations first by reviewing notes made during the project for each resource, created the draft evaluation, and met with the resource to discuss it with the person, allowing that resource to agree or disagree and suggest corrections if warranted. Once all performance evaluations were completed, and agreed to (signed) by each resource, the PM forwarded these to the company's Human Resource Dept., as a bulk spreadsheet report. Each functional manager (also a key stakeholder) was given a copy of the performance evaluations for the resources they provided to the project. Thank you, letters were given to each stakeholder, for providing human resources as well as equipment to the project.

The PM met with the team to tell them the project was completed on time, within budget, scope, and quality. The PM asked for any lessons learned, but none were offered. Despite that, it can be revealed that the company's network security team identified a few external cybercrime risks, and as per the mitigation activities, two new tasks were undertaken. First, the PM had to use a VPN to access the project when he was off-site. Second, the network security was boosted using a new firewall configuration. These were two crucial facts to document and pass on to future teams. This is ultimately the purpose of the lessons learned activity in the close phase. In other words, the lessons learned from closing activity need not be to document risks occurring. Still, instead, the risk mitigation plan needs to be cited to executives and shared with future teams as organizational learning.

4 Conclusions and Next Steps

The current Fintech industry case study was a simple example of an America-based project where ISO was applied. This was described retrospectively, meaning that the UPM developed here was interpreted using ISO 21502. The PMI PMBOK fifth edition was being applied by the PM, which is what ISO 21502 was based upon. This simple example demonstrates that UPM does not have to be complicated; it contains only four phases grouped into 10 PM functions (as compared to the five PMBOK or ISO processes grouped into ten knowledge areas). The UPM holds 29 suggested PM activities as contrasted to the ISO 21502 (2021), which had 41, and the PMBOK sixth edition, which had 47 and Agile examples. The PMBOK seventh edition [3] replaced the ten knowledge areas with 12 principles and eight performance domains (see Fig. 3).

This is a stark contrast to the ISO and PMI PMBOK. In most PM frameworks, risk management is considered a different phase, cycle, or process group, done in parallel but separately from managing the core project functions. In the UPM model proposed here, risk management takes place in the execute phase—no separate phase exists. The PM must concurrently manage the project plan, resources, and risks. If an internal or external risk manifests, action is needed, but in parallel, the scheduled tasks ought to continue as planned. Some risks require a change in scope, time, cost, quality, resources, and purchase goods, which would require the PM to document a change request, present that to the sponsor(s), and, if approved, the project management plan and schedule may change. In short, risk management is not an activity to be undertaken in a separate project phase or process group.

As a research-driven scholarly book chapter, we must review the original research questions (and any hypotheses, if created) to determine if they have been answered and supported. In this chapter, we did not specify a formal methodology because the nature of a book is often to apply critical analysis to the literature. In the current

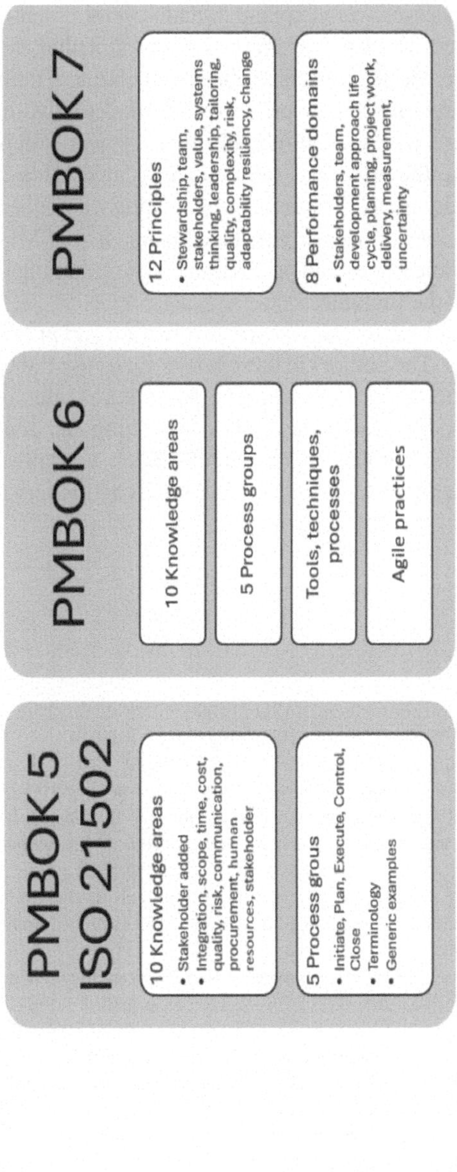

Fig. 3 Evolution of ISO 21502/PMBOK version 5 to PMBOK 6 and 7 (adapted from [3, 12])

chapter, we also used several general analytical techniques, including PERT and diagramming.

The research question appears to have been answered. This chapter was a simplified conceptual project management model, shown in a multi-dimensional visual diagram and transformed into a tabular two-dimensional matrix. The matrix was applied to an abbreviated case study from the financial industry (cited in the literature). A few key PM processes were explained, and several essential and valuable techniques were mentioned. There was an implied assumption that the PM techniques would be described in greater detail during subsequent chapters in these books.

In summary, this introductory chapter primarily introduced a two-headed rationale for both books: that there was a longitudinally high project failure rate and that the existing PM guidance literature was overly complicated. This chapter was the introduction to the generic model. The remaining chapters will discuss what needs to be added to the bodies of knowledge guidance books (from PMI and ISO) since there are no examples for industries and professions. The remaining chapters in both books will fill that gap in the literature.

The ideology of the book editors is to share information and promote the reuse of concepts and components. The authors in these books can reuse the UPM model and matrix as needed or create their own constructs. All cultures, ideologies, and ideas are presented in these two books to achieve our overarching end goal of developing a better simplified universal project management model, with practical examples from selected industries and professions explained by practitioners across different cultures.

References

1. Čabarkapa, J.: Analysis and comparison of ISO 21500—guidance on project management and PMBOK 6th guide (2019). https://doi.org/10.2991/senet-19.2019.44
2. Strang, K.D.: Man versus math: Behaviorist exploration of post-crisis non-banking asset management. J. Asset Manag. **13**(5), 348–367 (2012). https://doi.org/10.1057/jam.2012.14
3. PMI.: The Standard for Project Management and A Guide to the Project Management Body of Knowledge (PMBOK guide). ISBN: 9781628256673. Project Management Institute (PMI), PA (2021)
4. Borbath, M., Blessner, P., Olson, B.: An empirical approach to the evaluation of defence contractor performance. Int J Log Res Appl **22**(2), 138–153 (2019). https://doi.org/10.1080/13675567.2018.1497148
5. Eckerd, A., Snider, K.: Does the program manager matter? New public management and defense acquisition. Am. Rev. Public Adm. **47**(1), 36–57 (2017). https://doi.org/10.1177/0275074015596376
6. Pace, M.: A correlational study on project management methodology and project success. J. Eng. Project Prod. Manage. **9**(2), 56–65 (2019). https://doi.org/10.2478/jeppm-2019-0007
7. Shoar, S., Yiu, T.W., Payan, S., Parchamijalal, M.: Modeling cost overrun in building construction projects using the interpretive structural modeling approach: a developing country perspective. Eng. Constr. Archit. Manag. **30**(2), 365–392 (2023). https://doi.org/10.1108/ECAM-08-2021-0732

8. Masticola, S.P.: A simple estimate of the cost of software project failures and the breakeven effectiveness of project risk management. In: Paper Presented at the 2007 First International Workshop on the Economics of Software and Computation, May 10–14 (2007)
9. Kurek, E., Johnson, J., Mulder, H.: Measuring the value of enterprise architecture on IT projects with CHAOS research. Syst. Cybern. Inform. **15**(7), 13–18 (2017)
10. Strang, K.D.: Which organizational and individual factors predict success versus failure in procurement projects. Int. J. Inf. Technol. Project Manage. **12**(3), 19–39 (2021). https://doi.org/10.4018/IJITPM.2021070102
11. Strang, K.D., Vajjhala, N.R.: Mining project failure factors from big data using machine learning. Int. J. Inf. Technol. Project Manage. **14**(1), 1–24 (2023). https://doi.org/10.4018/IJITPM.317221
12. ISO 21502:2020(en). Project, programme and portfolio management—guidance on project management (2023). Retrieved Nov 24, 2023, from https://www.iso.org/obp/ui/en/#iso:std:iso:21502:ed-1:v1:en
13. Strang, K.D., Symonds, R.J.: Analyzing research activity duration and uncertainty in business doctorate degrees. Int. J. Risk Contingency Manage. (IJRCM) **1**(1), 29–48 (2012). https://doi.org/10.4018/ijrcm.2012010103

PMBOK® Challenges and Lessons Learned in the Quebec Public Administration: A Comparative Analysis with ISO 21502 Standard

France Desjardins

Abstract Since the early 1990s, managers of public organizations in Quebec have integrated and improved their project management processes and tools using the different PMBOK® editions. However, little research has focused on whether this framework meets public sector employees' needs. In this context, little research has also focused on whether ISO 21502 could be a more appropriate reference for the needs of public project managers. Despite theoretical and managerial advances in project management, public sector managers and project team members still face challenges. This qualitative and descriptive case study attempts to answer three questions: What are public sector organizations' current project management challenges? Does the Project Management Body of Knowledge meet the needs of these public managers? Could the ISO 21502 standard be a more appropriate reference to address the needs and challenges of public managers? The discussion of the results will outline the needs, challenges, and lessons learned by the employees and project managers of the public organization. Finally, we suggest that the PMBOK® is useful for managers and the ISO 21502 standard for project team members.

Keywords PMBOK® · ISO 21502 · Public sector project management · Quebec public organizations · Project management challenges · Project team needs · Project management frameworks · Qualitative case study · Public sector employees · Lessons learned

1 Introduction

For public sector organizations that are gradually integrating project management into their activities, the Project Management Institute's Project Management Body of Knowledge is an attractive guide that focuses on controlling cost, time and scope/

F. Desjardins (✉)
University of Chicoutimi, Quebec, Canada
e-mail: France1_desjardins@uqac.ca

K. D. Strang and N. R. Vajjhala (eds.), *International Program and Project Management—Best Practices in Selected Industries*, Information Systems Engineering and Management 31, https://doi.org/10.1007/978-3-031-80275-1_5

quality. The seventh edition of the PMBOK® [1] proposes eight performance domains and tips, accompanied by tools and techniques. These tips, presented in a single book, may have certain advantages for public managers new to project management, as they can be easily integrated gradually into an organizational context where everything is documented and standardized to deliver equal service to all citizens [2]. However, to date, little research has been done to determine whether the PMBOK® [1] effectively meets the needs of managers and project team members in Quebec public organizations, despite its importance as a reference [3–5]. In addition, the ISO 21502 standard as a potential reference has received limited research attention [6]. Although the primary mission of public organizations is to apply laws and regulations by providing services to the public, today's public managers are called upon to mobilize various stakeholders in society to participate in the design or improvement of these services [7, 8]. To this end, public managers must not only focus on the effectiveness of public services and the efficiency of resource management [9, 10], but also engage in an inclusive and collaborative dialogue with private and community organizations and civil society. This approach reflects the evolution from New Public Management towards New Public Governance [8, 11]. Similarly, project management involves the development of a new public service or its improvement with the participation of stakeholders [8].

In Quebec, the proportion of public spending on project management is difficult to estimate because some public organizations use their operating budgets to finance their projects. For example, infrastructure projects represent about 10% of public spending [12]. This low representation of projects compared to routine public activities means that managers do not develop their project knowledge and competencies to a great extent. This illustration of managing a project compared to managing routine activities is a good illustration of the lightweight matrix structure. It should be noted that the construction or rehabilitation of certain infrastructure is a shared responsibility among municipal (cities), provincial (Quebec) and federal (Canada) levels of government.

Considering the choice of reference in project management, the changing role of public managers and the low proportion of projects carried out by certain public organizations in Quebec, we pose the following questions: What are the current project management challenges facing public sector organizations? Does the Project Management Body of Knowledge meet the needs of these public managers? Could ISO 21502 be a more appropriate reference to address the needs and challenges of public managers? This chapter aims to answer these questions, which are important given the role of the public sector in our societies. To achieve this, we will describe a conceptual framework that relates to the specificities of the context of public organizations in Quebec, the performance domains of the 7th edition of the Project Management Body of Knowledge [1] and the ISO 21502 standard [6]. Next, the research methodology is detailed: a single case study of employees and managers with at least one year of project management experience in a large Quebec public organization with a lightweight matrix structure [13]. The reflections of the people involved in the data collection and the analysis of the administrative documents on the project management processes and tools used daily and the management challenges experienced will be described. Finally, the discussion of the results will bring together the needs and challenges of employees and project managers in the public organization to reflect on whether the PMBOK® [1] is still relevant or whether managers could more easily refer to the ISO 21502 standard [6].

2 Conceptual Framework

2.1 *Context of the Public Organizations of Quebec*

For almost thirty years, public organizations in Quebec have been progressively integrating project management to plan the construction or renovation of infrastructure that meets the needs of the population, improves the quality of services, creates new services and more [4]. This application of project management coincides with the modernization of public administration through results-based management [14]. This organizational transformation, specified in the Public Administration Act (chapter A-6.01), represents a shift from Weber's bureaucratic model (1864–1920), which was characterized by the imposition of written rules and procedures to standardize the delivery of public services. However, public organizations continue to adhere to standardized procedures inherited from the past because they must implement laws through administrative activities and create satisfactory services with and for the population [14, 15].

To achieve this, public managers must continue to optimize the use of resources to minimize the cost of providing public services and engage in dialogue with various stakeholders to create or improve them [16]. To accomplish this, public managers have changed their organizational structure to integrate project management into their day-to-day operations. Some public sector organizations have opted for a lightweight matrix organizational structure, where certain employees have a dual role of managing both routine operations and project activities [1, 15, 17–19]. This focus on collaboration with other stakeholders reflects the values of New Public Governance [8, 11].

This shift in organizational structure, from a functional to a lightweight matrix, offers advantages such as access to human resources already employed by the organization who have a good understanding of strategic and operational objectives. Employees are familiar with communication channels and reporting lines and the organization preserves the knowledge and competencies developed during the projects [13, 20]. However, there are disadvantages. Management, communication and decision-making processes are designed to support routine operations and must be followed in project management situations without regard to planning deadlines. As a result, coordinating human resources under the functional responsibility of the managers of the various administrative units/functions and securing approval for each decision from all levels of the hierarchy requires a great deal of time and energy from the project management team [13, 21]. To manage the integration of project management in public organizations, managers often refer to the PMBOK®, the characteristics of which are presented in the subsequent section.

2.2 Project Management Body of Knowledge (PMBOK®)

The predictive approach to project management refers to a reference guide for managing all projects, regardless of context or industry [22]. This universal perspective is beneficial to public organizations because it enables stable planning and execution of deliverables, as well as standardization of project management phases, techniques, and tools.

The 7th Edition of the PMBOK® [1] includes performance domains that provide tips to help project managers tailor their management style to the specifics of their organizational context and industry. These performance domains are: Stakeholder, Team, Development Approach and Life Cycle, Planning, Project Work, Delivery, Measurement and Uncertainty. The Stakeholder performance domain offers advice on identifying stakeholders, understanding their interests and power and engaging and communicating with them to encourage their involvement. The Team performance domain provides information on cross-functional teams, composed of employees from different administrative functions and how project management integrates into the organization. It also addresses the positioning of the project manager within the hierarchical levels. From the project team's perspective, this involves developing a productive working environment, maintaining motivation and avoiding conflicts. The Development Approach and Life Cycle performance domain considers the project management methodology (e.g., predictive, hybrid, iterative, or incremental) in relation to the degree of innovation, ease of change, project requirements, delivery options, risks and regulations. It also includes elements such as stakeholder involvement, schedules, funding and organizational structure and culture. The project lifecycle can be linear, from feasibility to completion, or involve several iterations of Concept, Feasibility, Plan and Develop in the phases. The Planning performance domain relates to the business case, incorporating stakeholder requirements and project scope. This scope defines the functionalities or features of the project deliverables. It involves estimating resources, time and costs using a breakdown structure. The Project Work performance domain includes optimizing workflows using Lean techniques, lessons learned, communication, resource management and procurement, including the preparation of tender and contract documents. This domain also encompasses knowledge management.

Since public sector organizations don't manufacture products, their project management often stops at the planning stage. Only project managers continue to supervise the progress of the Delivery performance domain with the business. They oversee compliance with requirements, deliverables, quality and stakeholder satisfaction. This supervision is covered by the Measurement performance domain, which gathers information to monitor project progress. Finally, the Uncertainty performance domain relates to risk management during project execution.

Overall, the project management processes by performance domain suggested by PMBOK® 7th edition [1] are shown in the Table 1.

2.3 ISO 21502 Standard

The ISO 21052 standard [6] outlines 17 integrated project management practices and 51 project management practices aimed at facilitating the attainment of planned results. Its applicability extends to organizations across all industries. Table 2 provides a summary of these practices.

These standardized project management practices align with the advice, techniques and tools offered by the PMBOK® [1]. The next section describes this potential integration between the Performance Domains of the PMBOK® [1] and the Integrated Project Management Practices or the ISO 21502 standard [6].

2.4 Conceptual Framework

In this chapter, the conceptual framework show the Performance Domains of the PMBOK® [1] and the Integrated Project Management Practices of the ISO 21502 [6]. Table 3 proposes an integration to facilitate comparison and address the research questions.

Following this presentation of the conceptual analysis framework, the research methodology will now be explained.

3 Methodology

To describe the project management needs and challenges of public managers, the present research adopts an interpretivist epistemological stance. It involves a personalized and open interpretation of the experiences and reflections of individuals working in the public sector regarding their project management practices[23]. The methodological approach is abductive, as project management practices are identified and defined using the PMBOK® as a reference guide. The abductive approach enables the capture of practices applied within each performance domain, resembling a form of field survey as suggested by Lorino [24]. The aim is not to identify new practices, but simply to classify those implemented by employees and managers.

Given the aim to capture the plurality of project management experiences and practices, this research is qualitative. The objective is not to quantify the number of practices implemented but to identify opportunities [25] and challenges encountered. These are articulated with the PMBOK® Performance Domains [1] and then evaluated for relevance against ISO 21502 [6]. To achieve this, the strategy employed

Table 1 Description of performance domains and tips

Project performance domains	Elements of the performance domain and tips
Stakeholder	Stakeholder engagement Interactions with other performance domains Checking results
Team	Project team management and leadership Project team culture High-performing project teams Leadership skills Adapting leadership styles Interaction with other performance domains Checking results
Development approach and life cycle	Development, cadence and life cycle relationship Delivery cadence Development approaches Considerations for selecting a development approach Life cycle and phases definitions Alignment of delivery cadence, development approach and life cycle Interaction with other performance domains Measuring outcomes
Planning	Planning variables Project team composition and structure Communication Physical resources Procurement Changes Metrics Alignment Interaction with other performance domains Checking results
Project work	Project processes Balancing competing constraints Maintaining project team focus Project communication and engagement Managing physical resources Working with procurement Monitoring new work and changes Continuous learning throughout the project Interaction with other performance domains Checking results
Delivery	Deliverables Quality assurance Suboptimal outcomes Interaction with other performance domains Checking results

(continued)

Table 1 (continued)

Project performance domains	Elements of the performance domain and tips
Measurement	Establishing effective measures What to measure Presenting information Measurement pitfalls Troubleshooting performance Continuous growth and improvement Interaction with other performance domains Checking results
Uncertainty	Uncertainty Ambiguity Complexity Volatility Risk Interaction with other performance domains Checking results

is a case study, wherein the actual context, the public organization and the project management practices used by employees and managers are intrinsically linked [26]. To account for the complexity of the case study, Yin [26] suggests the inclusion of multiple information sources, facilitating data triangulation to enrich subsequent reflections. This case study is described in the subsequent section.

3.1 Data Collection and Analysis Methods

The case is a public organization with a lightweight project matrix structure, as it is gradually integrating project management into its activities. To understand this context and the project management practices promoted and applied, the data collected came from administrative documents published between 2004 and 2018 and from semi-structured interviews with employees and managers who manage and carry out routine tasks and project activities.

To understand the organizational context in which the informants operate, the minutes of the meetings of the highest authority were selected and examined. Additionally, the organization's Auditor General's reports from 2018 to 2023 were analyzed, as they relate to the strengths and weaknesses of their project management. Semi-structured interviews were conducted with 24 employees who had been working in a dual function for more than one year in three administrative units active in projects, divided into four groups [13]. The composition of the groups was randomized according to people's availability. The choice of the group interview as a data collection tool is relevant because project management practices are carried out collectively. Topics covered included their organizational context, project management processes and practices and difficulties experienced. To validate the

Table 2 Description of the integrated project management practices

Integrated project management practices	Project management practices
Planning	Developing the plan Monitoring the plan
Benefit management	Identifying and analyzing benefits Monitoring benefits Maintaining benefits
Scope management	Defining the scope Controlling the scope Confirming scope delivery
Resource management	Planning the project organization Establishing the team Developing the team Managing the team Planning, managing and controlling the physical and material resources
Schedule management	Estimating activity durations Developing the schedule Controlling the schedule
Cost management	Estimating costs Developing the budget Controlling costs
Risk management	Identifying risks Assessing risks Treating risks Controlling risks
Issues management	Identifying issues Resolving issues
Change control	Establishing a change control framework Identifying and assessing change requests Planning the implementation of change requests Implementing and closing the change requests
Quality management	Planning quality Assuring quality Controlling quality
Stakeholder engagement	Identifying stakeholders Engaging stakeholders
Communications management	Planning communication Distributing information Monitoring the impact of communications
Managing organizational and societal change	Identifying the need for change Implementing organizational and societal change
Reporting	Planning reporting Managing reporting Delivering reporting

(continued)

Table 2 (continued)

Integrated project management practices	Project management practices
Information and documentation management	Identifying information management requests Storing and retrieving information and documentation
Procurement	Planning procurement Evaluating and selecting suppliers Administering contracts Closing contacts
Lessons learned	Identifying the lessons Disseminating the lessons

Table 3 Conceptual framework

PMBOK® 7th edition and performance domains	ISO 21502 standard and integrated project management practices
Stakeholder	Stakeholder engagement
Team	Resource management
Development approach and life cycle	–
Planning and delivery	Planning Benefit management (managing organizational and societal change) Schedule management Cost management Scope management Quality management
Project work	Procurement Resource management (physical) Communication management Lessons learned Information and documentation management. Issues management Change control
Measurement	Reporting
Uncertainty	Risk management

interpretation of the results, a group interview was conducted with six people who had participated in one of the four previous interviews. The preliminary results of this interview were presented to a group of six managers to provide a basis for understanding the organizational context and project management practices. Finally, the information-gathering process concluded with a meeting between the four employees who had participated in the previous interviews and the six managers to present all the preliminary results and to discuss the differences between perceptions and experiences within the projects. The aim was not to reach a consensus but to share different project management practices.

A careful transcription of the information obtained from all group interviews was made to analyze each sentence. As suggested by and to facilitate the classification of sentences from the group interview recordings, a coding framework was used with the eight PMBOK® performance domains. Six hundred and ninety-three (693) documents were downloaded from the public organization's website. A search was executed using the keyword "project," and one hundred and thirty (130) documents were selected. This made it possible to code phrases related to organizational context and project management practices in the same framework. This single coding framework made it possible to triangulate meaningful data. Given the wealth of information contained in the eight PMBOK® performance domains related to project management practices and the challenges experienced by team members and managers, they are used in the presentation of results.

4 Presentation of Results

This presentation of results will demonstrate the various performance domains alongside the project management practices and challenges encountered during data collection. This approach will facilitate reflection on the use of the PMBOK® [1], which serves as the reference framework utilized by public managers. It will also aid in determining whether its application is appropriate and thus address the questions raised.

Overall, participants in the research indicate that their organization primarily emphasizes the project management phases of feasibility, justification and planning. Subsequently, tenders are drafted and published. They also note a gradual integration of project management within their organization, with managers progressively standardizing project processes and activities. Participants perceive both organizational and individual project management maturity to be low. Managers underscore the importance of project management practices aligning with established processes for managing routine activities as well as human, material and financial resources.

Following this general presentation of the case, the eight PMBOK® performance domains are [1]: Stakeholder, Team, Development Approach and Life Cycle, Planning, Project Work, Delivery, Measurement and Uncertainty.

4.1 Stakeholder

The first performance domain of the PMBOK® [1] is the Stakeholder. Depending on the project, different public, private, community and civil society stakeholders may be involved. They may have claims or be contracted to carry out a mandate. To help identify stakeholders, one informant (Employee 22) stated:

> Stakeholders are a little less forgotten with the project management tool because we must do a project charter. We must identify all our stakeholders, including internal and external stakeholders. In short, when you present your project to the steering committee, your stakeholders need to be identified and you need to have names of internal people from the beginning, so that people feel involved.

In terms of external stakeholders, many other public and private organizations may be involved in public infrastructure projects. For example, some public infrastructure and utilities (e.g., telecommunications, gas, heating and electricity networks) are managed by various government and private entities. Utilities such as power lines are managed by the provincial government (Quebec), while telecommunications lines are managed by private companies. For projects involving infrastructure or utilities, informants report that they must submit their requests to the relevant authorities and wait for a response according to the authorities' schedules. This situation is very common and creates some tension because project team members have no control over the deadlines for these tasks. Another element of managing deadlines with multiple public stakeholders is respecting the end dates of different grants. According to the research participants, public funding sources come from all three levels of government.

In addition, for projects that revitalize neighbourhoods by upgrading or constructing public infrastructure, affected residents and businesses may be asked to comment on whether they accept the project. To encourage their participation, the public organization involved conducts consultations. According to several research participants, these consultations are a challenging activity to undertake. One informant in this research shared this (Employee 16):

> Yes, we put ourselves forward as a panel and people come and throw tomatoes at us because the happy people don't come. In fact, it's the negative people who come. In Project X, especially with the parking lot issue, people were afraid and the area around their homes was invaded. Today they're happy, they're calm, but it's not easy, we don't know who we're going to meet, what they're going to tell us.

The public can also influence project approval by lobbying elected officials. One employee who participated in the research said (Employee 46):

> When I'm on the executive committee, people say, '46, we need to tackle such and such a project, we need you to make a business case for it,' but we don't need it, 'but I promised to do it during the campaign.' Oh, well, that's what motivates them.

For the Stakeholder performance domain, the elements revealed are identification, tensions due to delays caused by external stakeholders, public consultation and public influence. The subsequent performance domain is Team.

4.2 Team

To locate the Team performance domain in this research organization, it is necessary to consider the organizational structure. The participants refer mainly to the respect of hierarchical levels and the simultaneous management of routine operations and project activities. One informant described the tension he felt (Employee 41):

> There you have the notion of hierarchy, the project is not above the line manager who is in control of the person's work, or the coordinator of the person's work and that's the problem. We see it a lot. You have the project manager and you have the division manager, you have two directions, two sources of direction that don't necessarily complement each other because the goals are not the same. There's one that's more operational and then there's one that's more project-oriented. And that's what's difficult. You can't necessarily ask the same person to do two different things because afterwards they're not going to do the two things correctly or completely. And that's the reality of what I see and what we experience.

In addition to these tensions between operational and project activities, in projects where several administrative units are mobilized, one informant adds (Employee 26): "Regarding priorities, each department gives more priority to its own priorities than to projects spread over several departments." As a result of this management of priorities, some respondents believe there is a certain degree of competition in terms of access and availability of staff. One informant explains (Employee 22):

> I don't think that we're, we're in a healthy competition, I'm not going to prevent a fellow employee's project, but it's certain that if you prioritize your own when you're under pressure from management, you're going to wait for the pressure to come from management to almost help your fellow employee, I don't think that's creating synergy, I don't know, maybe it can slow down collaboration a little bit.

In addition to these tensions, in projects where different administrative units are involved, the members must adapt to different ways of doing things. Some people mention that during a project, certain processes can change without warning, which becomes a new source of tension.

Regarding the creation of project teams, the interviewees say that they must document the profiles they are looking for and present them to the business unit managers, with one saying (Employee 22): "I think you also need to have good organizational knowledge; it helps with project management. I think you also need to know what the other departments are doing and what their involvement is […]." The choice of each project team member is up to the managers, as they have functional responsibility for their staff. For another informant, although he knows his organization well, he says (Employee 43):

> It's not always obvious what their background is, you don't... you don't necessarily get their resume. When somebody joins your team, you don't necessarily have their whole pedigree. Me, I'm in project management, sometimes I'm in water and environment service... the technicians that are loaned to me for projects, they come to our meetings. We think they've got it; we try to get as much out of them as we can, but we don't know what else they can do, we don't necessarily know everything...

An informant manager suggests that the validation of targeted team members could be done by the project manager's manager, since they have a broader vision of the entire organization. Then, once future project team members have been targeted, a request for their release must be made to their manager. This manager can refuse and target another person, or accept the request. However, one informant (Employee 11) says:

> My experience was that sometimes I was told that only 50% of the resource was available. So, in my management, I didn't have access to a full resource, I had access to half. That wasn't just a random estimate, it was based on statistics that we were able to collect. Sometimes the availability could be as low as 75% because there were too many operations that week and I was running late. That caused stress. Other times, the manager would come to see me because there were production problems, which happened quite often.

In such circumstances, project teams must respect the manager's decisions and reevaluate their project planning based on the resources allocated. If the project manager wants to hire a person from outside the organization to fill the gap between the project's human resource needs and the resources obtained internally, they must request an additional budget allocation from the six levels of hierarchy. It's important to note that project team leaders target human resources and estimate the time needed to complete each task; however, if that resource exists within the organization, there is no estimated budget allocated for it. According to the organization's administrative reports, this approach of either estimating or not estimating the cost of human resources partly explains the significant discrepancies in project costs between the estimated and actual expenditures.

One interviewee (Employee 15) says that when forming the team:

> It's important to communicate the project clearly, to have the idea, the structure, the vision of the project, the elements of the project, how we approach the project. It's not everyone who starts with a project, how to make sure people understand what's at stake, how they're going to be approached to provide information, the needs of the project…

Some informants describe this as an ideal situation for starting a project. However, others mentioned that the vision and stakes of projects are not always precise because the customer may not fully understand their requirements. Additionally, project managers are professionals who must adhere to the demands of higher hierarchical levels, even if they don't fully comprehend a project request. Nevertheless, several employee informants emphasized the importance of swiftly bringing team members together to ensure their ownership and involvement throughout the project.

The Team performance domain encompasses elements related to respecting the hierarchical levels of the lightweight matrix structure, managing tensions between routine and project activities and identifying and allocating team members. The subsequent performance domain is Development Approach and Life Cycle.

4.3 Development Approach and Life Cycle

Based on the project management administrative documents of the organization involved in the research, the official development approach is predictive. During the pre-project phase, various aspects such as identifying needs, stakeholders, existing data, the situation, expected benefits, preliminary budget, potential funding sources, risks, complexity, schedule and alignment with the organization's strategic planning are considered. One informant mentions (Employee 14):

> Good practice starts with identifying the need and the whole business case. It's true that this is rarely the case in real life - maybe it's not a tragedy, maybe it's natural for there to be trial and error and people don't necessarily know what they want. However, this should be included in the process.

Subsequently, if the project is included in the investment plan by the highest authority, preparations commence, including the creation of a project charter containing details about the governance structure, grant applications, communication plan, public consultations and marketing research.

Once the necessary documents receive approval from the designated authority, such as the administrative unit management, general management, or steering committee, the project team progresses to the planning phase. In this phase, the focus is on designing the project management plan, business case, environmental assessment, real estate appraisal, public consultations, agreements and protocols.

After completing these initial documents, the project teams advance to the plans and specifications phase. Here, they develop formal business cases, detailed environmental assessments, plans and acquisition requests, site monitoring plans, utility relocations, cost and schedule analyses, stakeholder management strategies and communication plans.

The subsequent phase is execution, which includes bidding and contracting.

Finally, the project culminates in the evaluation phase, followed by knowledge transfer activities. Notably, the organization utilizes a project management tool serving as a repository for various documents and as a reporting tool for team members.

Based on the annual audit reports, it's evident that some project teams utilize the adaptive technique, despite it not being formally outlined in a governance framework. One research participant (Employee 46) elaborated on this:

> That's true. But I'm telling you, I don't want to manage buildings or shoes in Agile mode. When we talk about projects, what kind of projects are we talking about? Because I'm dealing with software projects, IT projects and it's clear that the agile method works very well. I wouldn't think of building an aquatic center tomorrow morning in Agile mode.

Other informants' express confusion about this unofficially described approach because they lack clarity on when and under what circumstances they can use it, as well as how to maintain accountability within the project management tool. One informant (Employee 33) highlighted the lack of understanding among employees regarding the adaptive or agile approach:

> A bit like what was said at the start, the justification and the context are poorly defined at the beginning. In the sense that we've done a little bit of it, we're doing it and as we go along, we realize that maybe the analysis hasn't been done in its entirety. We're in agile mode, so there's no doubt about it, but it's more of a buzzword because we should have a backlog that's already prioritized. But we're just adding and adding and expecting to get it all done, but that's not what was planned.

For other research participants, the adaptive approach allows for day-to-day planning, which they find more relevant due to the unpredictable nature of operations management.

The Development Approach performance domain encompasses both formally predictive and informally adaptive approaches, leading to confusion among employees. The project life cycle includes phases such as justification, planning, implementation, execution and closure. The subsequent performance domain is Planning.

4.4 Planning

In the administrative documents of the participating organization, planning specifically encompasses the feasibility study, the functional and technical plan of the project results and the analysis of options for the results. However, informants emphasize that managers often fail to assess the availability of human resources needed for these project activities. Despite being integrated into employees' routine work, they struggle to meet these demands. As one participant noted (Employee 12):

> It's true, when there's a project mode, the operation is never considered. It's as if people magically tell themselves that it's going to continue its own... As project managers, they hope that people are there full-time for them, but that's not the case, everything continues as before, these people aren't replaced.

In cases where internal expertise is lacking, such as technical studies, a request for proposals may be issued.

Various aspects of the Planning performance domain have been discussed, highlighting the need for comprehensive planning encompassing assessments, deadlines, costs and scope. The subsequent performance domain is Project Work.

4.5 Project Work

In the Project Work performance domain, the aspect most frequently discussed by informants is communication. Throughout project management, team members communicate via email and meetings. However, some participants express frustration about being invited to meetings excessively, as one mentions (Employee 22):

> We receive numerous invitations to meetings, but it's essential to question whether it's relevant for us to attend. If you have to ask yourself that question, it's because we tend to invite too many people all the time. Even our meetings consume a lot of our time.

Other participants highlight that communication between the project team and other stakeholders presents challenges in terms of comprehension. While team members perceive themselves as competent in their field, engaging with the public and facilitating public consultations requires different competencies. To enhance communication with diverse stakeholders, participants express the need for professional support from communication experts.

Regarding the transfer of knowledge acquired during a project, respondents mention that employees in their functional unit can participate in workshops to share professional practices or new insights with others.

This performance domain primarily revolves around communication and information management. The subsequent performance domain is Delivery.

4.6 Delivery

For the informants, this performance domain, which also refers to a phase of the project, represents a management challenge. The steering committee or managers may ask to make up for accumulated delays. Considering that some deliverables have fixed dates set due to grants received or election campaigns, the pressure on the project manager and companies is substantial. It's worth noting that project delivery is often achieved through bidding. At this point, the project team stops meeting, and the project manager follows up with the companies. If the project manager needs clarification or explanation, they go directly to their colleague. Project managers are also responsible for resolving any planning issues raised by contractors.

Some informants express regret over the perception that project implementation marks its completion, whereas IT systems established require ongoing updates and explanations to other staff. Concerning the public infrastructures erected, participants believe that attention should also be given to devising procedures for their maintenance.

This performance domain is less prevalent in public sector organizations due to the reliance on bidding processes. The subsequent performance domain is Measurement.

4.7 Measurement

For managers, project activities align with the job descriptions of the organization's permanent employees. Consequently, when estimating the cost of each project, the salaries of internally mobilized employees are typically not factored in. One research participant encapsulated this sentiment (Employee 15):

> Often these are things that can happen. What we're talking about here is quite new, the project management tool. We're starting to use it and we're planning projects in advance, even very, very, very far in advance. Often, time is so unpredictable that you lose control, you say to yourself, 'Well, we're not going to be able to meet our deadline.' Often, I put it on my board, I follow it and it slips a little bit. We don't always have the resources.

Another measurement management tool is the control board. It summarizes the project and its key indicators, such as cost and schedule. Each indicator is assigned a color: green indicates the plan is being met, yellow signifies a variance and red denotes a significant variance. Some managers express a preference to see only green on the control board, fearing that any other color might prompt inquiries from the media or elected representatives about the project's progress. As one individual (Employee 51) articulated: "Well, that's it. What's the point of a control board when it's green and you can't see red? Otherwise, your control board is useless."

This performance domain highlights the presence of accountability albeit underutilized. The subsequent performance domain is Uncertainty.

4.8 Uncertainty

The final performance domain is Uncertainty. One respondent mentioned that many employees are not familiar with risk management because it is not commonly practiced in the management of routine tasks. Regarding digital projects, one person stated (Employee 33):

> Often, a technology is selected before the project begins. Someone has done an analysis. We're given a little bit of a schedule, but when we realize afterwards... when we get a little bit more into the functional analysis, well, 'it's not feasible.' That's when we become the bad guys... it's hard to say, but…

This last performance domain highlights the lack of competence in risk identification and management. Overall, these performance domains are condensed into two questions about the challenges and needs of public managers. These responses will lead to the third question, which compares the ISO 21502 [6].

5 Discussion of the Results

5.1 Question 1: What are Public Sector Organizations' Current Project Management Challenges?

Analysis of the results shows that despite the detailed descriptions of processes, techniques and management tools in the various editions of the PMBOK®, public managers and project team members continue to face numerous challenges. Table 4 summarizes these challenges and categorizes them according to the PMBOK® 7th Edition, Performance Domains [1] to track their evolution.

The analysis of the results indicates that two performance domains of the PMBOK® [1] aggregate a multitude of challenges for managers and team members of public projects: Stakeholder and Team. Within the Stakeholder domain, communication challenges arise due to the multitude of individuals and organizations involved in public projects. They often harbor divergent interests and needs, and project team members may lack the necessary competencies and resources to effectively manage them and facilitate public consultation meetings.

Table 4 Summary of public project management challenges

PMBOK® 7th edition and performance domains	Managerial challenges encountered in public projects
Stakeholder	✓ Lack of control by external public stakeholders in carrying out their tasks ✓ Tensions in public consultations ✓ Numerous communications with external stakeholders ✓ Difficult communication and tensions between administrative and political circles ✓ Multiple grant requirements
Team	✓ Different priorities between routine and project activities ✓ Different processes are established in each functional unit ✓ Lack of knowledge of employee job descriptions when selecting team members ✓ Gaps between human resource requests and allocations ✓ Lack of financial resources allocated to internal human resources ✓ Lack of project definition ✓ Difficulty in managing the individual aspect of routine activities and the collective aspect of projects
Development approach and life cycle	✓ Gap between the official project development approach and the one used ✓ Use of adaptive project approach without formalized description ✓ Lack of knowledge about project management techniques and practices
Planning	✓ Work overload for employees with dual functions
Project work	✓ Difficulty managing multiple internal communications
Delivery	✓ Difficulty managing fixed project deadlines
Measurement	✓ No cost allocation for internal human resources ✓ Poor management of the board of control
Uncertainty	✓ A lack of knowledge and competence in risk management

The first lesson learned from the analysis of the results is that the persistent challenges for public project managers and team members lie in stakeholder and team management, specifically the relational aspects of project management.

Furthermore, certain challenges identified by Engwall and Källqvist [27] endure, including the concurrent management of routine tasks alongside project activities, leading to potential conflicts in priorities and constrained access to internal human resources.

> The second lesson learned from the analysis is the importance of integrating management tools like a control board for both routine tasks and project activities to ensure the optimal utilization of human resources.

Another persistent management challenge is estimating the cost of internal human resources. Lack of cost estimates and uncertainty about staffing capacity can impede project completion or lead to increased requests for additional funding. It's important to note that such requests can be time-consuming as they must adhere to the approval processes within the organizational structure [21].

> A third lesson learned from the analysis is that the processes put in place to manage routine tasks can hinder the timely completion of project activities. This factor should be considered when planning project schedules.

5.2 *Question 2: Does the Project Management Body of Knowledge Meet the Needs of These Public Managers?*

Analysis of the results indicates that the PMBOK® adequately fulfills the technical requirements for managing public projects. The participating organization offers standardized project management processes and tools to project teams. However, on a day-to-day basis, team members adjust these processes based on project dynamics and circumstances. While the guidance provided in the 7th edition of the PMBOK® [1] could support this adaptation, challenges persist in integrating project processes with those for routine activities.

> A fourth lesson learned from the analysis: While the PMBOK® 7th edition can fulfill the knowledge requirements of public managers; it appears underutilized among the surveyed individuals.

Considering that employees of public organizations are selected for their professional expertise, grasping this reference might pose a challenge for newcomers, as Gottlieb [28] pointed out. Bloom's taxonomy of competency development delineates three levels: knowledge, understanding and application [29]. While the PMBOK®

imparts knowledge of project management, its comprehension and implementation are observed in real project scenarios. Alternatively, internal training sessions, as suggested by the research organization and Lee et al. [30], could bridge this gap. Such sessions would facilitate discussions on understanding and applying the processes, thereby nurturing project management competencies, as emphasized by Loufrani-Fedida [31].

> A fifth lesson learned from our analysis of the results is that merely reading the PMBOK® isn't adequate for cultivating project competencies.

5.3 Question 3: Could ISO 21502 Be a More Appropriate Reference to Address the Needs and Challenges of Public Managers?

Analysis of the results reveals that the lack of project knowledge and competencies among team members poses a significant management challenge, potentially impeding project performance [32–35]. Particularly highlighted are deficiencies in project definition and risk management skills. Moreover, it was observed that team members are primarily selected for their professional expertise or specialized knowledge within the organization. In this regard, the ISO 21502 standard [6] could enable public organizations to address certain management challenges. Table 5 presents the Integrated Project Management practices related to these challenges.

Implementing these few practices would help focus on managing stakeholder engagement, project team development, communication, schedules, reporting, and risk. Better identification and a mobilization plan can avoid tensions and communication difficulties in stakeholder engagement. Resource management refers to the integration of the project into the organization and the establishment of the project team. Thus, in the context of project management, a better understanding of certain human resource practices could help members to better manage their priorities. What's more, this special attention to human resources management could enable public organizations to better integrate them into project planning. The challenges in managing communication, scheduling, reporting, and risk are due to a lack of knowledge.

> A sixth lesson learned from analyzing the results is that the integrated project management practices outlined in ISO 21502 could be offered to project team members.

Table 5 Public project management challenges that can be addressed using the ISO 21502 standard

PMBOK® 7th edition and performance domains	ISO 21502 standard and integrated project management practices	Public project management challenges that can be addressed using the ISO 21502 standard
Stakeholder	Stakeholder engagement	✓ Tensions in public consultations ✓ Numerous communications with external stakeholders ✓ Difficult communication and tensions between administrative and political circles
Team	Resource management	✓ Different priorities between routine and project activities ✓ Lack of knowledge of employee job descriptions when selecting team members ✓ Lack of financial resources allocated to internal human resources ✓ Lack of project definition ✓ Difficulty in managing the individual aspect of routine activities and the collective aspect of projects
Project work	Communication management	✓ Difficulty managing multiple internal communications
Delivery	Schedule management	✓ Difficulty managing fixed project deadlines
Measurement	Reporting	✓ No cost allocation for internal human resources ✓ Poor management of the board of control
Uncertainty	Risk management	✓ A lack of knowledge and competence in risk management

However, there remain certain management challenges that are identified by the PMBOK® [1] Performance Domains, but are not well addressed by the ISO 21502 standard [6], as described in Table 6.

These persistent management challenges relate to the management of external stakeholders and dual roles. As for the development approach and the project life cycle, the ISO standard [6] touches on them briefly in its section on project management concepts, but not in practice. This general aspect of the context provides few answers to the challenges faced by public organizations.

Table 6 Managerial challenges encountered in public projects

PMBOK® 7th edition and performance domains	Managerial challenges encountered in public projects
Stakeholder	✓ Lack of control by external public stakeholders in carrying out their tasks ✓ Multiple grant requirements
Team	✓ Lack of knowledge of employee job descriptions when selecting team members ✓ Gaps between human resource requests and allocations
Development approach and life cycle	✓ Gap between the official project development approach and the one used ✓ Use of adaptive project approach without formalized description ✓ Lack of knowledge about project management techniques and practices
Planning	✓ Work overload for employees with dual functions

6 Conclusion

Public organizations in Quebec are gradually integrating project management to reflect New Public Governance. This chapter outlines the evolution from a functional organizational structure to a lightweight project matrix and the associated challenges of balancing project activities alongside routine tasks. Building upon the insights of Blixt and Kirytopoulos [32] and Pakarinen and Virtanen [15], this chapter examines the specific challenges encountered by public organizations operating within a lightweight matrix structure when managing their projects. By identifying these challenges, public managers can tailor training initiatives more effectively. A preliminary training activity could be the promotion of the ISO 21502 standard [6] for team members who do few projects and the PMBOK® for managers and project leaders. The second training activity could facilitate learning through project management scenarios, where team members are encouraged to apply knowledge acquired from repositories to real-life situations, as recommended by Desjardins et al. [18], Lee et al. [30], Eltigani et al. [33] and Desjardins and Horincq Detournay [36]. Suggested avenues for future research include continuing these reflections on the development of project management competencies. Continued focus on the challenges encountered by managers and team members in public project teams will aid in identifying pertinent training objectives, considering their dual roles.

References

1. Project Management Institute: A Guide to the Project Management Body of Knowledge (PMBOK® Guide)—The Standard for Project Management. Project Management Institute, Chicago (2021)
2. Marion, J., Richardson, T.: Managing Projects with PMBOK 7: Connecting New Principles with Old Standards. Business Expert Press, New York (2022)
3. Bredillet, C.: Bureau de projets en contexte organisationnel public. In: Mazouz, B. (eds.) Gestion de projets en contexte public. Presses de l'Université du Québec, Québec (2017)
4. Mazouz, B.: Gestion de projets en contexte public. Presses de l'Université du Québec, Québec (Québec) (2017)
5. Romero-Torres, A., Brunet, M., Lalonde, B., Aubry, M.: La Maturité en Gestion de Projet Améliorer la Performance Organisationnelle. Presses de l'Université du Québec, Quebec (2024)
6. International Organization for Standardization. Project, Programme and Portfolio Management. Guidance on Project Management, in ISO 21502, p. 52 (2020)
7. Moore, M.H.: Public value accounting: establishing the philosophical basis. Public Adm. Rev. **74**(4), 465–477 (2014)
8. Osborne, S.: Public Service Logic: Creating Value for Public Service Users, Citizens, and Society Through Public Service Delivery. Routledge (2021)
9. Lane, J.-E.: New Public Management. Routledge, London (2000)
10. Lorino, P.: In search of lost value: building processes that create value in the public sector. Polit. Public Adm. **17**(2), 21–34 (1999)
11. Ansell, C., Torfing, J.: Public Governance as Co-Creation A Strategy for Revitalizing the Public Sector and Rejuvenating Democracy. University of Cambridge ESOL Examinations, Cambridge (2021)
12. Government of Quebec, Quebec Infrastructure Plan 2022–2032, C.d. Trésor, Editor, p. 220. Quebec (2022)
13. Desjardins, F.: Proposition d'un cadre systémique en soutien à l'amélioration des compétences des membres des équipes projet: le cas d'une organisation municipale à structure matricielle faible, in Sciences économiques et administratives. Université du Québec à Chicoutimi, Canada (2020)
14. Government of Quebec, Guide de gestion axée sur les résultats, S.d.C.d. trésor, Editor (2014)
15. Pakarinen, M., Virtanen, P.J.: Matrix organizations and cross-functional teams in the public sector: a systematic review. Int. J. Public Sector Manage. **30**(3), 210–226 (2017)
16. Osborne, S.P.: Public Service Logic: Creating Value for Public Service Users, Citizens, and Society Through Public Service Delivery. Routledge, New York, NY (2021)
17. Kuprenas, J.A.: Implementation and performance of a matrix organization structure. Int. J. Project Manage. **21**(1), 51–62 (2003)
18. Desjardins, F., Jean, É., Bredillet, C.: Opening the black box of project team members' competencies improvement in a public sector organization for a successful transition to the project society. Proj. Leadersh. Soc. **3** (2022)
19. Project Management Institute. A guide to the project management body of knowledge: (PMBOK® guide), in Global standard. Project Management Institute, Newtown Square, Pennsylvania, USA (2017)
20. Sy, T., Beach, L., D'Annunzio, L.: Challenges and strategies of matrix organizations. Hum. Resour. Plan. **28**(1), 39–48 (2005)
21. Hobday, M.: The project-based organisation: an ideal form for managing complex products and systems? Res. Policy **29**(7–8), 871–893 (2000)
22. Turner, J.R., Anbari, F., Bredillet, C.: Perspectives on Projects. Routledge, London (2010)
23. Myers, M.D.: Qualitative Research in Business & Management, Third edition. SAGE Publications Ltd., London (2020)
24. Lorino, P.: Pragmatism and Organization Studies. Oxford University Press, Oxford, United Kingdom (2018)

25. Maxwell, J.A., Chmiel, M.: Generalization in and from qualitative analysis. In: The SAGE Handbook of Qualitative Data Analysis, vol. 7, pp. 540–553 (2014)
26. Yin, R.K.: Case Study Research and Application: Design and Methods, Sixth edition. SAGE, Los Angeles (2018)
27. Engwall, M., Källqvist, A.S.: Exploring the Multi-Project Matrix. Citeseer (2001)
28. Gottlieb, M.R.: The Matrix Organization Reloaded: Adventures in Team and Project Management. Bloomsbury Publishing, USA (2007)
29. Bloom, B., Engelhart, M., Furst, E., Hill, W., Krathwohl, D.: Taxonomy of Educational Objectives: The Classification of Educational Goals, Handbook I: Cognitive Domain. David McKay, New York (1956)
30. Lee, L., Reinicke, B., Sarkar, R., Anderson, R.: Learning through interactions: improving project management through communities of practice. Proj. Manag. J. **46**(1), 40–52 (2015)
31. Loufrani-Fedida, S.: Management des compétences et organisation par projets: une mise en évidence des leviers de gestion conjointe. La Revue des Sciences de Gestion: Direction et Gestion **43**(231/232), 73–83 (2008)
32. Blixt, C., Kirytopoulos, K.: Challenges and competencies for project management in the Australian public service. Int. J. Public Sect. Manag. **30**(3), 286–300 (2017)
33. Eltigani, A., Gardiner, P., Kirkham, R., Williams, T., Ou, L., Calabrese, A.: Learning in and from projects: the learning modes and a learning capability model. Prod. Plan. Control **31**(10), 786–798 (2020)
34. Loufrani-Fedida, S.: La gestion des ressources humaines à l'épreuve du management de projet: état des lieux et perspectives de recherche. @GRH **32**(3), 9 (2019)
35. Suikki, R., Tromstedt, R., Haapasalo, H.: Project management competence development framework in turbulent business environment. Technovation **26**(5–6), 723–738 (2006)
36. Desjardins, F., Horincq Detournay, R.: Rôles du coaching dans la gestion de projet: une réflexion andragogique basée sur une analyse inductive. Enjeux et société 9(2), 173–198 (2022)

Simplified Stakeholder Management in Project Management: Biopharmaceutical Case Study

Kenneth David Strang

Abstract A biopharmaceutical drug discovery project was leveraged as a case study to illustrate a proposed simplified stakeholder management approach. The driving force underlying this chapter was the lack of effective stakeholder management in projects thought to be a contributing factor to the approximate 50% project failure rate. Existing stakeholder management techniques in the project management body of knowledge were based on stakeholder theory from the strategic management discipline. This encourages decision-makers to consider the non-related interests of stock shareholders, customers, employees, suppliers, and the local community—such as ethics and social good. Those stakeholders are not necessarily as relevant in projects. In a project, the key stakeholders are those affected by WBS activities and deliverables. Therefore, a new simplified stakeholder management model was proposed by interviewing a panel of experienced project managers to document stakeholder identification best practices. Machine learning, nonparametric Kruskal Wallis, and Chi-Square contingency analysis statistical techniques were applied to analyze the results. The empirical literature was reviewed to inform the development of two models, the first for stakeholder analysis and the second for stakeholder engagement. Power and impact were argued to be the two key factors driving stakeholder analysis. A stakeholder engagement model was developed to leverage the stakeholder analysis.

Keywords Project stakeholder management · Stakeholder identification · Stakeholder analysis · Stakeholder engagement · Project management · ISO 21502 · Project management body of knowledge · PMBOK · Biopharmaceutical

K. D. Strang (✉)
W3-Research, New York, NY 11041, US
e-mail: professor@kennethstrang.com

University of the Cumberlands, Williamsburg, KY 40769, US

RMIT University, Melbourne, NSW 3001, Australia

K. D. Strang and N. R. Vajjhala (eds.), *International Program and Project Management—Best Practices in Selected Industries*, Information Systems Engineering and Management 31, https://doi.org/10.1007/978-3-031-80275-1_6

1 Introduction

This chapter summarizes the research-driven approaches for a project manager (PM) to apply stakeholder management. We present examples using a case study in the biopharmaceutical drug discovery industry (biopharma and biopharmaceutical).

1.1 Rationale for Biopharma Stakeholder Management Best Practices

The biopharmaceutical industry is important to study due to the high project costs, the large profits, as well as the potential for project outcomes to combat diseases and treat medical problems. A typical biopharma project costs $2.56 billion and often takes 10 years to reach the market [1]. The global market for the biopharmaceutical industry is estimated to be $516.79 USD Billion USD (at the time of writing), which is projected to grow 8–9% annually reaching USD 761.80 Billion USD by 2028 [2]. The largest geographical market share is the USA, at 40% [3]. Biopharma drug products include life-saving antibodies, hormone growth cells, regeneration proteins, disease vaccines, immune boosters, bad cell blockers, and other treatments. The global biopharma market leaders in approximate descending market size are Amgen, Eli Lilly & Company, Johnson & Johnson, Abbvie, and Bristol-Myers Squibb; while the other key players are: Pfizer, GlaxoSmithKline, Novartis, Hoffmann-La Roche, Merck. Sanofi, AstraZeneca, Novo Nordisk, Bayer AG, and Takeda Pharmaceutical [2]. The case study involved one of the above firms during a drug development project cycle.

The project failure rate in the biopharma industry is 99% when considering the research development ideation phase as the initiation point, which reduces only to 88% from the clinical trials phase to approval from the U.S. Food and Drug Administration (USFDA) or the International Council on Harmonization (ICH) in Europe [1]. High project failure rates are common for research & development industries. Nonetheless, stakeholder management becomes important after the biopharma ideation and planning phases are completed [1].

Stakeholder management in any industry is important for project managers to apply correctly because the longitudinal global project failure rate is approximately 50% [4–6]. To some extent, project failures may be attributed to unhappy stakeholders [7] or because the PM did not correctly identify, engage, and manage stakeholders [8]. Thus, stakeholder management best practices may be missing in the guidelines to help PMs.

1.2 Gap in Literature for Stakeholder Management Best Practices

There is a gap in stakeholder management best practice literature [8], especially for specialty sectors like the biopharma industry [1]. According to Martin [1], PMs in the pharmaceutical industry commonly lack project management training and certified PMs lack an understanding of the complexity involved in drug discovery projects. Strang and Vajjhala [4, 9] found empirical evidence that certified PM's may not be aware of stakeholder management guidelines because it is a relatively new knowledge area documented in the project management body of management (PMBOK) published by the Project Management Institute [10].

Another gap is that the PMBOK [10] and ISO 21502 [11, 12] standards do not provide relevant examples or tools to help identify, analyze, engage and manage project stakeholders. For example, the ISO standard is very basic—there are only two PM activities identified for the stakeholders: Stakeholder identification and stakeholder management [12]. The process details for identification, analysis, engagement, and management are missing. Furthermore, the primary output of "4.3.10 Manage stakeholders" [11] (p. 18) is a change request which is contrary to what is usually done in practice. Stakeholders could certainly provide the input for a change request, but it is the PM who determines if the change request is required and if so presents it to the project authority for approval. The output would be approval, modified change request conditional approval or rejection. The author argues the problem with the PMI and ISO standards is that they are missing accurate best practice details of how to perform stakeholder identification, analysis, and engagement. Additionally, PMI and ISO focus on stakeholder management. We assert the PM should instead be focused on stakeholder engagement, it is the team resources, not stakeholders, who need to be managed.

A final problem is the over reliance of speculative American Academy papers, where bias exists due to using university student samples, simulations, and self-cited literature reviews to propose best practices [8]. Many of the project stakeholder management guidance articles were based on literature reviews, speculation, or small sample sizes [5, 7, 13]. Stakeholder management theory was never properly credited to Mary Parker Follett who first proposed it 60 years before Freeman and other American Academy researchers popularized the theory without acknowledging her contributions [14]. Thus, the goal of this chapter is to provide research driven stakeholder management best practices, with a balance of theory and empirical evidence, along with examples about how to perform stakeholder identification, analysis, and engagement.

1.3 Research Question and Methods

The research question (RQ) underlying this chapter is what simple effective best-practices for stakeholder management are, with a focus on the biopharma industry as a case study example. An abductive literature review strategy was applied, which means only selected papers were selected for discussion [15]. A pragmatic ideology was taken, meaning customized techniques were used to collect, analyze, and interpret the data as well as to conduct the single case study [15, 16].

2 Project Stakeholder Management

This chapter starts by presenting empirical research analysis to show the important project stakeholder management facets. Then the facets are interpreted by the author, using literature, and examples are shown based on a case study. The focus is on demonstrating actual practices rather than discussing theories.

2.1 Research Driving the Gaps in Stakeholder Management Practices

Data were collected and analyzed from subject matter expert (SME) practitioners during three recent (2023–2024) project management conferences held internationally (outside the U.S.). Most of the common worldwide industries were covered, including manufacturing, utilities, service, agriculture, forestry, healthcare, information technology, software, public administration/government, and education. The average years of experience was 9.5 (standard deviation = 5.7), and all of the G20 nations were represented. The purpose was to answer the RQ based on the perspective of SME PM practitioners. Later in this chapter, research will be reviewed to correlate the SME findings. Audio recordings were captioned into transcripts and then analyzed using unsupervised machine learning (ML) techniques.

Figure 1 is a summary of the most important keywords, based on priority identified by the SME's highlighted in red, with the frequency shown by font size. To accomplish that, the ML tokenized the transcribed audio into lowercase full words without accents, dropping punctuation, whitespace, URLs, single letters and numbers. English work normalization was used based on the Lemmagen pre-trained model in Python. Unimportant stop words were removed such as prepositions. Topic modeling, similarity hashing, and keyword analysis were applied to create Fig. 1.

Figure 1 indicates stakeholder identification, engagement analysis, and group communication were the most important topics. Notice that stakeholder management is not a keyword. These findings show this chapter is focused on the correct topic, which keywords were relevant for project stakeholder management. This research

Fig. 1 Important stakeholder management keywords from SMEs (N = 69)

data will be used later to explain which techniques the SME practitioners recommend for stakeholder identification, analysis, and engagement.

2.2 Stakeholder Theory in Project Management

Stakeholder theory was originally discussed by Follett and then decades later it was extended by Freeman as well as other academic researchers [14]. Although stakeholder theory originally referred to addressing the interests of people beyond the shareholders of a corporation [14], it was enhanced in the PMBOK to include anyone impacted by, having interest in, or having the power to influence, a specific project [10, 17]. That is the definition applied in this chapter.

A simple explanation of project stakeholder management theory is that a PM must continually identify stakeholders (including phantom stakeholders who may be recognized later), document them, develop engagement strategies to win their support, and communicate with them [13].

In the project context, executive stakeholders are the sponsor, authority, and steering committee members if applicable. The project team—the resources doing the work—are a special stakeholder considered as a group but not individually. Internal stakeholders are those within the organization, while external stakeholders are typically regulators, customers, suppliers, or contracted specialists [10, 18, 19] Phantom stakeholders are unknown to the PM, but once identified, it is important to include them. Important stakeholders beyond the executive and project team can support or impede the project so their interests must be accommodated or at least skillfully appeased by the PM [10, 18, 19].

The literature and practices differ on how to perform stakeholder management. Project stakeholder management practices may differ across industries, such as manufacturing drugs versus manufacturing automobiles or consumer electronics. Generally, a PM goes through the process of identifying, analyzing, and engaging stakeholders, although the various bodies of knowledge or research differ on how many steps and which techniques are needed. Unfortunately, most of the literature is complicated or incomplete, which could leave a PM confused about stakeholder management. Those techniques are discussed individually in separate sections below.

2.3 Stakeholder Identification Methods in Project Stakeholder Management

We surveyed the literature for the generally accepted project management stakeholder identification methods across all industries, when applying stakeholder theory [7, 8, 13, 20–23]. The common methods were:

- ask sponsors;
- look at organization directories;
- ask team members (not other stakeholders);
- snowball by asking existing stakeholders to identify new ones;
- PM research by using current knowledge and project documentation;
- and surveying the organization's employees.

These above categories plus 'other' as a catch-all bucket was used to survey the SME's (N = 69). The ranking of each method for stakeholder identification was on a 1–5 scale where 5 was highest, for projects in a new context versus projects repeated in the same organizational context. The descriptive characteristics of the SME sample were the average age of 53 (SD = 4.8), the mean years of experience was 14.8 (SD = 4.1), 81% were male, and all were certified by PMI. There were several 'other' methods reported, but with low ranks so they were not included in the statistical analysis. For completeness, the other stakeholder identification methods were focus groups and regulatory agencies who maintain lists of approved specialists.

We conducted a chi square contingency analysis of the SME group interview results to determine if the proportions of ranks were different by stakeholder identification method conditioned on the project context of repeated versus new, and all were significantly different based on the 95% confidence level. Not all responses were provided so a pairwise test was not possible, and the subgroup sizes were too small for reliable applying parametric techniques. For a new project context, methods were ranked differently, $X^2[N = 33] = 31.441$, $p = 0.05$, likelihood ratio = 41.896, $p = 0.003$ (accepted as significant). For a repeated organizational project context, methods were also different, with $X^2[N = 36] = 40.879$, $p = 0.004$, likelihood ratio = 43.657, $p = 0.002$ (significant).

A Kruskall Wallis comparison was performed on ranks as the dependent variable, using the methods for nominal grouping and new versus repeat context as an independent factor. The results were: KW-Test for method [DF = 5] = 18.791, p = 0.002 (significant); and Context [DF = 1] = 1.353, p = 0.245 (ns). This can be interpreted that the use of some or all stakeholder identification methods were ranked significantly different, but they were not significantly different when contrasting existing versus new contexts. When considering the chi square test revealed different probabilities of 'applying' stakeholder identification method ranking across new versus repeated contexts, and the numerical ranking was also significantly different from one metho to the other although not different in a repeated versus new context, based on the Kruskal Wallis test, this suggests a few of the methods used differently, while most were similar in rank for any project context (new or repeated).

Indeed, this is what the data revealed when looking at the descriptive statistics, that is, one of the methods were being used differently by the SMEs for stakeholder identification. We used the descriptive statistics to create the plot shown in Fig. 1, to visually illustrate this. Figure 1. presents a line chart of the results for the mean rank (5 is highest) across the methods by the same repeated project context versus a new project context. There were no other methods mentioned by the SMEs.

Figure 2 reveals that the PM will use their own research knowledge if in a repeated context (RepeatProjectContext M = 4.429, SD = 0.535), which was the highest average rank in the data, as compared to a new situation (NewProjectContext M = 2.750, SD = 0.500). The remaining methods were all ranked lower for use in repeated projects as compared to a new situation. The next important stakeholder identification method was to ask the sponsor (NewProjectContext M = 3.846, SD = 1.345 versus RepeatProjectContext M = 3.231, SD = 0.725). Surprisingly, using a company directory was the next ranked method for identifying stakeholders, which was rated slightly higher for a new situation (NewProjectContext M = 3.333, Sd = 0.577) as compared to repeat context (RepeatProjectContext M = 1.667, SD = 0.577). Asking existing the team to identify stakeholders came next (NewProjectContext M = 2.667, SD = 0.577 versus RepeatProjectContext M = 2.333, SD = 0.577). The snowball method means to ask existing stakeholders to identify new ones, which was more likely to be used in existing situations (NewProjectContext M = 3.4, SD = 0.548 versus RepeatProjectContext M = 2.4, SD = 0.548). Finally, a survey was the least important approach to identify stakeholders (NewProjectContext M = 2.6, SD = 0.548 versus RepeatProjectContext M = 2.4, SD = 0.894).

An interesting insight from the above quantitative SME rank analysis was that the method for identifying stakeholders varies most according to the project organizational context. If a PM has already completed a project for the organization, they know some decision-makers and influencers, so they rely on their own research knowledge and secondly by asking the sponsor. For new situations, the PM first relies on the sponsor and then asks other stakeholders as well as searching company directories. A valuable fact derived from the SME interviews but not captured in the statistical analysis was that all PMs used at least three stakeholder identification techniques (sponsors and other stakeholders first), and then either their own research

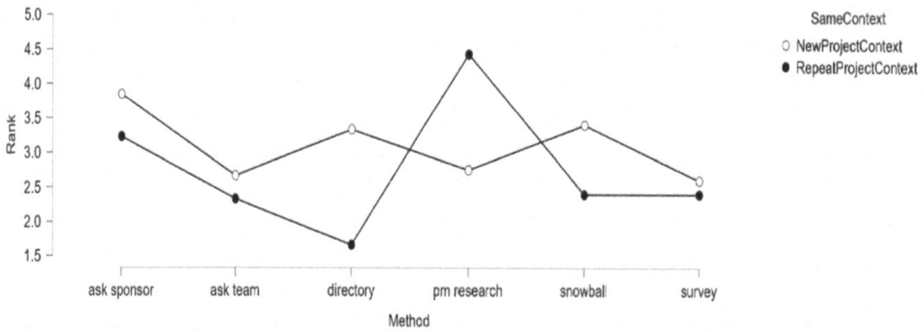

Fig. 2 Stakeholder identification methods by context using SME 1–5 rank (N = 69)

knowledge (for repeat contexts) or company directories (for new situations). A few SMEs used other stakeholder identification methods like surveys or asking the team.

Consequently, we assert a PM ought to use at least three stakeholder identification techniques, starting with asking the sponsor, then asking other stakeholders (snowball method). The third technique should be to review the company directory or organizational structure diagram, because that approach would likely uncover hidden or phantom stakeholders, people with authority and likely to be impacted, so when they are not engaged early, they could disrupt the project during the execution phase. Surveys and asking team members may not be as reliable as asking sponsors, asking other stakeholders and consulting the formal organizational structure governance.

The last step is to develop a stakeholder register database and enter all the stakeholders identified in the previous step. The stakeholder register database could be a simple spreadsheet (as used in this case study) or specialized project management software can be used. Each stakeholder should be entered as a new record or row in the database. The project team and PM should not be entered into the stakeholder register because the PM will be recorded in the charter and the team resources will be listed in the staffing pool database as part of project human resource management (see the project human resource management chapter in this book).

The next task is to determine which fields or attributes are needed for the stakeholder register. We argue there are too many fields described in the literature, but for completeness, all of them will be enumerated below, as adapted from PMI [10, 24] and ISO [11, 12]. We recommend including the bolded attributes.

1. **Stakeholder name**, **nickname**, and **contact details**;
2. **Functional job title** (from home department);
3. **Role** (what they contribute or perform in project, beyond the WBS tasks);
4. **Power** (formal legitimate legal power, in organizational chart, home position supervisor of certain project team resources or project equipment owner);
5. Interest (how much motivation to support or ignore the project);
6. Influence (informal power to support or interfere with the project);
7. **Impacts** (scale of how much their department are impacted by project outcomes or deliverables);

8. Knowledge (how much do they know about the project deliverables);
9. **Group** (category of executive, internal, or external);
10. **WBS** (which WBS work packages the stakeholder is involved with);
11. Priority (determined by PM, an ordinal scale of Low-Medium–High);
12. Status (some combination of custom coding to indicate interest and influence);
13. Update frequency (how often and by what method to update them);
14. **Communication date/notes** (for every important conversation with the stakeholder).

As mentioned above, the project team stakeholder is a group of all project team resources—they are not listed individually in the stakeholder register.

There are additional variations of the above attributes which call the stakeholder register the stakeholder priority matrix, engagement matrix, or sometimes the RACI matrix. One advantage of the RACI (responsible, accountable, consulted, informed) method is that it is well known in the software development industry, but we argue it is less useful in projects or other industries for stakeholder analysis because the attributes do not correspond to practical decision-making elements. Stakeholder maps, stakeholder circles, and stakeholder maps have been mentioned in the literature but without any convincing empirical evidence of their effectiveness.

Additional stakeholder attributes were suggested by some researchers, so we will mention that for completeness. The following stakeholder attributes were adapted from [4, 5, 11, 17–19, 22]:

1. Financial interests (ability to earn money from future project deliverables);
2. Moral ethical values (societal norms for expected project behavior);
3. Religious beliefs (these can impact worldviews and decisions);
4. Political opinions (these can impact opinions are shared);
5. Business interests (possibly outside the project or expected developments);
6. Demographics (age, gender, race -impacts interpretations and preferences);
7. Environmental stewardship (the ESG in stocks now arise in projects);
8. Asset value identification (ownership of projected revenues or benefits);
9. Communication mode (across demographics, old school or virtual).

We will skip explaining most of the attributes in the stakeholder register as they are either self-explanatory or unlikely to be needed in practice. Furthermore, project management software often provides alternatives to use for the stakeholder register, with or without the ability to add attribute fields to the database [8, 13, 20].

All these above attributes use various codes which can be complicated for a PM and their assistants to manage. We recommend keeping coding simple. The purpose of the stakeholder register is to have one place to record all key stakeholders, to find their contact information, and especially to keep notes about their engagement within the project. These involvement notes are the critical stakeholder register fields to analyze—these refer to the communication date/notes field. This would likely be a multiple dimensional field so that the PM could add notes about any important conversation with the stakeholder.

Notwithstanding the above advice to keep the stakeholder register simple, using only the above bolded fields and using simple low-medium–high codes, exceptions can be recognized. If the PM has the cognitive energy and time, or perhaps has assistants, it could improve the stakeholder analysis function by providing extensive factors to help understand and accommodate each stakeholder. Sometimes the organization collects extended demographic data including culture and other attributes cited above—if that data were available, statistical analysis could be performed to group stakeholders together with similar attributes, and to develop sociocultural profiles for every stakeholder, for use in stakeholder analysis and engagement. Demographic data can also be useful in the project closing phase, if it is used in conjunction with analyzing performance, to document the lessons learned. Having rich stakeholder socio-demographic data along with the traditional project performance metrics could provide valuable research insights about the current unknown factors impacting project success and failure. We do not know why projects are failing except that a few factors like competency, education, and certification can explain a very small effect size in robust empirical models to date [4, 6].

2.4 Stakeholder Analysis Methods in Project Stakeholder Management

Stakeholder analysis involves profiling the stakeholders to better understand how, when, why, where, how, and what the PM should do to engage the stakeholder. For practical purposes, most of that process is intangible rather than explicitly defined in the stakeholder register. That means the PM must have the skill and competency to know how to engage stakeholders without having a large panel of statistical variables to guide the process. The demographic attributes in the stakeholder register, names, titles, and contact data, are used for communication. The contact data should also contain the location because a busy PM can strategically schedule back to back visits to stakeholders who may be in close proximity, when projects include stakeholders sprawled throughout metropolitan areas, in other cities, states, or countries.

We recommend the analysis be focused on two attributes: Power and impact. If done correctly, these can be leveraged by the PM to provide as much descriptive value as using 10–20 additional factors for stakeholder analysis. The purpose of using power is that stakeholders often supply human or equipment resources to the PM for the project. The fact that they own the resources gives them high power over those team members, where the stakeholder would be their home position supervisor. In a similar sense, high power could refer to the stakeholder's formal rank as compared to other stakeholders, thus making this a relative power indicator. In either case, power tends to be a resource control factor. Impact on the other hand refers to deliverables or outputs from the project which will affect the stakeholder. Examples of high impact would be installing a new ERP which the stakeholder and their staff, in a department, will have to learn to use, and put a lot of effort into converting from the old to the new system.

We advise that it is not necessarily a scale for power and impact, but rather, the who and what which the PM must focus on. Pitting stakeholders against one another, sparring for decision making superiority, according to power and impact is rated, will not be useful for anyone. Instead, the PM should use power and impact for engagement, asking stakeholders with high power to help with the resources they control to avoid delays or to provide additional funding for team training or mentoring, and asking stakeholders with high impact to be involved in helping to develop solutions during the project to avoid negative impacts. In other words, power and impact are not rated attributes, but rather descriptive factors, listing resources under control and WBS deliverable impacts. An example data value for high power might be the names of resources and equipment controlled by the stakeholder. An example data value for impact would be the deliverables or WBS sections most likely to force the stakeholder's department (or themself) to change operational processes.

It should be noted that in the literature, interest and influence are used by PMI and ISO instead of impact and power. The first approach for stakeholder analysis is the influence versus interest matrix. The interest attribute as explained in the previous section means how much the stakeholder is motivated while the influence refers to informal power. The stakeholder theory underlying this model is that four quadrants are identified which would inform the PM how to use different communication or engagement techniques as shown below (adapted from: [10, 13, 24]):

- High interest + high influence = these are the key players to be proactively managed and consulted because their feelings can make or break the project;
- High interest + low influence = although they have slight influence, they can motivate others by satisfying their curiosity and keeping them informed;
- Low interest + high influence = dangerous people to keep on the radar, and satisfied because even though they are not currently interested, if they become dissatisfied, they could quickly become interested and take action;
- Low interest + low influence = stakeholder with no interest or influence so can be ignored, yet as above they may become more interested later on.

Another variation to the above is to substitute power instead of influence, with much the same interpretations for the action to take.

Wood et al. [23] presented an interesting stakeholder analysis typology based on three stakeholder register factors, namely:

1. power (formal authority as described above—but data type is unknown);
2. urgency (implying interest as explained above—unknown data type);
3. legitimacy (combination of formal power and interest—data type unknown).

The problem with the Wood et al. three-factor model is that the factors are likely to overlap, they would confound one another since they tend to have similar meanings. This means that a model uses three factors that measure nothing informative, or at best only one attribute—power. We argue that interest need not be measured because it is the PM's job to communicate and engage the stakeholders, motivating them, to create interest. A PM needs to know the formal power of each stakeholder, and the impact of the project deliverables on each stakeholder. Those two attributes are sufficient.

The problem with these stakeholder analysis matrix approaches is they generalize four discrete conditions to all stakeholders. Additionally, the matrix is limited to these two or three factors, and it is questionable whether the correct factors are being analyzed. Furthermore, there is no empirical evidence in the literature proving that these stakeholder analysis matrix approaches increase project success. Finally, it takes a lot of valuable cognitive overload from the PM to develop and continually maintain these stakeholder analysis matrices, an effort which may be better placed in the stakeholder engagement function or in other project management functions.

Instead of a matrix, we recommend putting effort into accurately recording the values for formal power and impacts. The stakeholder ought to be consulted and agree to your coding for both of those attributes. Power should indicate which resources (human and equipment) the stakeholder has control over, meaning which employees formally report back to the stakeholder outside of the project context. The impacts should indicate which deliverables, specifically which WBS work package activities or milestones (for deliverables) directly impact the stakeholder in their function. In this way power may be used to understand when to engage a stakeholder for any resource not performing, or equipment problems, to request training funding, mentoring, additional formal supervision (possibly reassignment or removal from the project subject to sponsor approval), or equipment replacement/maintenance. Impacts, as a plural attribute, allows the PM to know when and why a stakeholder must be consulted, based on the WBS item or deliverable. Engagement notes, also as a plural, allows the PM to record critical conversation results by date for evidence.

Statistically, we can argue the principle of parsimony would apply, which means that the best model is one with the fewest variables needed to explain most of the variation or effect [16, 25]. If we examine the intent and practical use of all the stakeholder register attributes, ultimately, power and impact are the most important factors, no other factor is needed, beyond the contact and notes information. Impact ought to drive the motivation and interest of the stakeholder. Power would refer to the resource involvement in any WBS activity and subsequently correlates with impact (since impact relates to the same deliverable work in the WBS). Note that we recommend including WBS in the stakeholder register as well as the impact field, because WBS could reveal which activities the stakeholder is involved with, while the impact field would pertain to outcomes or decision-making milestones where engagement will be required.

In summary, having many stakeholders register attributes could be useful for some situations, but likely having too many fields will complicate the process of stakeholder engagement. The more fields in the stakeholder register then the more administration and decision-making overhead the PM will face every time it is viewed. Table 1 is a simplified stakeholder register with the recommended attributes using a subset of the case study project (explained later).

In Table 1, there was only one stakeholder in the executive group and the project group was not used. In larger projects, where a Deputy PM or Assistant PM is helping

Table 1 Simplified stakeholder register for a PM

First name	Last name	Job title	Contact + Location	Project group	Project role	Functional power	WBS Involved	Change impacts	Engagement DATE/NOTEs
Ken	Strang	VP	xxx	Executive	Sponsor	PM boss, controls budget	1.1, 9.9	Final deliverable < year end	2024/1/1 approved charter
Joe	Doe	Mgr	xxx	Internal	ERP GL	Res. 1–3	2.3–2.8	GL changes	24/2/3 agreed
Jane	Doe	Mgr	xxx	Internal	ERP AP	Res. 4–5	3.1–3.8	AP updates	24/2/2 agreed
Sally	Doe	Mgr	xxx	Internal	ERP AR	Res. 7–8	4.1–4.9	Receivables	24/2/5 agreed
IM	Smart	ERP	xxx	External	Vendor 1	Authorizes free software + training	2.4, 3.4, 4.4, 5.6	Vacation blackout	24/3/3 briefed

Notes R1–3 are Auditing Dept staff; R4–5 are Accounting Dept staff.; R7–8 are Collections Dept. staff the PM, or if there are regional PMs, it is advisable to add them to the stakeholder register, because they are key stakeholders as well as key team members, and so that all contact information is stored in the same place.

The stakeholder identification and stakeholder analysis are recursive processes because they must be repeated every time a new stakeholder is identified, or when any of the stakeholder's attributes change.

Finally, we would recommend a project organizational chart be created using the stakeholder register and staffing pool database (from PM HRM). The project team members should be added to the chart. The major WBS deliverables should also be added if possible using contrasted notation, such as circles for WBS and squares for the stakeholders and resources. This share would visually represent the power lines, and implicitly show the impact relationships between stakeholders and deliverables. Figure 3 is an example project organizational chart based on Table 1. In Fig. 3, the dotted lines represent legal or formal power, while the solid lines indicate the reporting relationship back to the PM within the project contact. It is interesting that our approach uses the opposite conventions from an organizational structure chart, with solid lines to show PM informal power, and dotted lines to show formal power.

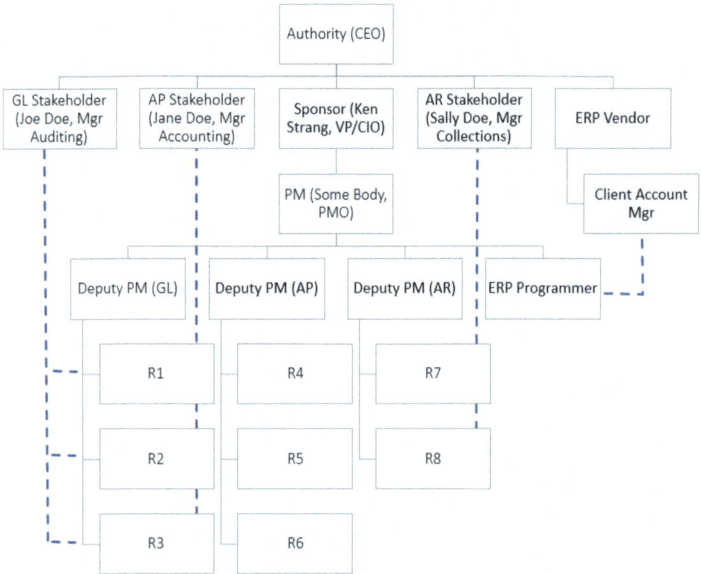

Fig. 3 Example project organization chart based on stakeholder register (in Table 1)

2.5 Stakeholder Engagement Methods in Project Stakeholder Management

We do not call this function stakeholder management as found in the literature from PMI or ISO, because the goal is engagement not management. The PM must manage the team resources, not the stakeholders. The PM engages the stakeholders and if they do not step up to the task then it must be escalated to their boss or to the project sponsor.

Here we depart from the literature to explain stakeholder engagement. This function requires the PM to communicate with the stakeholder either directly or in a group situation. The nature of the meeting, the reason for contacting the stakeholder, will determine what engagement approach is needed, and whether it ought to be a one-to-one or group session. The generally accepted managerial and leadership literature provides guidance for how to communicate with co-workers and superiors. In the literature, influence tactics focused on stakeholder buy-in are often recommended. Essentially, stakeholder engagement is a composite process of persuasion, influence, communication, and engagement. Sometimes engagement is merely communicating status, but at other times, engagement becomes problem solving with key stakeholders or even the sponsor. Handling team resources is not stakeholder engagement—instead that function is known as project human resource management (see other chapter focused on PM HRM).

Generally, stakeholder engagement is a function of the project phase in the life-cycle and the severity of the event impact on stakeholders. Figure 4 depicts a conceptual model of how a PM could engage stakeholders depending on the project phase and impact severity of the event. Although every situation can be different, a generic matrix can help make these decisions. Early in the project life cycle, during the initiation and planning phases, stakeholders who are strongly impacted by project deliverables, such as a new computer system replacing the GL software in the accounting department, the managerial stakeholders ought to want to be involved.

However, the staff of the managerial stakeholder may resist the new system. Therefore, it would fall back on the functional manager (stakeholder), not the PM, to motivate their own staff, with assistance from the PM if needed. Early in the project life cycle, managers are often the key stakeholders, and they should be met with first by the PM, on a one-to-one basis, to complete the stakeholder register. Then all key stakeholders, such as managers, could be brought together to complete the WBS, and perhaps identify which staff could be seconded as project team members. Key stakeholders could also identify when more resources will be needed when they do not have them. In those instances, the PM may have to meet with the sponsor to request funding for external resources if none exist internally.

Once the project has entered the execution phase, the PM needs to perform communication management. That means everyone needs to be updated at appropriate times depending on certain events. Communications management uses the stakeholder register to determine who, why, how, when, and where those communications messages are given. Email and newsletters can be used to update all stakeholders

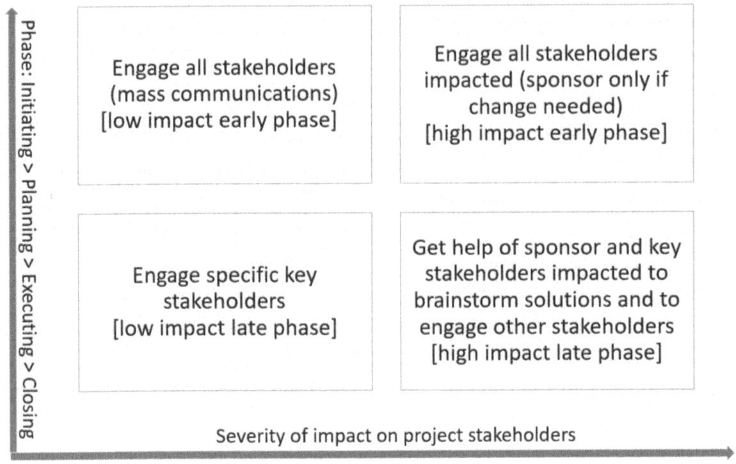

Fig. 4 Conceptual model of engagement approaches by phase and impact severity

and team members, as well as to make employees outside the project aware of the project progress, and when key milestones are reached.

The positive outcomes of communications management are easy for a PM. Usually everyone will accept a positive message regardless of the format, email, meeting, newsletter or whatever. The problem is how the PM needs to address the negative situations, or when risks manifest.

In negative message situations, the stakeholder register will indicate who should be contacted, based on the power and impact attributes. The PM must look at which WBS activities, which deliverables, stakeholders are associated with, when a risk occurs or a negative message is needed (e.g., deadlines slipped, and the entire schedule needs to be moved back). That is the starting point for a negative message, the key stakeholders impacted and those with the power over the resources in question. Those key stakeholders need to be contacted first on a one-to-one basis, and if possible, brought together in a small focus group, to strategize on or agree with whatever needs to be done, and to develop the strategy for messages to be sent to other stakeholders.

A possible outcome from a negative event during the stakeholder engagement function could trigger a change request—if so, the PM would evaluate if a change request were needed, and if so, create one and submit it to the sponsor—if approved or not approved, the outcome would flow back to the stakeholders based on power and impact first, as part of the iterative stakeholder engagement. It should be noted that a change request is a defined process in project management performed by the PM to gain approval from the sponsor for changing anything in the project plan which would impact cost, time, scope, quality, risk, procurement, human resource management, and possibly stakeholders or integration with the organizational governance. The change request process is different from stakeholder engagement (or communication), although the latter (stakeholder engagement) could be the impetus for initiating the change request process.

In essence, stakeholder engagement is communicating to the right stakeholder(s) at the right time, in the right way, about the right issues, and recursively performing this throughout the project when and where needed. The PM leverages the expertise and power of the key stakeholders impacted to help engage other key stakeholders. Engagement and communication could possibly extend outside the project context, such as business partners, suppliers, or important customers. Ultimately the PM decides if a message is needed, the content of the message, the channel to communicate the message (usually), and the PM carries the feedback from stakeholders back through the channel when needed. The PM makes these decisions about stakeholder engagement content/why, why, when, where, how, who—the PM must have (or develop) the skills, rather than rely on templates of que cards. That simple methodology tends to work well for most projects.

Stakeholder engagement differs for external versus internal stakeholders. Buy-in, persuasion, cooperation, empathy, understanding and agreement, are the underlying strategies for stakeholder engagement communication techniques. Most of the engagement techniques have applied to internal stakeholders. Email is often the most frequent technique, followed by direct conversations, either face-to-face, phone, or video (e.g., Teams, Zoom, Meet, sometimes even via Facebook Messenger or Apple iPhone). For generic dissemination, email or newsletters are useful, occasionally a recorded video or a group meeting can be helpful, the latter for special purpose activities like problem brainstorming.

The stakeholder engagement communication approaches for external stakeholders can vary from formal letter, cleared through your legal department, to email, phone, or personal meetings. For community groups, usually a prepared briefing is used in a press release format. For public disclosure and needs analysis, as is common in many public infrastructure projects like parks, bridges, roads, commercial land zoning, the public has a legal right to have a say, and the PM must manage that process as stakeholder engagement. If external community groups are to be consulted, the organizational often provides protocol and procedures to conduct this type of stakeholder engagement. Scientific information gathering or analytical meetings can often take the form of a conference for stakeholder engagement, with further mass dissemination of the results from those meetings to share with all stakeholders.

The details of how to write or verbally communicate to engage stakeholders is beyond the scope of this chapter. A professional with leadership soft skills will generally have the competency to know how to apply stakeholder engagement without having to develop extra fields in the stakeholder register or to create another matrix. If the PM is unskilled, then training in managerial communication and leadership soft skills is recommended, rather than putting time into developing engagement matrices or tools.

2.6 *Project Stakeholder Management in a Biopharma Case Study*

Now we present an applied example based on a biopharmaceutical industry case originally published with data collected for a different study [26, 27].

In the drug development industry, the project activities are practically driven by the regulations, to the extent that the USFDA (in the USA) and the ICS (in Europe) publish guidelines to inform projects. In other words, the PM must follow the network critical paths outlined in the regulatory documents to comply and to improve the odds that the final deliverable—the drug product—will be approved. It can also be noted that in the pharma industry, projects often span ten years from ideation through to product approval or application withdrawal.

As noted earlier, pharma projects can be expensive since many of the team resources are highly educated doctors and scientists, while external stakeholders sometimes must be compensated, including specialists, clinical doctors, and supply chain practitioners. The biopharma project management phases are shown in Fig. 5, subdivided by the six drug development stages which drive the WBS and schedule.

Biopharma project plans sometimes span the entire drug development cycle, all stages, as illustrated for the case study in Fig. 5, or a project may be constrained to one drug stage (e.g., ideation). There has been a long-standing custom in the industry to promote lead pharmaceutical scientists to perform the project management functions,

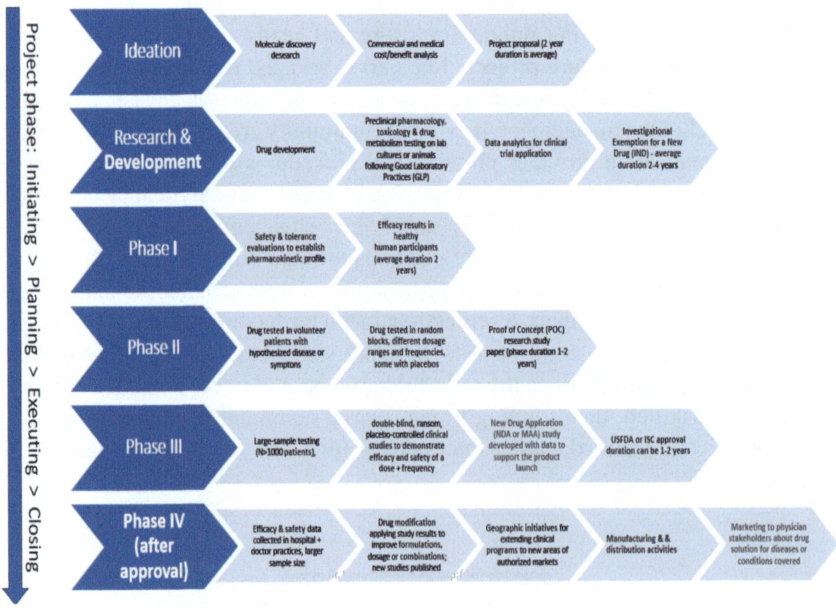

Fig. 5 Biopharmaceutical drug project phases superimposed on drug life cycle stages

which no doubt has contributed to the project failure rate, notwithstanding the low success rate in the entire biopharma industry. This industry may have skewed global project failure statistics. To combat that, in the last two decades most medium to large sized pharmaceutical companies have contracted certified and experienced project managers because the profits of a successful product are very significant, and the cost of hiring a high quality PM are relatively very small in comparison to the salaries paid for medical scientists and laboratories [1, 20, 28]. The problem is not many PMs have experience with biopharma projects or stakeholder management in these contexts.

Stakeholder identification, analysis and engagement take place iteratively, in a recursive process throughout both the project phases as well as during each new drug development stage. This is conceptually summarized in Fig. 6, which extends Fig. 5 by replacing the stage details with the three stakeholder management processes of identification, analysis, and engagement. The key stakeholders of pharma industry projects commonly include: doctors, pharmacists, patients, pharmaceutical traders, insurance agencies, academia, professional associations, regulatory authorities, consumer groups, and sometimes also specific government departments such as health, commerce, trade, and industry [20].

The biopharma drug development stages usually start with an ideation process, where subject matter specialists develop a conception of the product core chemicals. This is commonly known as drug discovery research, to articulate molecules which show potential for combating diseases or health problems, and which would be commercially profitable, as summarized in Fig. 6. This stage normally averages 2 years. The project initiation phase may start at the end of this or the next stage.

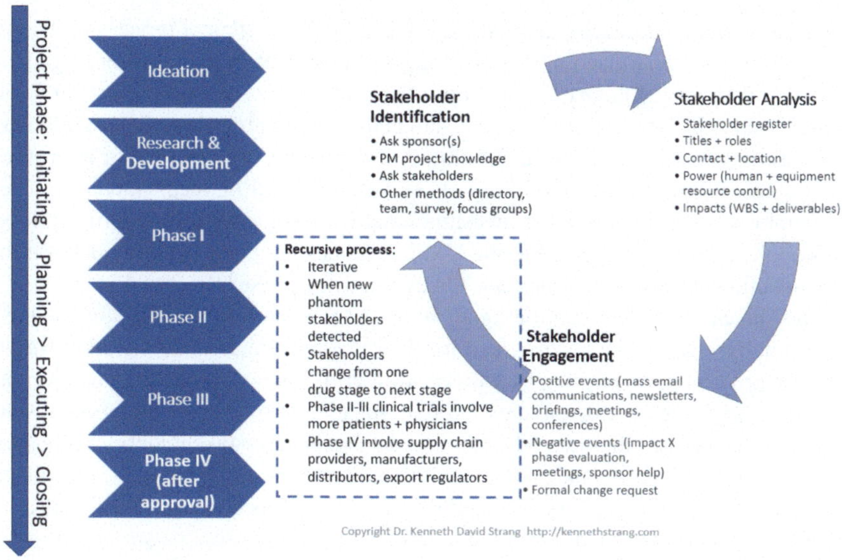

Fig. 6 Stakeholder identification, analysis, engagement in biopharma projects

If started in this phase the stakeholder register is created and a draft staffing pool database will be created to identify internal and external team resources based on templates of previous projects. This industry is high in the maturity life cycle, so templates of previously successful projects are always used if available.

Although the execution and controlling phase often starts with ideation, most of the work starts in the drug development stage/phase 1, where safety and tolerance are evaluated. In this stage the pharmacokinetic profile is developed further, and preliminary clinical efficacy of the drug tested with healthy human volunteers are analyzed. A stakeholder register is required at the start of this stage, which is iterative updated from the planning through the executing and closing project phases.

Stakeholder identification in the ideation stage commonly relied on sponsor and PM knowledge. The key stakeholders were research division directors who could provide research scientists, usually with doctorates in their respective fields such as chemistry or molecular analysis. We also included directors or managers from engineering and manufacturing so they could allocate two resources to provide a practical perspective. Finally, we included a stakeholder from the product division who could guide us in selecting appropriate medical and healthcare experts to confirm the characteristics of the problems being targeted. This stakeholder also provided the laboratory equipment and technical human resources. The legal research team was involved to provide some time who could help us avoid patent, trademark, license, or methodology conflict with existing market products. Clearly, the stakeholder identification strategy was to identify people who could provide human and equipment resources, along with sharing guidance on product design. Stakeholder analysis was simple and focused on contact details. Engagement was regular and periodic, following previous project templates because all stakeholders except for the scientists were involved in other projects—time was of the essence.

The research and development stage involved the pre-clinical processes. These focused on basic pharmacology, toxicology and drug metabolism testing to gather sufficient data for regulatory applications. The purpose was to secure approval for a clinical trial application using humans—but at this point only cultures or lab animals were used. As noted in Fig. 4, all laboratory experiments used in vitro or approved lab animals following the Good Laboratory Practice (GLP) ethical standards. The deliverable of this stage was the Investigational Exemption for a New Drug (IND). This application was presented to the USFDA requesting the necessary approval to conduct clinical testing with humans. On average this phase lasts 2 to 4 years.

The stakeholder identification again followed the sponsor and PM knowledge. Additionally, stakeholders were asked for their recommendations to include several friendly pro-company physicians who could recruit human volunteers. We do not consider the volunteers or patients explicit stakeholders or team members (even though conceptually they are very significant healthcare stakeholders). Here the emphasis was to identify stakeholders who could provide the volunteers, resources and lab for the testing as well as IND writing experience. Stakeholder analysis and engagement were focused on the external players to ensure power and impact aligned with where we intended to conduct this and phase II and III trials (the same stakeholders would be used while expanding to more).

The next project management activities in the biopharma drug development process was to complete the second approval stage (Phase II). This stage is where the drug product was tested on volunteer patients (or animals) showing the disease. Here the drug was applied in different dosages and frequencies, to calibrate the best efficacy configuration (dosage, application method, and frequency) for effective treatment. The major project deliverable in this stage was the proof of concept (POC) documenting the human testing evidence, which typically becomes a published study by our researchers to be used for credibility in the next stages. It could be noted that drug stages I and II in Fig. 4 take around 2–4 years in project duration. Occasionally drugs are developed for animals or fish, so the processes are different (beyond the scope of this chapter).

Stakeholder identification extended the register from the previous stage, adding more physicians and clinics, this time to recruit sick patients—those experiencing the problem. Analysis and engagement focused on the physicians once again, to ensure they were kept appraised of dosage and frequency changes through the feedback loop back to the design group. Stakeholders were again included from legal compliance as well as from the product division to provide resources to help with the POC study writing and publication. Production resources were provided to fine tune the design.

In the third drug discovery stage (Phase III in Fig. 4), larger samples are obtained using clinical trials with patients experiencing the disease or symptoms. In this stage, double-blind, placebo-controlled clinical trials are executed to prove the efficacy and safety of specific doses and frequencies. The PM also worked with team members (provided by the key stakeholders) to develop the new drug application or market application approval (NDA or MAA), citing results from our studies.

As with the previous stage, stakeholder identification extended the register by adding more physicians and clinics, this time to use more rigorous experimental trial designs, with more manipulated conditions. To ensure all scenarios were covered as well as the healthy patient situation. Analysis and engagement focused on the physicians once again, to ensure they were kept appraised of experiment results, through the feedback loop back to the design group. Stakeholders were again included from legal compliance as well as from the product division to provide resources to help with the NDA application writing. Production resources were requested from the stakeholders to fine tune the design.

Finally, in the fourth drug development stage (Phase IV in Fig. 4), started when approval has been received from USFDA or ISC. In this stage the PM organized more studies to collect efficacy and safety data from physician practices and hospitals. The medical researchers developed and published studies to document and disseminate the clinical results. Those studies were leveraged and cited for marketing purposes and to secure patents for the structure of the drug. The PM duties also included the marketing life cycle plan, to manufacture and distribute the new drug through supply chain partners. Additional pilot initiatives were included in the project plan to explore which diseases or conditions could be addressed by enhancing the product or changing the marketing message.

Stakeholder identification extended now involved many more players, from manufacturing, marketing, distribution, legal compliance as well as product design. The purpose was to request resources and to start the production process. While not much stakeholder analysis and engagement were required, beyond communications, this was a project where approval was granted. If approval were refused at any of the drug development stages, the stakeholder engagement would have been more analytical, following the negative event path identified in Fig. 1 where stakeholders would have been asked to help avoid project failure.

The closing phase of the biopharma project results in project team member evaluations, those results are commonly shared with the company human resource department as well as to the key functional stakeholders who provided the resources. Project success communications were done by the PM, with help from a few key stakeholders, and the sponsor, to share the success with the entire organization. The project plan, stakeholder register and staffing pool database (with resource evaluations completed) were cataloged in the PMO for use in future projects.

3 Conclusion

This chapter has summarized the most important and relevant literature underpinning contemporary project stakeholder management. While the ISO and PMI bodies of knowledge were consulted, several independent empirical research studies were reviewed to inform the problem addressed as well as to identify the key principles driving a proposed better solution for stakeholder management. A revised stakeholder management approach was developed and applied to a biopharmaceutical case study involving a successful new drug discovery project.

The RQ answered that effective best-practices for stakeholder management were developed and applied to a biopharma industry case study as an example. The traditional stakeholder management literature documented the functional requirement as two steps: Identify stakeholders and manage stakeholders. This chapter revealed that managing stakeholders is actually engaging stakeholders, and a third required step is analyzing stakeholders. Two key project stakeholder management artifacts were shown and explained: Stakeholder register and the project organizational chart.

Empirical research was embedded in the stakeholder identification section. An abductive literature review strategy was applied, which means only selected papers were selected for discussion, and a pragmatic ideology was taken, meaning statistical techniques were customized to meet the goals of the RQ. Subject matter experts (SMEs) were interviewed ($N = 69$), and their responses were analyzed using nonparametric statistics. The results revealed the best practices for stakeholder identification which were corroborated by citations to the literature. The highest ranked most common identification methods were, asking the sponsor, using PM research knowledge from previous projects, and asking stakeholders to identify others (snowball technique). Additional techniques included company directory searches, surveys, asking team resources, and sometimes using focus groups.

An interesting finding from the SME data analysis was that the method for identifying stakeholders varies most according to the project organizational context—whether it was a new situation or a project in a repeated organizational context. If a PM had already completed a project for the organization, they know some decision-makers and influencers, so they rely on their own research knowledge and secondly by asking the sponsor. For new situations, the PM first relied on the sponsor and then asked other stakeholders (snowball technique) as well as searching company directories. A valuable insight derived from the SME interviews was that all PMs used at least three stakeholder identification techniques (sponsors and other stakeholders first), and then either their own research knowledge (for repeat contexts) or company directories (for new situations). Consequently, it was recommended that a PM ought to use at least three stakeholder identification techniques, starting with asking the sponsor, then asking other stakeholders (snowball technique), followed by either reviewing the company directory or organizational structure diagram. Strategically, reviewing the company directory might uncover hidden or phantom stakeholders, people with authority and likely to be impacted by project deliverables.

Two models, based on stakeholder theory, were developed and explained, namely: Engagement based on severity and project phase, and power x impact engagement decision making. Two case study project plan overview models were shared. The first to illustrate how the biopharmaceutical drug development stages are sequenced over the project phases. The second overview model demonstrated how the stakeholder identification, analysis and engagement functions were applied in an iterative recursive fashion by the PM in the biopharmaceutical case study.

The limitations of this chapter are that the intended audience is emerging or inexperienced PMs. The recommendations were simplified. Nevertheless, alternative theories, techniques, and models were explained so that experienced practitioners and researchers could enhance this study. The research design was also explained in sufficient detail to allow researchers to understand and extend this study.

The second significant limitation of this chapter concerned the case study—which was focused on a successful drug discovery project in the biopharmaceutical industry. This industry is much different than most other manufacturing industries, and also different from practicing project management in the other highly populated industries including education, information technology and public administration. This constraint must be kept in mind for practitioners outside the biopharma industry.

Additionally, since a successful project was used for the case study examples, the negative events were not encountered to fully illustrate how the power-impact engagement model would be used when severe negative events occur in a project, such as a drug application being refused, or clinical trials failing. Perhaps another researcher would be able to extend this study by illustrating case studies of unsuccessful projects in this and other industries.

Finally, although international literature and empirical research studies from Europe were favored, much of the project management body of knowledge was developed in the USA and the case study organization was based in the USA. Consequently, other researchers ought to study project stakeholder management best practices by examining empirical studies and case study organizations, from outside the USA. Furthermore, there was some reliance on European literature. Additional global regions ought to be targeted for empirical evidence and research.

References

1. Martin, P.: Speeding time to market with better pharmaceutical project management. In: Pharmaceutical Technology Outsourcing Resources, no. Supplement. https://www.pharmtech.com/view/speeding-time-market-better-pharmaceutical-project-management (2017)
2. MordorIntelligence.: Biopharmaceutical Industry—Demand, Overview & Trends. Pharmaceutical Research, India. https://www.mordorintelligence.com/industry-reports/global-biopharmaceuticals-market-industry. Accessed 10 May 2024
3. Persistence, M.: Biopharmaceutical Market to Expand With Rise in Investment and Innovations; Protein Therapeutics Gain Traction. Persistence Market Research, London, UK (2024). https://www.persistencemarketresearch.com/market-research/biopharmaceutical-market.asp. Accessed 10 May 2024
4. Strang, K.D., Vajjhala, N.R.: Mining project failure indicators from big data using machine learning mixed methods. Int. J. Inf. Technol. Proj. Manag. **14**(1), 1–24 (2023). https://doi.org/10.4018/IJITPM.317221
5. Strang, K.D.: Which organizational and individual factors predict success vs. failure in procurement projects. Int. J. Inf. Technol. Proj. Manag. **12**(3), 19–39 (2021). https://doi.org/10.4018/IJITPM.2021070102
6. Strang, K.D., Vajjhala, N.R.: How bias impacted the project manager decision to not terminate a failing project. Int. J. Inf. Technol. Proj. Manag. **13**(1), 1–17 (2022). https://doi.org/10.4018/IJITPM.304059
7. Boutilier, R.G.: Chapter 19: Understanding stakeholders and their influence. In: Handbook of Social Impact Assessment and Management, pp. 292–307. Edward Elgar Publishing, Cheltenham, UK (2024). https://doi.org/10.4337/9781802208870.00029
8. Lim, S., Skaik, S.: Successful stakeholder engagement in not-for-profit projects: a systematic literature review. Int. J. Proj. Organ. Manag. **16**(1), 1–21 (2024). https://doi.org/10.1504/IJPOM.2024.136266
9. Vajjhala, N.R., Strang, K.D. (eds.): Global Risk and Contingency Management Research in Times of Crisis. In: Advances in Logistics, Operations, and Management Science. IGI Global (2022). https://doi.org/10.4018/978-1-6684-5279-0
10. PMI.: The Standard for Project Management and A guide to the Project Management Body of Knowledge (PMBOK Guide), 7th edn. Project Management Institute (PMI), PA, USA (2017)
11. ISO.: ISO 21502:2020(en), Project, Programme and Portfolio Management—Guidance on Project Management (2012). https://www.iso.org/obp/ui/en/#iso:std:iso:21502:ed-1:v1:en. Accessed 23 Nov 2023
12. ISO.: BS ISO 21500:2021 Project, Programme and Portfolio Management. Context and Concepts (2023). https://www.en-standard.eu/bs-iso-21500-2021-project-programme-and-portfolio-management-context-and-concepts/. Accessed 24 Nov 2023
13. Uribe, D.F., Ortiz-Marcos, I., Uruburu, Á.: What is going on with stakeholder theory in project management literature? A symbiotic relationship for sustainability. Sustainability **10**(4), Art. no. 4 (2018). https://doi.org/10.3390/su10041300
14. Schilling, M.A.: Decades ahead of her time: advancing stakeholder theory through the ideas of Mary Parker Follett. J. Manag. Hist. **6**(5), 224–242 (2000). https://doi.org/10.1108/13552520010348371
15. Strang, K.D.: Cybercrime risk found in employee behavior big data using semi-supervised machine learning with personality theories. Big Data Cogn. Comput. **8**(4), Art. no. 4 (2024). https://doi.org/10.3390/bdcc8040037

16. Strang, K.D. (ed.): Handbook of Research Design in Business and Management, 1st ed. In: Research Methods. Palgrave Macmillan US, New York (2015). https://doi.org/10.1057/9781137484956
17. PMI.: PMI Lexicon. In: PMI Lexicon of Project Management Terms—Version 3.2 [free public version]. Project Management Institute (PMI), PA, USA (2017). https://www.pmi.org/pmbok-guide-standards/lexicon. Accessed 2 Jan 2024
18. Strang, K.D.: Examining effective technology project leadership traits and behaviors. Comput. Hum. Behav. **23**(1), 424–462 (2007). https://doi.org/10.1016/j.chb.2004.10.041
19. Strang, K.D.: Examining effective and ineffective transformational project leadership. Team Perform. Manag. Int. J. **11**(3/4), 68–103 (2005). https://doi.org/10.1108/13527590510606299
20. Geer, M.I.: Stakeholders in pharmaceutical policy development. In: Pharmacovigilance—Volume 2. IntechOpen (2022). https://doi.org/10.5772/intechopen.105606
21. Vanclay, F., Esteves, A.M.: Handbook of Social Impact Assessment and Management. Edward Elgar Publishing, Cheltenham, UK (2024). https://doi.org/10.4337/9781802208870
22. Bulgarella, C.: Toward a Perspective of Stakeholder Culture in the Financial System. American College of Financial Services, King of Prussia, PA (2022). https://ethics.theamericancollege.edu/sites/ethics/files/Stakeholder-Whitepaper-ML_v4.pdf
23. Wood, D.J., Mitchell, R.K., Agle, B.R., Bryan, L.M.: Stakeholder identification and salience after 20 Years: progress, problems, and prospects. Bus. Soc. **60**(1), 196–245 (2021). https://doi.org/10.1177/0007650318816522
24. PMI.: The Standard for Program Management, 4th edn. Project Management Institute (PMI), PA, USA (2017). Accessed 2 Jan 2024. https://www.pmi.org/pmbok-guide-standards/foundational/program-management
25. Strang, K.D., Korstanje, M.E., Vajjhala, N. (eds.): Research, practices, and innovations in global risk and contingency management. In: Advances in Business Information Systems and Analytics. IGI Global (2018). https://doi.org/10.4018/978-1-5225-4754-9
26. Strang, K.D.: Resistance, diseconomies, and abnormal AI behavior in HRM: a real-time big data action research experiment at a pharmaceutical. Hum.-Intell. Syst. Integr. **4**(3–4), 35–52 (2022). https://doi.org/10.1007/s42454-022-00046-6
27. Strang, K.D., Sun, Z.: ERP staff versus AI recruitment with employment real-time big data. Discov. Artif. Intell. **2**(1), 21 (2022). https://doi.org/10.1007/s44163-022-00037-1
28. Peng, T.: Polysorbates: part(icle) of the problem?. Pharm. Technol. **2017**(3), 30–32 (2024). https://www.pharmtech.com/view/speeding-time-market-better-pharmaceutical-project-management. Accessed 10 May 2024
29. Strang, K.D.: Strategic analysis of CSF's for not-for-profit organizations. Meas. Bus. Excell. **22**(1), 42–63 (2018). https://doi.org/10.1108/MBE-07-2016-0035

Framework for Managing Artificial Intelligence (AI) Projects: Avoiding Harms, Losses, and Damages

Gloria J. Miller

Abstract Leveraging studies on artificial intelligence (AI) stakeholders and success factors, this article sets out to embed an AI perspective in a project management standard and center it around avoiding moral issues—harms, losses, and damages—in AI projects. The study provides an AI Project Framework that identifies the significant differences between AI projects and other information technology (IT) projects, including the AI development lifecycle, risks, issues, and challenges. The study creates a conceptual structure that combines aspects from the International Organization for Standardization (ISO) 21502:2020-12 Project Management standard and the AI project lifecycle. Finally, it weaves a practical framework of interdependencies and success factors for managing AI projects. The study uses an integrative literature review methodology that follows a hermeneutic framework. The study results should offer practical benefits to sponsoring organizations, project sponsors, and project managers in planning and governing AI projects.

Keywords Artificial intelligence (AI) · AI project framework · AI development lifecycle · AI projects versus IT projects · Project management standard · ISO 21502:2020-12 · Moral issues in AI projects · AI project risks and challenges · Success factors in AI projects · Integrative literature review

1 Introduction

This article defines a project management framework to prevent artificial intelligence (AI) systems developed in AI projects from causing harms, losses, and damages to individuals and society. AI is an "umbrella term covering a wide range of efforts to computationally reproduce complex human capabilities such as language use, vision and autonomous action" [12, p. 1]. It includes categories of systems such as chatbots (examples include Siri and Alexa), humanoid robots, behavioral software,

G. J. Miller (✉)
Maxmetrics, Heidelberg, Germany
e-mail: g.j.m@ieee.org

K. D. Strang and N. R. Vajjhala (eds.), *International Program and Project Management—Best Practices in Selected Industries*, Information Systems Engineering and Management 31, https://doi.org/10.1007/978-3-031-80275-1_7

and predictive systems. It has been argued that some AI systems are moral agents, which is to say that they are responsible for actions or decisions that can benefit or harm people, and the decisions involve choice where moral factors are balanced against other factors in the decision-making [53]. Further, some AI systems perform a complex series of actions without human control or guidance.

The algorithms that constitute AI systems are developed for and used in consumer platforms (e.g., online shopping, social media), enterprise systems (e.g., credit approval), and critical and life-saving systems (e.g., diagnosing Parkinson's disease, breast cancer, or COVID-19). Unfortunately, there are examples of AI system failures, such as job discrimination, plane crashes, emissions fraud, and false imprisonment [50]. Since the launch of ChatGPT in 2023, interest, enthusiasm, and concerns associated with AI have grown dramatically. ChatGPT is a chatbot that "is a language model based on reinforcement Learning from Human Feed-back and was created to produce conversational outputs" [37, p. 2].

Consequently, the AI project actors, including project sponsors, project managers, and project team members, are not mere participants in the AI project. They are moral agents; their in-project decisions have the potential to significantly impact the life, liberty, human, or civil rights of individuals and society. Their actions can lead to environmental, financial, reputation, and political damage. Therefore, they bear a moral responsibility to apply fair, ethical, and transparent processes to the AI systems which they develop.

Leveraging studies on AI stakeholders and success factors, this article sets out to embed an AI perspective in a project management standard and center it around avoiding moral issues—harms, losses, and damages—in AI projects. The Framework intends to overcome the limitations of applying existing AI ethical principles to a project management context, where such principles are either too broad or too difficult to translate into practice.

The Framework's architecture is purposefully designed. It comprises two main elements: an AI lifecycle model and success factor definitions. The lifecycle model is a strategic tool that aligns the problems, challenges, and failures of AI projects to the complex relationships between a project management standard, an AI project lifecycle, and AI success factors. On the other hand, the success factor definitions provide detailed descriptions of the deliverables, acts, or situations necessary for project actors to succeed in the interest of the sponsoring organization and impacted stakeholders.

The study results are designed to benefit sponsoring organizations, project sponsors, and project managers in planning and governing AI projects.

The remainder of the paper is structured as follows: Sect. 2 provides literature on AI systems, development lifecycle, risks, issues, challenges, and project management standards; Sect. 3 provides the research methodology and introduces the conceptual overview of the Framework; Sect. 4 presents details of the AI Project Framework, including the International Organization for Standardization (ISO) processes and AI project practices; finally, Sect. 5 introduces a procedure for using the framework.

2 Literature Review

2.1 AI Systems

The algorithms used in AI systems are developed using computer science and statistical methods, such as natural language processing, machine learning, deep learning, artificial neural networks, and Generative Pre-Trained Transformer (GPT) technology (as in ChatGPT) [3, 31].

A GPT is a language model that takes human inputs and uses machine learning techniques and sophisticated algorithms to provide natural language and conversational responses to human inquiries. The algorithms underlying ChatGPT have been trained with data from much of the internet, meaning they are built using billions of data sources. The systems are generative because there is a learning process that allows the machines to acquire new data, learn from that data, and continuously improve and make better predictions on how to respond to human queries [41]. Further, chatbots can be integrated with other systems to perform specific tasks; they are virtual assistants in such cases. While today, chatbots take input via text or voice, emerging technologies will allow two-way communication between humans and chatbots via thoughts. For example, Neuralink's brain implant has been used to direct a computer cursor using human thoughts [20].

Natural language processing (NLP) functions by manipulating human language. Machine learning uses various supervised and unsupervised methods to discover and model patterns and relationships in data, allowing the algorithm to make predictions. Deep learning uses machine-learning approaches to automatically learn and extract features from complex unverified data without human involvement. Artificial neural networks are models trained on data to make predictions.

An important element in the AI systems discussion is the degree of human intervention in decision-making, which varies according to the type, purpose, and method of integrating the algorithms into other systems or processes. For example, robots or other artificial agents perform a complex series of actions without the need for human control or guidance [56]. The decision-making autonomy of these artificial agents has provoked a review of civil liability, criminal liability, data protection, and security issues and raised ethical questions about how moral decisions are to be made when humans are not involved [49, 62, 78]. However, AI systems are not sentient at the current state of technology: "Sentience is the ability to have subjective experience, which includes perceiving and experiencing things…it differs from the ability to think and the ability to feel" [25, p. 2].

For the current study the definition of AI systems follows that in Miller [50, p. 2]: "AI systems include data-driven computer systems that incorporate algorithms that learn from data and algorithmic decision-making that uses computerized systems for autonomous or human decision-making and problem-solving."

2.2 AI Risks, Issues, and Challenges

Significant differences exist between AI projects and other information technology projects [50]. When not properly managed within the project context, these differences present unique challenges that can lead to issues for society or the environment or to legal or financial issues for the sponsoring organization. While some of these challenges may not directly impact project efficiency, they underscore the need for specialized governance and management in AI projects, and this is a point of particular concern for policymakers and professionals. Some of the differences and challenges that are specific to AI projects are listed and described below.

Violations of regulations and laws. Existing laws and regulations rely on organizations to comply with data protection regulations or other laws such as the Americans with Disabilities Act [11]. The proposed European Union Artificial Intelligence Act will require a risk assessment for high-risk AI systems before deployment [50]. Nevertheless, weak technology policies, inadequate enforcement, and lack of laws and regulations create situations where individuals are subjected to extensive personal data collection, intrusive models, privacy violations, and constant surveillance with no ability to influence or contest their treatment [14, 63].

Extreme environmental impacts. AI projects that train large data models have high energy and water consumption and produce carbon emissions that may affect the environment; the scale of carbon emissions for training some models is on the scale of a trans-American flight [6, 62, 69].

Harmful in-project decisions. The designers and developers of an AI system make important decisions, and their personal biases, blind spots, and design choices can influence the consequences of using the systems [35, 43]. For example, computing methods such as artificial neural networks produce models with hidden features, and the designers may not know or understand which parameters influence the model. The data used for training the models may be biased, incomplete, or influenced by past practices [14, 63]. Entire communities of people may be missed or misrepresented in data due to social infrastructure, systematic disparities, technology policies, or the costs of data collection [63].

The "black box" problem. Developers can provide systems for which the end users cannot explain or interpret the results. Consequently, AI systems may be "black boxes." The designers may not understand some model decisions, the end users may be unable to explain or interpret the results, or both [63]. This lack of transparency and explainability can be due to the method used or other reasons, such as the user interface design. Nevertheless, the developed systems are artificial agents [49, 68].

Poor deployment, lack of controls. Lack of quality controls in the commercial deployment and usage of AI systems can harm employees and customers and ultimately impact the financial performance of a business. For example, an American retail provider used facial recognition software for surveillance to track security risks. There were several mistakes in the usage and deployment of the system. First, the system produced thousands of false-positives. Then, there were problems with inadequate staff training, and with data collection and retention. Consequently, the system led employees to mistakenly accuse consumers of wrongdoing due to false positives. The Federal Trade Commission finally required the company to discontinue use of the system, citing unfair biometric surveillance and unfair data security practices [21].

Mistakes of usage. AI systems generate new content that is not necessarily based on facts. The large language models used in AI systems mimic human language and intelligence without understanding what the words or concepts mean. They assign weights and probabilities to stitch together convincing sentences. "Rather than any singular, definitive understanding, then, this socially generated text contains many different 'truths'…Language models, in a very tangible sense, have no knowledge of the facts and no integrated way to evaluate truth claims" [54, p. 4]. For example, in the court case *Mata v. Avianca, Inc.*, attorneys were sanctioned for submitting non-existent judicial opinions with fake quotes and citations that were created by ChatGPT [2].

Moral buffer in accountability. The developers provide a system that is an artificial agent embodying a moral code; that is, it makes decisions or recommendations where the outcomes can benefit or harm people or society. A moral buffer occurs when neither the human decision-makers who use the systems nor the engineers who build them are accountable for decisions that result from use of the AI system [26].

2.3 AI Lifecycle

Big data, predictive analytics, facial recognition, and machine learning are some of the project types that apply data science in the development of AI systems. The lifecycle stages of an AI system are (1) planning and design; (2) data collection and processing, model building and interpretation, verification and validation, and deployment; and (3) operation and monitoring [56]. Operation and monitoring and the impacts and consequence of usage accrue months or years after the development project have been completed. The individual roles and stakeholders involved in or impacted by projects vary depending on the lifecycle phase [74].

Figure 1 provides an overview of the AI project life cycle.

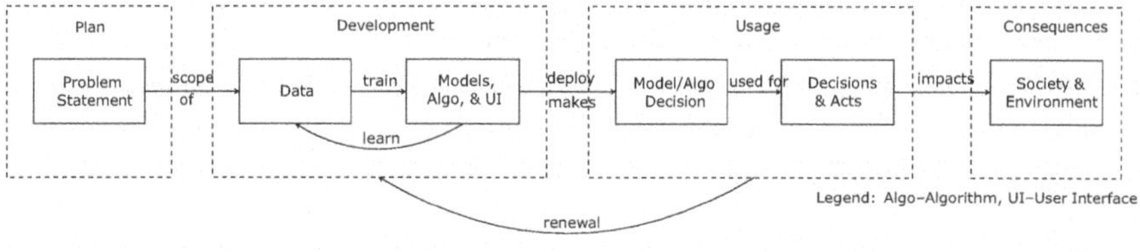

Fig. 1 AI lifecycle (author's representation)

Plan. The problem to be addressed by the AI system is defined in the planning and design stage. The problem statement defines the aims and rationale for the algorithmic system [15]; it should act as a contract, setting out the goal of the desired algorithm and the behavior that can be anticipated.

Development. The next stage is the development process, involving data collection and processing, model and user interface development, and validation [18, 56]. The source data come from original systems, such as enterprise resource planning systems, websites, or social media sites. The source data are collected, stored, and manipulated to make them fit for purpose. This manipulation may augment the data with labels, tags, identifiers, or metadata [53]. It may also anonymize personal data or use synthetic data. The volumes of the data may be vast, including millions of transactions and massive databases.

A subset of the data is then used to design the AI algorithms. These data enable the models to learn and are usually referred to as training data [6, 43]. Construction and interpretation of model and algorithm are based on the data science methods and techniques already described [43]. The algorithms are verified and validated against relevant measures such as bias and false positives [56].

A user interface is developed for inputting parameters to trigger the algorithm or for the AI system to make autonomous decisions and present the results. The form of the interface will vary depending on the type and purpose of the application. It may be an automated system, drones, chatbots, social media or e-commerce sites, or an appliance such as a smartphone or smart device [56, 67, 73].

The technical infrastructure for creating and deploying AI systems vary in technical scope and complexity. Robust technical infrastructure is required to collect, process, and store the data [6, 17]. High-performance computing systems and architecture are necessary in order to support computational needs and data loads [6, 17]. Finally, the system configuration, including the architecture and deployment, determines how the algorithms function.

Usage. End users input parameters to the user interface to trigger the algorithm. The model, algorithm, and user interface produce the output with or without the user's knowledge of the algorithm's existence [43, 73]. The system's performance, environment, and supporting staff are continually surveyed to identify problems and learn. From time to time model values or choices become obsolete, so monitoring and renewal processes during ongoing operations are essential. The renewal process refreshes obsolete model values or choices [56, 78].

Consequence. The model or algorithm output decisions or recommendations that impacts specific people, organizations, or groups. AI systems can impact human rights based on the design and nature of the system, its implementation, use, or all of these together. AI ethical principles such as for fairness, trustworthiness, transparency, explainability, accountability, and sustainability are key criteria for assessing the implications of an AI system for society and the environment [33, 62].

2.4 Project Management Practices

Project management bodies of knowledge are used worldwide as guiding frameworks for standard practice. The guidelines include, for example, *ISO 21502:2020-12, Project, Programme and portfolio management—Guidance on Project Management 1st Edition* [32], *APM Body of Knowledge 7th Edition* [5], *A Guide to the Project Management Body of Knowledge (PMBOK guide)—7th Edition* [57], and *Managing Successful Projects with Prince2* [1]. The standards represent the collective identity of project managers and have evolved to suit different delivery approaches, e.g., predictive, incremental, iterative, agile [32].

Each standard uses various names to describe similar content, processes, and artifacts, but they all cover similar topics, which are, namely: planning, initiating, directing, monitoring, controlling, and closing projects. We used the subject areas from the ISO standard as the governing framework. It provides a holistic and structured way of describing project activities and subjects. Table 1 gives an overview of the ISO activity areas.

Table 1 Project management processes (ISO 21502:2020)

Role	Processes	Description
Sponsoring org	Pre-project	Identifying and formalizing the needs and opportunities to realize benefits before authorizing the project
	Overseeing	Monitoring that organizational needs and stakeholder expectations are met at an acceptable risk level
	Post-project	Verifying the outcomes are sustainable and expected benefits have been realized
Project spr	Directing	Engaging directly and through boards to confirm organizational resources are used as expected or to terminate when support is no longer justified
Project mgr	Initiating	Planning the project, organization, governance, and management structure and mobilizing the project team
	Controlling	Monitoring and measuring performance against agreed plans or authorizing changes
	Closing or terminating	Confirming the completion, enabling post-project benefits realization, and demobilizing resources and facilities
Work package leader	Managing delivery	Defining and planning required outputs and outcomes and delivering outcomes to achieve the expected benefit

Abbreviations: Mgr-Manager, Org-Organization, Spr-Sponsor

3 Research Methodology

The aim of the article is to define a project management framework that could prevent AI systems developed in AI projects from causing harm, losses, and damage to individuals and society. AI is an emerging topic, and there are limited studies that address the practical aspects of managing AI projects. Therefore, an integrative literature review methodology was chosen as the best approach, as described by Torraco [70, p. 356]: "The integrative literature review is a form of research that reviews, critiques, and synthesizes representative literature on a topic in an integrated way such that new frameworks and perspectives on the topic are generated."

3.1 Hermeneutic Framework

We used a hermeneutic framework for the literature review process [8]. The framework includes two recursive processes for understanding a subject based on a body of literature relevant to a particular problem. The first process is the search and acquisition of literature to develop an understanding. The second process involves analysis and interpretation to map the literature to the research problem through classification, critical assessment, and development of an argument. Reading is the key activity that links the two processes.

We chose a hermeneutic framework because it treats the literature review as an ongoing process, achieving a "point of saturation" where the researchers' understanding and purpose become an important contributing factor: "The deeper a researcher's understanding of the relevant literature, the more convincing the argument for comprehensiveness and insightfulness of the literature review" [8, p. 273].

The project, a "temporary endeavour to achieve one or more defined objectives" [32, p. 3], was the unit of analysis for the study, and the themes of AI and project management were the main concepts used in the literature review. The details of the literature review are described in the following sections.

Search and Acquisition. In the search and acquisition process, we started from previous literature reviews on artificial intelligence in project management. Based on those reviews, we selected a small set of relevant documents to read, which is the usual review strategy for a hermeneutic framework. We selected three papers as the starting point for our review: [49, 50, 62]. We selected those papers because they were published in the three years prior to this study, were on AI and project management, were written in English, and were peer-reviewed. Each document was read in detail and relevant articles were selected for further study. The latter articles were used in the analysis and interpretation process.

Analysis and Interpretation. In the analysis and interpretation process, the first step was to read the relevant literature thoroughly to extract relevant subjects. The next step was to map and classify the literature under the headings of ISO project management, AI life cycle, project success factors, and AI ethical concepts. This involved systematic analysis of relevant concepts and their organization according to relevance to the research question. For purposes of the critical assessment, connections were made between the concepts, hidden explanations were exposed, and white spots and gaps were resolved. Arguments were then put forward to support the connections that were established during the critical assessment. The arguments are supported by explanations and justifications discovered during the literature review.

3.2 Conceptual Structure

The conceptual structure of the study combines aspects from *ISO 21502:2020-12, Project, Programme and portfolio management—Guidance on Project Management 1st Edition* [32] and the AI project lifecycle. Figure 2 provides an overview of the structure, which produces the AI Project Framework. The Framework is a conceptual model derived from critical analysis and synthesis. It weaves the different streams of research together to provide practical guidelines for visualizing interdependencies and managing AI projects.

First, the gray process steps in Fig. 2 represent the ISO activities that are the "recommended project management practices that should be used when undertaking a project" [32, p. 6]. Next, the yellow processes relate to the AI project lifecycle. The AI project lifecycle covers the project and post-project periods, taking the system's impact on society and the environment into consideration. For each ISO process, the Framework describes the AI success factors related to that process for each stage in the AI project lifecycle.

The success factors are the deliverables, acts, or situations that are necessary to avoid harm or ensure the benefits of AI projects and systems. These factors were originally described in Miller [49] and have been refined and updated as part of this study. That study was judged appropriate because it took a project perspective on AI, provided a practical approach to AI ethical principles, and was based on a review of 55 peer-reviewed articles.

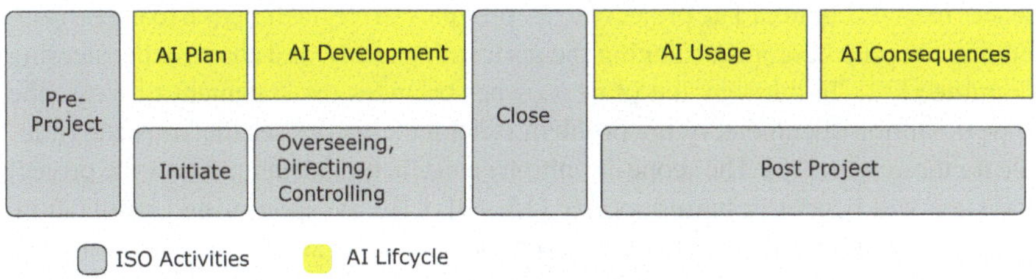

Fig. 2 AI project framework (author's representation)

4 AI Project Framework

The AI Project Framework interweaves the ISO activities and subjects with the AI project lifecycle and success factors.

4.1 ISO Processes

Pre-project. In the pre-project process, the sponsoring organization recognizes the strategic needs of the project [32]. A set of target benefits are defined for the project investment before the project starts. In addition to target benefits, firms expect other tactical goals, such as project efficiency, to be met. In the context of AI projects, developing or using an AI system may impact society, the environment, and workers. Individuals and communities have fundamental rights and freedoms, such as the right to privacy, and some types of decisions are made by humans and not systems [11, 62]. Further, AI systems may introduce new job structures and patterns, eliminate specific types of jobs, or change the way of working. Programs may need to address employee and employment-related issues [60, 62]. Therefore, the target benefits must consider public interests and address the need to protect individual civil rights and liberties, workers' rights, and environmental sustainability.

Overseeing. In the overseeing process, the sponsoring organization monitors the project to ensure that organizational needs and stakeholder expectations are being met at an acceptable level of risk. The system's designers and developers make important decisions, and their personal biases and design choices can influence the consequences of using the systems [35, 43]. Thus, sponsoring organizations need deep insights into the in-project decisions that may impact individual, environmental, or worker rights.

Directing. In the directing process, the sponsor directs the project through ongoing personal engagement with it and through boards to confirm that organizational resources are used to deliver outputs, outcomes, and benefits as expected or to terminate when organizational support is no longer justified. The project sponsor is the agent for the sponsoring organization and is, therefore, also known as the "project owner" in an AI context. The project owner provides strategic direction to the project, defining the project scope, arranging the performing entity, and chairing the steering committee [32]. In this way the project owner provides the statement of work, the scope definition document, or the problem statement that defines the aims and rationale for the system [15]. The scope definition is one factor that defines how the project or system will benefit or harms society [15, 49]. Like the sponsoring organization, the project owner needs insights into in-project decisions and methods to influence the project deliverables.

Initiating. After management approval, the initiating process consists of formal planning to reach the goals. The project manager defines the organization, governance, and management frameworks, mobilizes the project team, and identifies the stakeholders. In addition, the project manager builds on the preliminary project justification by further developing and documenting the business case for the project.

Controlling. The project manager monitors and measures project performance against agreed plans or authorizes changes in the controlling process. In the AI context, the project manager is accountable for the project's outputs as per an agreed statement of work provided by the project sponsor [32]. The sponsoring organization establishes the project scope, and provides funding, strategic direction, and operational guidance through the project owner. The organization's strategic goals are imposed on the project, and policies flow down from the organization to the project level [16]. Thus, the project manager ensures that the team adheres to ethical, privacy, and security norms, as well as to the expectations and engagement of the relevant stakeholders [50]. Consequently, the project manager is able to manage approval processes, but not as the party responsible for approval itself [59]; this latter responsibility stays with the project sponsor as the corporate agent for the sponsoring organization. However, the project manager is responsible for record-keeping and managing the expectations and engagement of relevant stakeholders.

Managing delivery. In the managing delivery process, the project work takes place and is managed by workstream leaders and managers. This stage includes producing the required output and outcomes and realizing the project's expected benefits. The project team members are decision-makers and designers who directly influence the system's models, data, and trustworthiness. Membership of the project team is context-specific and determined by the project's scope. The team may include members with various specializations depending on the type of technology, industry, and business function involved. Furthermore, the people responsible for labeling, annotating, tagging, or curating training datasets or moderating content may be temporary, contract, or gig workers engaged through platforms such as Mechanical Turk [53].

In addition to responsibilities for quality, the team is accountable for activities that secure the legal and financial benefits expected of the project sponsor. There is limited overlap between these responsibilities and the usage qualities; team members are accountable for some quality controls, providing interpretable models, and supporting knowledge transfer with stakeholder-centric communications and onboarding procedures. Specifically, the team is responsible for avoiding the "black box;" that is, they must avoid building systems that they do not understand how the model makes its inferences [63].

If the project team is responsible for data manipulation of offensive topics (e.g., pornography, hate speech, violence), its members may face psychological harm in developing the systems [55]. In addition, team members may face other harm when they find themselves in conflict with the project direction on moral grounds [36, 62]. Nevertheless, the project team members are decision-makers and designers who directly influence the project output. The team may include the following specializations: business users, data analysts, data scientists or AI practitioners, data engineers, software architects, and software developers [48, 50].

Closing or terminating. The terminating process involves completing the project and disbanding the team. Termination is organized by the project manager. In AI, closing the project also means transitioning into operational mode, which varies vastly depending on the type of system. For example, robots, chatbots, and social media platforms sold to consumers require a different transitional model compared with systems used for enterprise decision-making. In both cases, addressing usage quality is a prerequisite before closing the project.

Post-project. In the post-project process, the sponsoring organization will have realized the expected benefits and outcomes will be sustainable. In the context of AI, the post-project phase is the transition to an operational model that requires monitoring and tuning of system performance on an ongoing basis. Usage processes and controls established during the project are operationalized in the post-project phase.

The operating organization is the purchaser or consumer, including the end users and decision makers, and the organization may have expectations of success that are different from those of the end users. The organization provides usage policies and practices, monitors the systems and staff, and engages with the end users and people impacted by the decisions. Systems and staff are monitored to prevent the system's decision process from becoming ineffective [26]. In addition to carrying out operations, the operators perform due diligence to ensure the system's appropriateness and quality.

4.2 AI Lifecycle, ISO Subject, AI Success Factors

AI Plan. The AI plan process is where the sponsoring organization formulates the AI problem statement. The problem statement is usually created in the pre-project stage, handed down to the project sponsor, and then to the project manager and workstream leaders. Further activities occur at the initiating stage and throughout the project. The following sections use the ISO subjects to describe the project artifacts which the project manager creates to initiate, further manage, and monitor an AI project.

Planning. Projects need practical guidelines and organizational structures which address multiple aspects of the AI life cycle—system engineering, management strategies, and independent oversight. Consequently, frameworks should be established for value, ethics, and data governance.

Ethics framework. Eleven categories of ethical principles are applicable to AI projects: beneficence, dignity, freedom and autonomy (including protection of human rights, contestability), justice and fairness (including diversity), non-maleficence (including safety, security, reliability), privacy, accountability/responsibility, solidarity, sustainability, transparency (including audibility), and trust [39, 62]. These principles need to be transformed into practical guidance and organizational roles and processes must be established to support them. Thus, the ethics framework should be multi-faceted, including ethics policies, ethics training, ombudsman, conflict of interest rules, and professional standards [4, 23, 52, 62].

Data governance framework. Data availability and use are foundational for AI systems. However, data protection is regulated in many jurisdictions. For example, in the European Union, the General Data Protection Regulations (GDPR) protect the data privacy rights of citizens. Further, the GDPR address data profiling (i.e., algorithm development), use, and contestation [60]. Therefore, data governance should consider methods for managing the data life cycle [30], including processes for allocating authority and control over data, its collection, management, use, and destruction. The data retention policy should specify time periods and obligations for retaining and destroying data.

Legal framework. AI systems can impact human rights through the design and nature of the systems, their implementation, use, or all of these. Here, human rights are understood to refer to "all the rights and obligations that people should enjoy to survive and develop, and to live in harmony with others and society under a certain social and historical condition and under a relatively stable social and natural environment" [47, p. 922]. Many adverse effects of AI on human rights are addressed by implementing a robust ethical framework. However, additional legal safeguards and privacy safeguards are required during project execution to ensure regulatory compliance, protect privacy and confidentiality, and limit liabilities [60].

Benefits management. AI systems create value using data and algorithms. They can benefit the value proposition of a firm or be used to monetize data: "Value propositions define an organization's position in a market, its core skills and how these are applied for the benefit of others" [10, p. 168]. Consequently, introducing AI systems is a business model and technical problem.

A value framework should identify the ethical implications of transforming a firm's data and algorithms into value and the guardrails to constrain unethical behaviors. The value framework that is needed varies depending on the business model behind the system. For example, insurance companies reduce costs by monetizing risks, "so any data-driven means to reduce or manage risk will be of benefit" [10, p. 176]. Thus, sourcing data from consumers that helps rate an individual's risk level is of interest to an insurance organization. Conversely, the individuals providing such data would have ethical concerns about consent, privacy, and discrimination.

As discussed, financial benefits will arise from the intended use of the AI system, e.g., cost reductions from making faster, less expensive, or better decisions or improved efficiency from reducing or eliminating tasks [77]. However, other benefits may arise from licensing or reselling the AI artifacts or attracting investment funds. Additionally, the ownership of AI artifacts (e.g., data, models, algorithms) is needed in order to retain intellectual property rights, and in-project actions are needed for purposes of intellectual property protection [19, 60].

Scope Management. The AI problem statement, containing the aims and rationale for the algorithmic system, feeds into the project scope definition document [15].

The scope should act as a contract between the project manager, the project sponsor, the sponsoring organization, and the team regarding the goal of the algorithm and the behavior that can be anticipated. It should be used as the starting point to build trust on how moral and ethical issues will be addressed. A well-defined scope document protects the sponsoring organization against spurious claims and the misapplication or misuse of the system. It is therefore one of the most important project artifacts, providing legal protection to the firm that uses the AI system.

Based on AI ethical principles, AI systems should be developed to benefit society as a whole (beneficence), or at least they should avoid harming others (non-maleficence) [40, 62]. The scope document should reflect how the proposed AI system will address these ethical principles. An algorithmic impact assessment investigates how an AI system may impact society and anticipates and proposes steps to address potential deficiencies or harms [60, 78]. It can be a useful tool in defining and refining the project scope and in risk mitigation.

Resource Management. The project team plays a significant role in how and to what extent AI systems impact individuals and society. Therefore, responsible development requires the engagement of developers, operators, and representatives for passive stakeholders. The project organization should include people from numerous backgrounds, disciplines, and specializations to protect the system from inbuilt bias.

A diverse working environment promotes exchange and cooperation across regions and among organizations [62]. Furthermore, the project team needs members with specialized skills and knowledge to process data and accomplish design and development of the algorithms, since the level of the team's skills and expertise impacts usability of the algorithms and governance policies [71]. Further, passive stakeholders should actively participate in the project through representation; "[p]assive stakeholders may be affected by the project but do not actively contribute to the project outcome" [50, p. 2]. For example, if the system is meant to affect a community of women, then a representative from a community of women should be given the opportunity to judge whether the system will benefit the community or make it more vulnerable [9, 50] (Tables 2 and 3).

A responsibility assignment matrix, also known as a responsibility accountability consultation information matrix (RACI), is an essential project management tool to avoid moral hazards in AI projects. The matrix maps persons and organizations to accountabilities and responsibilities [45]. Accountability ensures that a task is done satisfactorily and accountabilities cannot be delegated. The other classifications relate to responsibilities for taking action, consultation, or giving or receiving information. Responsibility and accountability assume a proper understanding of subject matter and the possession of other relevant knowledge [62].

Moral hazard can arise from inefficiency or abuse of trust due to opportunistic behavior between a principal and an agent [75], for example, between supplier (i.e., developer) and client. The issue is relevant in AI projects because of the high degree of specialization required in the field and the "black-box" problem. A transparency gap caused by concealing information or engaging specialists with little competence and experience can create legal issues for the sponsoring organization. The responsibility

Table 2 Decision tradeoffs

Product quality—versus—societal impacts
Even larger language models—versus—high environmental costs
Product quality—versus—usage quality
Automation—versus—human autonomy
High number of choices—versus—limiting human intervention
High transparency—versus—ability to manipulate algorithms
Societal impacts—versus—financial benefits
Benefits to society—versus—financial gains
Usage quality—versus—legal safeguards
Alternative decision choices—versus—limiting choices to limit liabilities
Correct erroneous/biased decisions—versus—limit liabilities in justifying different decisions
High flexibility for end users—versus—low flexibility to limit liabilities
Product quality—versus—legal protections
Biased models with no errors—versus—unbiased models with high costs
High explainability algorithms—versus—preserving IP rights
High system flexibility—versus—low legal safeguards
High transparency—versus—protecting IP rights
Higher accuracy model—versus—high development costs
Financial benefits—versus—legal protections
High project efficiency—versus—low regulatory compliance

Abbreviations: IP-Intellectual property

assignment matrix is therefore vital when assigning tasks and roles; it should be used to mitigate the risk of morally hazardous behaviors.

An ethical framework, policy guidelines, and standards should be used when aligning people around the project through the scope document, in order to build consensus and provide an understanding of the relevant issues, such as algorithmic transparency, accountability, fairness, and ethical use.

Schedule Management. The life cycle of an AI system involves multiple periods: plan, development, consequences, and usage. In a typical IT project, the plan and development stages are part of the project, while consequences and usage are post-project. This is also the case in AI projects, but in AI projects ethical considerations for the post-project stages must be addressed during the development stage to avoid harmful impacts.

Cost Management. Project costs are an important contributor to value as investments are needed to ensure the available human, material, and financial resources.

Model development has environmental impacts and energy costs, which can be high when training large language models. Energy costs of computing power, including electricity and water consumption (for on-premises or cloud-based services), are substantial when training models [6]. For example, "the well-known GPT-3 model was trained on 300 billion training tokens with 175 billion parameters, using computing power of 10,000 V100 GPUs, which took weeks and costs up to multiple millions of dollars" [17, pp. 251–252]. A token is a unit of information

Table 3 Stakeholders and project roles

ISO role	Role	Description
Sponsoring org	Developing org	Sponsors the project and establishes the project scope
	Operate org	Consumers or enterprise that purchases or license the system
	Platform owners	Organizations that own the technical infrastructure and architecture upon which the system resides
	End users	System users working in the operating organization or consumers
	Prj funder	A funding entity that pays for and governs the project
	Data controller	Responsible for data governance and its use in system development
	Data custodian	Manages, assures, secures, and preserves the data
Prj Spr	Prj owner	The sponsor that provides strategic direction to the project
Prj mgr	Prj/Prog mgrs	Accountable for the project's outputs per an agreed SOW
Prj team members	Prj team	Context specific collection of people that is determined by the by the project organization
	AI practitioner	Specialist (e.g., computer vision engineer) that may be relevant, depending on the technology and context of the project
	Architects	Specialist in the technical infrastructure and may include architects, the IT team, the process manager, and security specialists
	Business users	Responsible for specific content development, editing, or coaching the team or users
	Data analyst	Experts within the context needed to interpret raw data, ask relevant questions, generate relevant hypotheses, and interpret the results
	Data engineer	Prepares data, including extracting, cleaning, enriching, and transforming the data
	Data scientist	An interdisciplinary role that aims to transform data into knowledge by finding patterns and trends in the data
	Model mtn	Reviews and refreshes models for fitness and refreshed over time as model values or choices can become obsolete
	Software dev	Usually computer scientists who design, develop, and optimize the processing of large data sets or develop user interfaces
Prj stakeholders	Workers	Employees and contractors that face new job structures and patterns eliminate specific jobs or change their way of working

Abbreviations: AI-artificial intelligence; dev-development; mtn-maintenance, org-organization; prj-project; prog-program; spr-sponsor from one character to one word and parameters are the variables that govern how the model behaves and its results. Further, in a mock scenario [17], estimated cloud computing costs and model training costs for a large language model for radiology practice would vary from US$ 27,000 per month for 5100 radiology studies to about US$ 77,568 for 17,000 studies. Reduction of environmental impact and financial costs when training models can be achieved by building smaller models or using a method that is less energy-intense.

Data representativeness, where one set of data stands for another (e.g., a sample for a population, an instance for a category), has power, value, and cost elements which must be treated with caution. Data cost efficiency occurs when acquiring and using information is less expensive than the costs involved if the data were absent [22]. "Only by considering the ways we translate the world into data in the first place can we fully examine the realities data represent and potentially, perpetuate" [14, p. 78]. Data costs are impacted by the answer to questions such as: what methods were used to collect the data? Do we need to collect more data if the data are unrepresentative or insufficient? Will collecting the data require undue exploitation or surveillance? What ethical judgments determined who was included or excluded from the data set? How much are the license fees for this intellectual property?

Stakeholder engagement is another cost factor of AI projects: inter-organizational solutions such as committee meetings, auditor, and outreach programs may be necessary to engage third parties or community members.

Finally, project efficiency evaluates the project management's success in meeting stakeholder requirements for quality, schedule, and budget [49].

Risks Management. Risk management is the project management practice of identifying, evaluating, and mitigating risks to improve the likelihood of project success. Multiple approaches are available for identifying and assessing risks; at a macro-level, each approach compares the goals with a value baseline for benefits on the one hand and ethics on the other. The assessment of the project's impacts on the system organization, individuals, and society depends on frameworks for ethics and value that are established as part of the project's governance. If risks are not properly managed, they can lead to project failure [79]. Projects are affected by uncertainty that is internal to the project organization and external uncertainty produced by market conditions and competitive environments.

AI is a fast-moving area with external risks from extensive market and competitive uncertainties. Internal project-specific risks are of a private nature, separate from external risks, and can impact project costs and the market value of product output if they lead to creation of models with bias, errors, or poor accuracy, or if they create potential for misuse of the AI system. Specific methods for mitigating AI risks include retraining and fine-tuning models, adding new components (wrappers) to original solutions, and building functionally equivalent models (copies) [72]. A model risk assessment is an active risk management process for managing technical risks associated with AI models [72].

Quality Management. The quality aspects of AI systems are discussed in the AI development and AI usage sections of this article. There are trade-offs between many quality aspects and management decisions. Table 2 provides an overview of the trade-offs identified in the study.

Stakeholder Engagement. The AI stakeholder groups associated with the AI lifecycle include development, usage, and external stakeholders. Development stakeholders are involved during the project, while usage stakeholders are involved post-project. "Consequently, they [the development stakeholders] are the only stakeholders who can consider the concerns of all stakeholders" [50, p. 13]. External stakeholders are not part of the project but may affect or be affected by the project. As noted in the resource management discussion, communities impacted by the AI system should be engaged in the project. Table 3 provides a list of AI stakeholders.

Miller [50] recommended that power, legitimacy, urgency, and harm attributes should be considered when identifying stakeholders. The harm attribute includes an ethical dimension that involves economic, social, and environmental sustainability considerations in the analysis of stakeholders [50]. To qualify as an AI project stakeholder, a party should have at least one of the following attributes:

- Power: defines the ability of the stakeholder to negatively or positively impose their will on the system's development or use. Such powers may include: the power to punish for non-compliance (coercive); the power to provide a reward for complying (reward); formal authority to control (legitimate); ability to provide or withhold special knowledge, skill, or expertise (expert); and ability to be a role model (referent).
- Legitimacy: evaluates the existing contractual relationships or societal agreements. Aspects include terms of services (terms), project statement of work (SOW), employment contract (employ), organizational agreements and rules (rules), regulations or laws (laws), or organizational positions (hierarchy).
- Urgency: assesses time-sensitive relationships between the act of development or usage and the consequences. Urgency occurs during development (D), during usage (U), and associated with consequences of usage (C).
- Harm evaluates the harms, damages, or losses caused by the development or use of the system. Aspects include: bodily harm, loss of life, limitation of rights or freedom of movement, surveillance, and psychological impacts (cases of harm); violations of human or civil rights such as loss of privacy, security, freedom, funds, and employment (cases of loss); and trust, reputation, and environment (cases of damage).

Procurement Management. As mentioned in the section on resource management, AI projects require specialized skills and knowledge, so one or more suppliers are usually engaged in projects. In addition to the need for a clear RACI matrix, agreements with specialized suppliers should avoid setting conditions that would create a moral hazard.

Moral hazards can occur when the supplier and client are from different organizations, and where the supplier benefits from intentional misbehavior in the current project [75]. Examples include intentional poor completion of tasks to create a competency gap or underpricing development work to win operational work.

Alternative procurement models for infrastructure procurement of computing resources, data, storage, and model training are relevant to AI projects. They involve new components and terminology not relevant to other information technology (IT) projects. As discussed in the section on cost management, the costs of training in a large language model include token, computing, and storage costs. Infrastructure (cloud versus on-premises) and administrative costs are also important [17]. Intellectual property procurement through licensing or buying rights to copyrighted or proprietary data as a data source may also enter the supplier planning process.

Finally, people who tag or apply metadata to the training datasets may be temporary workers —gig or contract—from specialized organizations or procured platforms such as Mechanical Turk [53].

Document Management. Systematic record keeping is the process for retaining logs and other documents with contextual information about the process, decisions, and decision-making, from project inception through to system operation [15]; the list of records that should be retained are listed Table 4 even though they extend across many subject areas.

Most of the listed records were proposed by the reviewability framework, which is an approach to improving the accountability of algorithmic decision-making by drawing on administrative law, understanding "decision-making as a process beginning before the decisions are made and extending to their consequences and effects" [15, p. 598]. Record keeping is the key accountability for a project manager in AI projects.

AI Development. As discussed in the AI lifecycle section, the AI development process is where data collection and processing, model and user interface development, and validation are completed. Based on ISO project management activities, this stage would be a managing delivery activity, governed by workstream leaders. The following sections describe the project artifacts identified as necessary to ensure success and avoid harms, losses, and damages caused by use of the resulting systems.

Source Data. Source data are the data from original systems or sites that are used in the development process. Depending on the geographical region, there are regulations and laws that constrain how data may be accessed, processed, and used in analytical processes. Furthermore, there may be internal rules from the sponsoring organization that govern data privacy, retention, accessibility, and confidentiality. Factors related to source data include: data accessibility, data transparency, and data confidentiality [7]. The reviewability framework recommends maintaining data collection records

Table 4 AI records and documents

Audit finding records	Metadata versioning	Procurement records
Audit response records	Model training records	Risk assessment records
Consequence records	Model validation records	Technical deployment records
Disclosure records	Model versioning	Technical logging
Data collection records	Process deployment records	Usage records

and the datasheet for datasets provides guidance on the content of such a document [15, 24].

Data protection secures data in various states during storage and transportation or protects content from intrusion by unauthorized persons. Data encryption, data anonymization, privacy notices, informed consent, and personal data controls are methods for ensuring data protection [28].

Training Data. Training data is a subset of the source data used to train models and algorithms to learn. The data quality and data relevance determine if the data are fit for purpose. As discussed in the resource and cost management sections, decisions on the data significantly impact how representative and equitable the resulting system will be. Relevant factors include equitable representation in the data and the team performing the work. In this sphere, people working with the data may come into contact with violent or offensive data, so they may need assurance of physical and psychological safety when interacting with and processing data (interaction safety). The datasheet for datasets Gebru et al. [24] and model cards Mitchell et al. [51] provide frameworks for documenting facts about the training data.

Model and Algorithm. The data science techniques used vary the development process and also influence explainability, algorithmic transparency, interpretability, and auditability of the resulting system [42, 65]. For example, models can be interpretable by design, provide explanations of their internals, or provide post-hoc explanations of their outputs. Alternatively, the design can be completely hidden, i.e., a "black box". In any case, considerations of accuracy and consistency should determine suitability for a specific use case [27, 65]. The model validation checks for adherence to defined principles and standards, effectiveness, adverse situations, sensitivity, and bias. Further, it is important at this stage to eliminate non-equitable treatment, whereby similarly situated people are treated differently [22]. The reviewability framework recommends maintaining model validation records and model cards that provide detailed guidance on content [15, 51].

User Interface (UI). The user interface interacts with the model or algorithm. It should meet front-end transparency needs, consider usability by people with differing abilities [61, 76], and permit human intervention to override default decisions [28]. Furthermore, it should be noted that default choices may introduce bias in decision-making.

System Configuration. The availability, robustness, cost, and safety capabilities of the software, hardware, and storage are essential for algorithm development and use [38]. Quality of the system and architecture may impact the algorithm's outcomes, result in indeterminate behavior or cause system behavior that does not always produce the same results when given the same inputs. The degree and type of automation may limit the user's choices.

Accidental, unlawful, or malicious actions compromising data availability, authenticity, integrity, and confidentiality should be mitigated by providing technology, processes, and people to establish security safeguards [61].

AI Usage. The AI usage process is where the sponsoring organization and end users put the AI system to productive use. This stage occurs after the development project is complete, so it would be a post-project activity under the governance of the sponsoring organization. The following sections describe project artifacts which the sponsoring organization creates in the development or post-project stages.

Risk Migration. Most harm from AI systems to individuals and society occurs in the course of AI system usage. Steps should therefore be taken to mitigate the risk of putting a harmful AI system into use. First, algorithm testing checks for unexpected issues or makes inferences from the data. It is usually carried out as part of validation during the project and is conducted by the project team.

Other mitigation strategies engage people or organizations who are independent of the project or the sponsoring organization. An algorithmic impact assessment investigates the social impact of the system [13, 46] and algorithmic auditing reveals how algorithms work [66]. An audit produces audit finding records for the audit results and audit response records for resolutions [15, 23]. Other possibilities include obtaining certification from a third-party institution that rewards compliance and punishes for breaches by the application of an expected performance standard [66].

Usage. The audience for the AI system depends on its design and purpose, and the level of explanations, domain knowledge, and expertise needed to use or explain the system varies accordingly. First, there should be stakeholder-centric communication that considers the profile of the target audience: executives and managers, regulators, auditors, end users, domain-experts, model maintainers, etc. [64]. Next, for enterprise systems, there should be on-boarding procedures to orient the users and to explain important parameters [29].

Success in usage will depend on interpretable models provided by the developers [27, 62] and allowing users to choose whether they use the model results, thus maintaining a human-in-the-loop control mechanism [62]. However, specialized skills and knowledge may be required to choose between alternatives. Therefore, professional expertise, staff training and supervision, and coaching may be necessary to ensure appropriate use and decision quality [34, 38]. Here again, interaction safety should protect users who are confronted with offensive data. Finally, problem reporting mechanisms should allow users to report concerns such as bugs or algorithmic biases [19].

Usage Controls. Several processes are needed to detect improper system usage or under-performance, including a complaint process, quality controls, staff monitoring, and system monitoring [34, 40, 76]. Since model performance degrades overtime and model values or choices become obsolete, an algorithmic renewal process is necessary to review, replace or refresh values [78].

Decision Quality. From the perspective of the individuals who are impacted by system usage, decision quality depends on having people and processes in place to provide awareness of the existence and capabilities of the AI system, to take responsibility for decision outcomes, and to provide access and redress for user issues [44]. Finally, privacy and confidentiality are bedrock principles throughout the whole process, and they include any activities, procedures, policies, or people that are necessary to protect and maintain confidential information of identified or identifiable natural persons [60].

AI Consequences. The last stage of the AI lifecycle is Consequences. In this post-project stage, the AI system is operational, and the sponsoring organization is interested in sustainability and realizing the expected benefit.

At this point, the stakeholders, whose role is pivotal in addressing AI's adverse effects, should have been identified and their interests addressed. As discussed in previous sections, robust ethical practices play a significant role in mitigating AI's adverse effects on human rights. Further actions at this stage include steps to assess the AI system's impact and evaluate its performance.

Similar to AI usage, risk mitigation strategies that involve independent third parties can play a role at this stage in assessing and mitigating risks of harm. These strategies include algorithmic impact assessments and algorithmic auditing [15, 23]. However, it is important to note that such evaluations may be initiated by regulators or the media due to issues in the market.

In sum, individual rights should be protected by activities enumerated in the respective sections of this article (e.g., equitable treatment, accessibility, choices, privacy, and confidentiality) [11, 40, 62]. Decision quality, legal safeguards and privacy safeguards are critical success factors in protecting human rights.

Environmental sustainability is considered in the sizing of the computing environments and the training of the models [6, 62]. The development and renewal process should consume less energy by training smaller models or by using an alternative model-building method to reduce environmental impact [6].

In-project workers' rights can be protected by providing a safe space for dealing with offensive data and a whistleblower process to address conflicts [7, 22, 39, 55, 62].

5 Conclusions

The chapter's aims were to define a project management framework to prevent new AI systems from causing harm, losses, and damage to individuals and society. First, the article points out significant differences between AI projects and other IT projects, including the AI development lifecycle, risks, issues, and challenges. The article then creates a conceptual structure that combines aspects from the ISO 21502:2020-12 project management standard and the AI project lifecycle. Within that context, it weaves together a practical framework of interdependencies and success factors for managing AI projects.

The framework presents several methods for avoiding violations of regulations and laws. First, in the AI plan stage, projects should establish ethical, data governance, and legal frameworks. Next, at the AI usage stage, a proactive risk mitigation strategy and usage controls should be considered. Finally, privacy and security safeguards should be planned early on and should be put in place across all project stages.

The framework points to several places where transparency is needed to avoid the system remaining a "black box." Testing, auditing, and risk mitigation factors are aligned at the relevant points in the AI lifecycle. Specifically, the model risk assessment factor is proposed before putting the system into productive use. This recommendation is consistent with the principles of the European Union AI Act that require assessment of high-risk AI systems before their productive use; the Act also contemplates transparency rules for all but low-risk AI systems [58].

More than a dozen situations involve tradeoffs affecting the quality of the system. It is almost impossible to eliminate all bias or quality issues. Nevertheless, Table 2 in the quality section provides some guidance on the tradeoffs to consider.

The reasons mistakes in usage occur and why they occur vary depending on the type and purpose of the AI system. The framework proposes some factors that ensure transparency for end users and prepare the AI for proper use. Such factors, together with identification of who is accountable for decision outcomes, play a key role in avoiding a moral buffer where no one is accountable to society for the use of an AI system.

Finally, one of the most important issues is to identify project actions that affect the environment. Energy consumption in developing large language models is a real concern for the environment and is very expensive. Here, while we identify the source of the problem, the study does not offer any silver bullets for avoiding or mitigating the impact. As has been suggested by Bender et al. [6], we should question the value of larger and larger language models.

In sum, the AI Project Framework could be a useful tool in the project planning process for team assignment and stakeholder engagement. A project sponsor and manager could use the procedure proposed in the ISO guidelines to make effective use of the Framework.

The study does not fully address all possible risks from AI systems. For example, it does not provide any metrics or measures for assessing the quality of the model. How many false positives are too many? Where should the threshold be? These are important questions, but they were not in the scope of the study. Furthermore, the study does not offer strategies to mitigate the environmental impacts of AI systems. These topics would be an interesting area for future research.

References

1. Managing Successful Projects with Prince2. AXELOS Limited (2017)
2. Mata v. Avianca, Inc. (PKC), vol. U.S. Dist. (No. 22-cv-1461), pp. *1. S.D.N.Y. (2023)
3. Aggarwal, J., Kumar, S.: A survey on artificial intelligence. Int. J. Res. Eng. Sci. Manage. **1**(12), 244–245 (2018). https://doi.org/10.31224/osf.io/47a85
4. Aguirre, A., Dempsey, G., Surden, H., Reiner, P.B.: AI loyalty: a new paradigm for aligning stakeholder interests. IEEE Trans. Technol. Soc. **1**(3), 128–137 (2020). https://doi.org/10.1109/TTS.2020.3013490

5. APM: APM Body of Knowledge 7th Edition. Association for Project Management, Buckinghamshire, United Kingdom (2019)
6. Bender, E.M., Gebru, T., McMillan-Major, A., Mitchell, M.: On the dangers of stochastic parrots: can language models be too big? In: FAccT 2021: Proceedings of the 2021 ACM Conference on Fairness, Accountability, and Transparency, pp. 610–623. ACM (2021). https://doi.org/10.1145/3442188.3445922
7. Bertino, E., Kundu, A., Sura, Z.: Data transparency with blockchain and AI ethics. ACM J. Data Inf. Qual. **11**(4), 1–8 (2019). https://doi.org/10.1145/3312750
8. Boell, S., Cecez-Kecmanovic, D.: A hermeneutic approach for conducting literature reviews and literature searches. Commun. Assoc. Inf. Syst. **32**, 257–286 (2014). https://doi.org/10.17705/1CAIS.03412
9. Bondi, E., Xu, L., Acosta-Navas, D., Killian, J.A.: Envisioning communities: a participatory approach towards AI for social good. In: AIES'18: Proceedings of the 2018 AAAI/ACM Conference on AI, Ethics, and Society, pp. 425–436. ACM (2021). https://doi.org/10.1145/3461702.3462612
10. Breidbach, C.F., Maglio, P.: Accountable algorithms? The ethical implications of data-driven business models. J. Serv. Manag. **31**(2), 163–185 (2020). https://doi.org/10.1108/JOSM-03-2019-0073
11. Büchi, M., Fosch-Villaronga, E., Lutz, C., Tamò-Larrieux, A., Velidi, S., Viljoen, S.: The chilling effects of algorithmic profiling: mapping the issues. Comput. Law Secur. Rev. **36**, 1–15 (2020). https://doi.org/10.1016/j.clsr.2019.105367
12. Caldwell, M., Andrews, J.T.A., Tanay, T., Griffin, L.D.: AI-enabled future crime. Crime Sci. **9**(1), 14 (2020). https://doi.org/10.1186/s40163-020-00123-8
13. Chan, L., Doyle, K., McElfresh, D., Conitzer, V., Dickerson, J.P., Borg, J.S., Sinnott-Armstrong, W.: Artificial intelligence: measuring influence of AI 'assessments' on moral decision-making. In: Proceedings of the AAAI/ACM Conference on AI, Ethics, and Society, pp. 214–220. ACM, New York, NY, USA (2020). https://doi.org/10.1145/3375627.3375870
14. Chasalow, K., Levy, K.: Representativeness in statistics, politics, and machine learning. In: FAccT'21: Proceedings of the 2021 ACM Conference on Fairness, Accountability, and Transparency, pp. 77–89. ACM, (2021). https://doi.org/10.1145/3442188.3445872
15. Cobbe, J., Lee, M.S.A., Singh, J.: Reviewable automated decision-making: a framework for accountable algorithmic systems. In: FAccT'21: Proceedings of the 2021 ACM Conference on Fairness, Accountability, and Transparency, pp. 598–609. ACM (2021). https://doi.org/10.1145/3442188.3445921
16. Derakhshan, R., Turner, R., Mancini, M.: Project governance and stakeholders: a literature review. Int. J. Project Manage. **37**(1), 98–116 (2019). https://doi.org/10.1016/j.ijproman.2018.10.007
17. Doo, F.X., Kulkarni, P., Siegel, E.L., Toland, M., Yi, P.H., Carlos, R.C., Parekh, V.S.: Economic and environmental costs of cloud technologies for medical imaging and radiology artificial intelligence. J. Am. Coll. Radiol. **21**(2), 248–256 (2024). https://doi.org/10.1016/j.jacr.2023.11.011
18. Dumrak, J., Zarghami, S.A.: Reimaging stakeholder analysis in project management: network theory and fuzzy logic applications. Eng. Constr. Archit. Manag. (2020). https://doi.org/10.1108/ECAM-06-2020-0391
19. Eslami, M., Vaccaro, K., Lee, M.K., On, A.E.B., Gilbert, E., Karahalios, K.: User Attitudes towards algorithmic opacity and transparency in online reviewing platforms. In: CHI 2019: Proceedings of the 2019 CHI Conference on Human Factors in Computing Systems, pp. 1–14. ACM (2019). https://doi.org/10.1145/3290605.3300724
20. Fiani, B., Reardon, T., Ayres, B., Cline, D., Sitto, S.R.: An examination of prospective uses and future directions of neuralink: the brain-machine interface. Cureus **13**(3), e14192 (2021). https://doi.org/10.7759/cureus.14192
21. Rite Aid Banned from Using AI Facial Recognition After FTC Says Retailer Deployed Technology without Reasonable Safeguards. https://www.ftc.gov/

22. Gandy, O.H.: Engaging rational discrimination: exploring reasons for placing regulatory constraints on decision support systems. Ethics Inf. Technol. **12**(1), 29–42 (2010). https://doi.org/10.1007/s10676-009-9198-6
23. Garfinkel, S., Matthews, J., Shapiro, S.S., Smith, J.M.: Toward algorithmic transparency and accountability. Commun. ACM **60**(9), 5 (2017). https://doi.org/10.1145/3125780
24. Gebru, T., Morgenstern, J., Vecchione, B., Vaughan, J.W., Wallach, H., Daumé III, H., Crawford, K.: Datasheets for datasets (2018). https://arxiv.org/abs/1803.09010v7
25. Gibert, M., Martin, D.: In search of the moral status of AI: why sentience is a strong argument. AI & Soc. **37**, 319–330 (2022). https://doi.org/10.1007/s00146-021-01179-z
26. Green, B., Chen, Y.: Disparate Interactions: an algorithm-in-the-loop analysis of fairness in risk assessments. In: FAT*'19: Proceedings of the Conference on Fairness, Accountability, and Transparency, pp. 90–99. ACM (2019). https://doi.org/10.1145/3287560.3287563
27. Hamon, R., Junklewitz, H., Malgieri, G., De Hert, P., Beslay, L., Sanchez, I.: Impossible explanations? Beyond explainable AI in the GDPR from a COVID-19 use case scenario. In: FAccT 2021: Proceedings of the 2021 ACM Conference on Fairness, Accountability, and Transparency, pp. 549−559. ACM (2021). https://doi.org/10.1145/3442188.3445917
28. Helberger, N., Araujo, T., de Vreese, C.H.: Who is the fairest of them all? Public attitudes and expectations regarding automated decision-making. Comput. Law Secur. Rev. **39**, 1–16 (2020). https://doi.org/10.1016/j.clsr.2020.105456
29. Hopkins, A., Booth, S.: Machine learning practices outside big tech: how resource constraints challenge responsible development. In: AIES'18: Proceedings 2018 AAAI/ACM Conference on AI, Ethics, and Society, pp. 134–145. ACM (2021). https://doi.org/10.1145/3461702.3462527
30. Hutchinson, B., Smart, A., Hanna, A., Denton, E., Greer, C., Kjartansson, O., Barnes, P., Mitchell, M.: Towards accountability for machine learning datasets: practices from software engineering and infrastructure. In: FAccT 2021: Proceedings of the 2021 ACM Conference on Fairness, Accountability, and Transparency, pp. 560–575. ACM (2021). https://doi.org/10.1145/3442188.3445918
31. Iqbal, R., Doctor, F., More, B., Mahmud, S., Yousuf, U.: Big data analytics and computational intelligence for cyber-physical systems: recent trends and state of the art applications. Future Gener. Comput. Syst. 766–778 (2017). https://doi.org/10.1016/j.future.2017.10.021
32. ISO: ISO 21502: 2020-12 Project, Programme and Portfolio Management—Guidance on Project Management, First Edition. International Standards Organization, Geneva, Switzerland (2020)
33. Jobin, A., Ienca, M., Vayena, E.: The global landscape of AI ethics guidelines. Nat. Mach. Intell. **1**(9), 389–399 (2019). https://doi.org/10.1038/s42256-019-0088-2
34. Joerin, A., Rauws, M., Fulmer, R., Black, V.: Ethical artificial intelligence for digital health organizations. Cureus **12**(3), e7202 (2020). https://doi.org/10.7759/cureus.7202
35. Kasinidou, M., Kleanthous, S., Barlas, P., Otterbacher, J.: I agree with the decision, but they didn't deserve this: future developers' perception of fairness in algorithmic decisions. In: FAccT'21: Proceedings of the 2021 ACM Conference on Fairness, Accountability, and Transparency, pp. 690–700. ACM (2021). https://doi.org/10.1145/3442188.3445931
36. Kerr, A., Barry, M., Kelleher, J.D.: Expectations of artificial intelligence and the performativity of ethics: implications for communication governance. Big Data Soc. **7**(1) (2020). https://doi.org/10.1177/2053951720915939
37. Khosravi, H., Shafie, M.R., Hajiabadi, M., Raihan, A.S., Ahmed, I.: Chatbots and ChatGPT: a bibliometric analysis and systematic review of publications in web of science and scopus databases (2023). arXiv:2304.05436
38. Langer, M., Landers, R.N.: The future of artificial intelligence at work: a review on effects of decision automation and augmentation on workers targeted by algorithms and third-party observers. Comput. Hum. Behav. **123**, 106878 (2021). https://doi.org/10.1016/j.chb.2021.106878
39. Lim, J.H., Kwon, H.Y.: A study on the modeling of major factors for the principles of AI ethics. DG.O2021: 22nd Annual International Conference on Digital Government Research, pp. 208–218. ACM, Omaha, NE, USA (2021). https://doi.org/10.1145/3463677.3463733

40. Loi, M., Heitz, C., Christen, M.: A comparative assessment and synthesis of twenty ethics codes on AI and big data. In: 7th Swiss Conference on Data Science, pp. 41–46. IEEE (2020). https://doi.org/10.1109/SDS49233.2020.00015
41. Lund, B., Wang, T., Mannuru, N.R., Nie, B., Shimray, S., Wang, Z.: ChatGPT and a new academic reality: artificial intelligence-written research papers and the ethics of the large language models in scholarly publishing. J. Assoc. Inf. Sci. Technol.https://doi.org/10.1002/asi.24750
42. Mariotti, E., Alonso, J.M., Confalonieri, R.: A framework for analyzing fairness, accountability, transparency and ethics: a use-case in banking services. In: 2021 IEEE International Conference on Fuzzy Systems (FUZZ-IEEE), pp. 1–6. IEEE (2021). https://doi.org/10.1109/FUZZ45933.2021.9494481
43. Martin, K.: Ethical implications and accountability of algorithms. J. Bus. Ethics **160**(4), 835–850 (2019). https://doi.org/10.1007/s10551-018-3921-3
44. Matthews, J.: Patterns and antipatterns, principles, and pitfalls: accountability and transparency in artificial intelligence. AI Mag. **41**(1), 82–89 (2020). https://doi.org/10.1609/aimag.v41i1.5204
45. McGrath, S.K., Whitty, S.J.: Accountability and responsibility defined. Int. J. Manag. Proj. Bus. **11**(3), 687–707 (2018). https://doi.org/10.1108/IJMPB-06-2017-0058
46. Metcalf, J., Moss, E., Watkins, E.A., Singh, R., Elish, M.C.: Algorithmic impact assessments and accountability: the co-construction of impacts. In: FAccT 2021: Proceedings of the 2021 ACM Conference on Fairness, Accountability, and Transparency, pp. 735–746. ACM (2021). https://doi.org/10.1145/3442188.3445935
47. Miao, Z.: Investigation on human rights ethics in artificial intelligence researches with library literature analysis method. Electron. Libr. **37**(5), 914–926 (2018). https://doi.org/10.1108/EL-04-2019-0089
48. Michalczyk, S., Nadj, M., Mädche, A., Gröger, C.: Demystifying job roles in data science: a text mining approach. In: Twenty-Ninth European Conference on Information Systems (ECIS 2021), pp. 1–7 (2021)
49. Miller, G.J.: Artificial intelligence project success factors—beyond the ethical principles. In: Ziemba, E., Chmielarz, W. (eds.) Information Technology for Management: Business and Social Issues. FedCSIS-AIST 2021/ISM 2021. Lecture Notes in Business Information Processing, vol. 442, pp. 65–96. Springer International Publishing, Cham (2022) https://doi.org/10.1007/978-3-030-98997-2_4
50. Miller, G.J.: Stakeholder roles in artificial intelligence projects. Proj. Leadersh. Soc. **3**, 100068 (2022). https://doi.org/10.1016/j.plas.2022.100068
51. Mitchell, M., Wu, S., Zaldivar, A., Barnes, P., Vasserman, L., Hutchinson, B., Spitzer, E., Raji, I.D., Gebru, T.: Model cards for model reporting. In: FAT* 2019— Proceedings of the 2019 Conference on Fairness, Accountability, and Transparency, pp. 220–229. ACM (2019). https://doi.org/10.1145/3287560.3287596
52. Mittelstadt, B.: Principles alone cannot guarantee ethical AI. Nat. Mach. Intell. **1**(11), 501–507 (2019). https://doi.org/10.1038/s42256-019-0114-4
53. Moser, D.C., Hond, P.F.d., Lindebaum, P.D.: Morality in the age of artificially intelligent algorithms. Acad. Manag. Learn. Edu. **21**(1), 139–155 (2022). https://doi.org/10.5465/amle.2020.0287
54. Munn, L., Magee, L., Arora, V.: Truth machines: synthesizing veracity in AI language models. AI & SOCIETY, 1–15 (2023). https://doi.org/10.1007/s00146-023-01756-4
55. Munoko, I., Brown-Liburd, H.L., Vasarhelyi, M.: The ethical implications of using artificial intelligence in auditing. J. Bus. Ethics **167**(2), 209–234 (2020). https://doi.org/10.1007/s10551-019-04407-1
56. OECD: Artificial Intelligence in Society, pp. 152. OECD Publishing, Paris, France (2019). https://doi.org/10.1787/eedfee77-en
57. PMI: A Guide to the Project Management Body of Knowledge (PMBOK® Guide)—Seventh Edition. Project Management Institute, Newtown Square, PA (2021)

58. Resseguier, A., Ufert, F.: AI research ethics is in its infancy: the EU's AI Act can make it a grown-up. Res. Ethics **20** (2023). https://doi.org/10.1177/17470161231220946
59. Rezania, D., Baker, R., Nixon, A.: Exploring project managers' accountability. Int. J. Manag. Proj. Bus. **12**(4), 919–937 (2019). https://doi.org/10.1108/IJMPB-03-2018-0037
60. Rodrigues, R.: Legal and human rights issues of AI: gaps, challenges and vulnerabilities. J. Responsib. Tech. **4**, 100005 (2020). https://doi.org/10.1016/j.jrt.2020.100005
61. Rossi, A., Lenzini, G.: Transparency by design in data-informed research: a collection of information design patterns. Comput. Law Secur. Rev. **37**, 1–22 (2020). https://doi.org/10.1016/j.clsr.2020.105402
62. Ryan, M., Stahl, B.C.: Artificial intelligence ethics guidelines for developers and users: clarifying their content and normative implications. J. Inf. Commun. Ethics Soc. **19**(1), 61–86 (2021). https://doi.org/10.1108/JICES-12-2019-0138
63. Sambasivan, N., Arnesen, E., Hutchinson, B., Doshi, T., Prabhakaran, V.: Re-imagining algorithmic fairness in india and beyond. In: FAccT '21: Proceedings of the 2021 ACM Conference on Fairness, Accountability, and Transparency, pp. 315–328. (2021). https://doi.org/10.1145/3442188.3445896
64. Scoleze Ferrer Paulo, S., Galvão Graziela Darla, A., de Carvalho Marly, M.: Tensions between compliance, internal controls and ethics in the domain of project governance. Int. J. Manag. Proj. Bus. **13**(4), 845–865 (2020). https://doi.org/10.1108/IJMPB-07-2019-0171
65. Shin, D., Park, Y.J.: Role of fairness, accountability, and transparency in algorithmic affordance. Comput. Hum. Behav. **98**, 277–284 (2019). https://doi.org/10.1016/j.chb.2019.04.019
66. Shneiderman, B.: Bridging the gap between ethics and practice: guidelines for reliable, safe, and trustworthy human-centered AI systems. ACM Trans. Interact. Intell. Syst. **10**(4), 1–31 (2020). https://doi.org/10.1145/3419764
67. Simon, J.P.: Artificial intelligence: scope, players, markets and geography. Digit. Policy, Regul. Gov. **21**(3), 208–237 (2019). https://doi.org/10.1108/DPRG-08-2018-0039
68. Singh, J., Cobbe, J., Norval, C.: Decision provenance: harnessing data flow for accountable systems. IEEE Access **7**, 6562–6574 (2019). https://doi.org/10.1109/ACCESS.2018.2887201
69. Strubell, E., Ganesh, A., McCallum, A.: Energy and policy considerations for deep learning in NLP (2019)
70. Torraco, R.J.: Writing integrative literature reviews: guidelines and examples. Hum. Resour. Dev. Rev. **4**(3), 356–367 (2005)
71. Umar Bashir, M., Sharma, S., Kar, A.K., Manmohan Prasad, G.: Critical success factors for integrating artificial intelligence and robotics. Digit. Policy, Regul. Gov. **22**(4), 307–331 (2020). https://doi.org/10.1108/DPRG-03-2020-0032
72. Unceta, I., Nin, J., Pujol, O.: Risk mitigation in algorithmic accountability: the role of machine learning copies. PLoS ONE **15**(11), e0241286 (2020). https://doi.org/10.1371/journal.pone.0241286
73. Vesa, M., Tienari, J.: Artificial intelligence and rationalized unaccountability: ideology of the elites? Organization (2020). https://doi.org/10.1177/1350508420963872
74. Vos Janita, F.J., Achterkamp Marjolein, C.: Stakeholder identification in innovation projects: going beyond classification. Eur. J. Innov. Manag. **9**(2), 161–178 (2006). https://doi.org/10.1108/14601060610663550
75. Wachnik, B.: Moral hazard in IT project completion. An analysis of supplier and client behavior in Polish and German enterprises. In: Ziemba, E. (ed.) Information Technology for Management. LNBIP, vol. 243, pp. 77–90. Springer, Cham (2016) https://doi.org/10.1007/978-3-319-30528-8_5
76. Wagner, B., Rozgonyi, K., Sekwenz, M.-T., Cobbe, J., Singh, J.: Regulating transparency? Facebook, Twitter and the German network enforcement act. In: FAT* 2020—Proceedings of the 2020 Conference on Fairness, Accountability, and Transparency, pp. 261–271. ACM (2020). https://doi.org/10.1145/3351095.3372856
77. Watson, H.J., Conner, N.: Addressing the growing need for algorithmic transparency. Commun. Assoc. Inf. Syst. **45**, 488–510 (2019). https://doi.org/10.17705/1CAIS.04526

78. Wieringa, M.: What to account for when accounting for algorithms: a systematic literature review on algorithmic accountability. In: FAT* 2020—Proceedings of the 2020 Conference on Fairness, Accountability, and Transparency, pp. 1–18. ACM (2020). https://doi.org/10.1145/3351095.3372833
79. Yim, R., Castaneda, J., Doolen, T., Tumer, I., Malak, R.: A study of the impact of project classification on project risk indicators. Int. J. Project Manage. **33**(4), 863–876 (2015). https://doi.org/10.1016/j.ijproman.2014.10.005

An Exploratory Study of Project Team Agility in Business Intelligence and Analytics Projects Using a Complex Adaptive Systems Lens

Mikhail Tsoy and D. Sandy Staples

Abstract Currently, few empirical studies of agile project management methods used in analytics projects exist. That, combined with a lack of theoretical conceptualization of Business Intelligence and Analytics (BI&A) project team agility, motivates this study. We aim to address these research gaps by conducting an exploratory case study of project team agility in the context of analytics projects. First, we introduce the research context and the Complex Adaptive Systems (CAS) theory as a theoretical framework for the study. Next, we describe the research methodology and data analysis used. Finally, we introduce and discuss the findings.

Keywords Agile projects · Project team agility · Agile project management · Analytics projects

1 Introduction

Agile project management methods are widely adopted in IT projects [1]. The methodology originated largely from the "Manifesto for Agile Software Development" (Agile Manifesto 2001), published online by a group of ISD engineers [2]. The authors of the manifesto promoted core values and principles of agility (e.g., individual empowerment, incremental and iterative development cycles, and customer involvement throughout the development process). Since the publication of the manifesto and the consequent increasing popularity, some researchers have emphasized the challenges [3] and benefits [4] that characterize the adoption of agile methodologies.

M. Tsoy (✉)
Brandon University, Brandon, MB, Canada
e-mail: tsoym@brandonu.ca

D. S. Staples
Queen's University, Kingston, Canada
e-mail: sandy.staples@queensu.ca

K. D. Strang and N. R. Vajjhala (eds.), *International Program and Project Management—Best Practices in Selected Industries*, Information Systems Engineering and Management 31, https://doi.org/10.1007/978-3-031-80275-1_8

Existing research on agile methods also has limited theoretical grounding [5]. Baham and Hirschheim [6] point that while some of the more recent Agile software development (ASD) studies applied theoretical frameworks, they cite a low number (11 vs. 291 between 2017 and 2020, IS Top 8 journals versus Software Engineering and Computer Science journals) and a very low proportion of ASD studies (less than 1%) studies published in Top 8 IS journals. The authors cite the lack of theoretical core as the main reason to explain both numbers. Our search of recent project agility studies (2014–2024) shows a similar trend: there are only a few studies that study project agility [7–12]. Out of these six studies only three used a theoretical framework. We aim to contribute to further development of this theoretical core as proposed by Baham and Hirschheim [6].

We aim address the need for further theoretical grounding of project team agility research by using Complex Adaptive Systems theory (CAS). Vidgen and Wang [13] posit that there have been very few empirical studies using this theory in IS due to the challenging nature of adopting abstract ideas of complex adaptive systems theory to empirical research. Ueland et al. [14] posit that the abstract nature of CAS may appear to be challenging when trying to understand actual events. However, the authors also recognize CAS theory potential to guide further understanding through abstract view using the theory. One contribution of our paper is an illustration of how CAS can be used to study projects.

Furthermore, Marchand and Pepper [15] postulate that both researchers and practitioners focus too much on technical challenges and less so on institutional and social aspects of BI&A projects. They argue that it has resulted in an imbalance between the socio- and technical sides of the process. In a more recent meta-analysis study by Oesterreich et al. [16], the authors discovered that institutional and social aspects such as human resources, management capabilities, and organizational culture have significant influence of how BI&A initiatives are being translated into business value. Our study hopes to address these needs by applying complex adaptive systems theory to the context of BI&A projects to better understand the nature of project team agility and explain the underlying mechanisms of project agility. The research question for our study is:

- *How do various factors conceptualized by CAS theory enable project team agility in BI&A projects?*

In study, we follow Conforto's [17, p. 667] conceptualization of project team agility where it is defined as "the project team's ability to quickly change the project plan (or product) as a response to customer or stakeholders needs, market or technology demands in order to achieve better project and product performance in an innovative and dynamic project environment." This definition focuses on project teams as a focal entity and emerging nature of project context.

This study of project team agility focuses on analytics projects for several reasons. Business Intelligence and Analytics (BI&A) projects are often associated with the IT function and sometimes are run by IT departments. However, treating BI&A projects similarly to a project to upgrade a CRM system may lead to failure [15]. Marchand and Peppard [15] suggest that BI&A projects represent special cases of IT projects

[15], e.g., Eckerson [18] and require different management approaches. They note that a typical ISD project operates around "defined outcomes, required tasks, and detailed plans for carrying them out," and a standard BI&A project "frames questions to which the data might provide answers, develops hypotheses, and then iteratively experiments to gain knowledge and understanding" [15, p. 106]. We see that while Agile methods are adopted across various industries, not all of them equally benefit from such adoption. BI&A appears to represent a case that could benefit more due to the nature of the product and higher uncertainty about the final outcome due to the exploratory nature of some BI&A projects. That, combined with the rapid growth of analytics initiatives in organizations, provides strong motivation to study how to manage these types of projects effectively.

BI&A projects may also pursue overarching goals that are different from IT project goals (e.g., IT project: improved efficiency, productivity vs. BI&A: challenge existing business strategy). IT projects are concerned with the optimization of existing business processes, and BI&A ones are focused on finding alternative business opportunities. Consequently, BI&A projects may benefit from different project structures and agile methodologies applied to reflect the iterative nature of some components of BI&A projects. Also, in comparison to some of the IT department led initiatives, requirements for BI&A projects are often challenging to be fully defined at the start of the project. Although business users and analytics teams may be able to define some initial requirements, the process of extracting, consolidating, cleaning, and testing the data may result in the emergence of new ideas or force the search for and pursuit of alternative routes. Such tasks would require certain degrees of flexibility from project team members, suggesting an agile approach may be well-suited to these types of projects. Studying how project team agility is enabled in BI&A projects will potentially contribute to both practice and research knowledge.

To address the proposed research questions, we conducted multiple case studies of analytics projects. The rest of this chapter will describe our study's theoretical background, methodology, and findings. We conclude by explaining potential contributions to the existing body of knowledge.

2 Theoretical Background

Dingsøyr et al. [5, p. 1217] point out that while "agile development evolved from the personal experiences and collective wisdom of the consultants and thought leaders of the software community" and "most individual agile practices have intuitive appeal," theoretical foundations are certainly deficient and only a few manuscripts that applied theoretical perspectives. Among them, knowledge management, personality, and organizational learning perspectives were used the most. The use of knowledge management and organizational learning perspectives is based on the view that the software development process is a knowledge development effort and, thus, represents a logical choice. Similarly, the use of personality perspectives (e.g., Big Five personality theory) appears to be a good fit when explaining the in-group dynamics of project teams.

Complex adaptive systems (CAS) theory has been used in several organizational studies [19], e.g., [20–23] and Anderson [20] suggests that this theory may no longer be counted as a new theory in organizational studies. Supporting our choice of Complex Adaptive Systems (CAS) theory for our study, proponents of agile project management methodology posit that the CAS perspective fits appropriately in the explanation of ISD team agility e.g., [24, 25] and is "the only way to make sense of the world" [26, p. 48]. As we mentioned earlier, our search of recent project agility literature identified three studies that used theoretical framework [9, 10, 12] in 2014–2024. Interestingly, all three used the same theoretical lens—Complex Adaptive Systems theory.

McConnon and Chowdury [9] used three principles of CAS (self-organization, interaction, feedback) as guiding lens for their single case study of cloud computing. Radhakrishnan et al. [10] uses CAS theory as a foundation for Adaptive Project Performance as a mediator between project agility and project success in online survey of 292 project managers. Finally, Werder and Maedche [12], in multi-case study, develop a theoretical model and propositions based on CAS and ASD literature, use three case study findings to support developed model and propositions. Werder and Maedech [12] theorize that factors such as individual experience, job clarity, technology access, goal interdependence, team task complexity, user research, management support will have positive influence on autonomy which in turn will have positive influence on emerging state of team agility. The authors also theorized that development length will have negative influence on autonomy.

In our study, we apply a theoretical perspective (Complex Adaptive System theory) as an analytical guide to code the qualitative data and potentially enable us to explain how adherence to some agile practices leads to the achievement of higher levels of project team agility. CAS principles and sub-dimensions are explained below.

2.1 Complex Adaptive Systems

A complex adaptive system is characterized by the presence of loosely interconnected autonomous parts (agents). Agents possess the capacity to intervene meaningfully over the course of events [27] as they operate based on their local principles (schemata—"changeable cognitive structures used to make sense of the environment and determine what actions to take": [13, p. 357]. The consequential interaction of agents in complex adaptive systems may result in complex behaviors.

The main principles of CAS theory relevant to academic and practitioner literature are described as follows. Volberda and Lewin [28] broke it down into three principles: match the co-evolutionary change rate, optimize self-organization, and synchronize exploitation and exploration. These three principles provide a platform for key CAS concepts such as co-evolution, the edge of chaos, interconnected autonomous agents, self-organization, and the edge of time [19–21]

Principle 1. Match Co-evolutionary Change Rate

The first principle states that an organizational entity should match or exceed the co-evolutionary rate of the system of which they are a part of [29]. Mittleton-Kelly [22] describes matching the co-evolution rate as sensing changes in the external environment and within the entity and responding at a pace that matches or exceeds that of internal or external change. For instance, a project team that manages to keep up with internal and external change has a better chance to deliver higher business value to its customers by delivering a product that closely matches changes in requirements. Thus, the principle of the coevolutionary change rate is important for project success.

Similarly, following the principle of matching the co-evolutionary change rate, Vidgen and Wang [13, p. 357] describe the edge of chaos as a "zone between total order and complete disorder. At the edge of chaos, a system shows bounded instability; that is, it is stable and unstable at the same time. It is stable in the short term as it follows the plan to address the change that occurred in the most recent iteration. It is unstable in the longer term in that the path which the system will follow is uncertain and unpredictable". The edge of the chaos is characterized by sufficient freedom to cope with change but enough structure to prevent the project from failure. Vidgen and Wang [13] described user requirements as "the most significant source of change" that a project team would face. For instance, the way planning is done in agile projects demonstrates short-term stability in the form of a more detailed plan for the length of the current iteration and long-term flexibility achieved via only high-level planning for a project. Thus, for principle 1, there are two sub-dimensions: poise at the edge of chaos, and co-evolution.

Principle 2. Optimizing Self-Organization

Self-organization is defined by Vidgen and Wang [13, p. 358] as "the ability of interconnected autonomous agents of a complex adaptive system to evolve into an organized form without external force." Similar to Principle 1, there is a certain balance between individual autonomy to sense and respond to change and local rules that are defined and followed by the individuals and teams [13]. Such a form of organization matches the system's focus on keeping up with internal and external changes. Moreover, this form of organization relies on knowledge sharing and team learning. Furthermore, local rules evolve over time in order to match or exceed the co-evolutionary change rate (Principle 1). Therefore, self-organization in complex adaptive systems facilitates shared responsibility among team members, knowledge sharing, and team learning.

Collective mindfulness is defined by Vidgen and Wang [13, p. 370] as "openness to novelty, alertness to distinction, sensitivity to different contexts, awareness to multiple perspectives, and an orientation in the present." All of those characteristics are important for project team self-organization, continuous awareness to change, better ability to work in cross-functional team due to awareness to other perspectives and focus on the present resembles the focus on near-future results.

However, self-organization requires team members with relevant skills and experiences. For instance, some team members might be overwhelmed by the amount of autonomy, communication, and collaboration that is required in self-organized teams. To be effective, it is important to have sufficient levels of relevant experience and expertise. Thus, principle 2 has three sub-dimensions: collective mindfulness, sharing and team learning, and individual skills.

Principle 3. Synchronize Exploration and Exploitation

This principle emphasizes the importance of a balance between simultaneous exploration and exploitation. More specifically, Vidgen and Wang [13, p. 359] define exploration as innovation and knowledge creation and exploitation as improvements in productivity, process efficiency, product extensions, and enhancements. Brown and Eisenhardt [19, p. 12] explained the balance of exploration and exploitation in terms of the CAS concept of balancing at the edge of time where you are "rooted in the present, yet aware of the past and future". An example of that can be a procedure of daily stand-up meetings in agile projects: the main focus is on the near-term future (e.g., what was accomplished yesterday, what is to be done today, and what are the immediate concerns, "roadblocks").

Benbya and Shelvey [30], in the conceptual paper on CAS, posit that modularity of design enables higher rate of adaptive response. More specifically, the authors hypothesize citing Parnas [31, p. 24] that modularity has three benefits: (1) shortened development time due to possibility of parallel work, (2) product flexibility due to the ability to change only one module at a time, (3) comprehensibility due to gradual design and understanding of the whole product or system. Other potential benefits of design modularity is the ability to reuse or repurpose some of the modules developed, and ability to deliver testable product increments for customers to test and review e.g., Sun et al. [32]. Thus principle 3 has two sub-dimensions: the edge of time and the modularity of the product and process.

In our study, we adopted a case study methodology (explained further below) as it allowed us to explore the three dynamics identified by Choi. Our study aimed to explore project team agility and the theoretical CAS mechanisms that potentially influence project team agility. Furthermore, we adopted the complex adaptive systems theory as the theoretical lens for our project team agility study. Compared to the existing studies, mentioned above, we took a more exploratory approach in our study, and tried to identify and explain how different factors enable project team agility using CAS principles as a theoretical lens.

As there is "no definitive account of CAS theory" [13, p. 357], we follow Volberda and Lewin's [28] three principles of CAS: 1. Match coevolutionary change rate, 2. Optimize self-organization, 3. Synchronize exploitation and exploration. Furthermore, we rely on subdimensions based on three principles, as theorized by Vidgen and Wang [13], Benbya and Shelvey [30], Kautz [33], and Meso and Jain [34]. We used the three principles and respective subdimensions as a guide to qualitative coding in order to find codes that either support, contest, or expand our view of CAS theory-based project team agility.

3 Methodology

This study uses a case-based research strategy to study projects throughout their life cycle. This strategy is appropriate as it fits our research questions and allows us to examine contemporary events that have complex contextual conditions, as almost all projects do. Within any project, there are many practices and factors that interact with the setting of the project. For example, the stakeholders involved and their power within the organization could affect the outcomes, the core team skills and their interaction patterns could affect productivity, and the organizational (or departmental) culture could affect the support received for the project and the role of the project leadership. As described above, the case-based research strategy allows us to examine the dynamics suggested by CAS theory. A multiple-case design is also appropriate for the following reasons. The case study approach allows us to study project team agility in "a natural setting, learn about state of the art, and generate theories from practice" [35, p. 370]. In this situation, it furthers our understanding of the nature of the relationship among the factors, discovers new factors, or sees that concepts that emerged from literature-based theorization are irrelevant in a given context. The case study approach also allows us to understand better the "nature and complexity of processes taking place." Therefore, a small number of cases will be able to provide rich insights. Furthermore, multiple-case designs allow for "cross-case analysis and theory extension" [35, p. 373]. As much as possible, Yin's [36] suggestions for establishing reliability and validity and analyzing case study evidence are followed (e.g., developing a detailed case study protocol, case study database, interview script, coding, pattern matching, etc.). Our unit of analysis is the project.

3.1 Sample

We used convenience sampling at companies using existing contact information. Companies presented in the data sample are those who agreed to participate in the study and offered several ongoing projects to choose from for the study. We aimed to pick projects with varying degrees of project team agility, as well as uncertainty in terms of project requirements and project management practices employed. Data collection for each project was done by interviewing key project stakeholders: members of the project team (all if possible), the project leadership (e.g., sponsors, senior executives affected by the project, people involved in project selection and approval), and the project clients (those who ends up using the outcome of the project). Brief details of each case project follow, along with our assessments of the level of project team agility found in each project.

3.2 Data Collection and Case Details

Initial data collection was completed at the end of August 2018 for four projects. All four projects are analytics projects in large financial institutions. Interviews lasted about 1.5 h with each participant, and 29 interviews were completed in total for the four projects. In both participating organizations, we had a few of projects that we could select from. We aimed to study projects with varying degrees of agility and uncertainty of final product. That would provide necessary variance across the sample. We also looked to study projects that were on-going which would provide the opportunity to study how they unfolded over the time.

All four projects were from two large financial institutions. Interviews with Project C and Project P participants were conducted on-site. Interviews with for Project M and Project W were conducted via conference calls.

Further, we conducted follow-up interviews with several participants from Project C and one participant from Project P. Participants from Projects M and W were unresponsive and, hence, we were not able to collect follow-up data.

In terms of project performance, we observed quite a range of results. Two projects struggled to meet project requirements, deliver results on time, and deliver business value. Project P finished model development with significant delays (i.e., several months). The developed model did not pass the test by the customer (business) team and was not implemented. Similarly, Project W experienced significant delays and did not pass customer team requirement test. However, the Project W team decided to salvage business value by pivoting towards an alternative application of the product, albeit at business value below what was originally planned.

Two of our case study projects met project requirements and deliver business value, albeit with slight delays. Project C met business requirements albeit slightly behind the initial timeline (14 weeks instead of planned 12). The developed BI&A model passed the validation stage. At the time of follow-up interview, the Project C model was being tested in preparation for implementation. With promising results at the time, it was expected that the model would be implemented after six months.

Project M was able to meet project requirements with slight delays and over delivered in terms of produced business value. The developed product was deployed and was undergoing further optimization.

3.3 Construct Measurement and Analysis

The semi-structured interview protocol asked questions covering the following topic areas: project team agility activities based on [17], project objectives, project performance to date, project selection activities, project initiation and planning activities, project execution and monitoring/controlling activities, project closing, and product transition activities (if appropriate—depending on the stage of the project), the analytics environment (using Eckerson's [18] framework), project leadership activities (based on Cross and Brohman's [37] Project Leadership model), and views on the challenges and rewards of analytics projects. The interview protocol was pilot tested with an experienced project manager and a doctoral student.

All interviews were recorded, transcribed, and entered into NVIVO 12 for analysis. Initial coding was done on interview question topics, as well as open coding to capture the dimensions of agility and the dynamics suggested by CAS theory. Axial coding was done to identify relationships among the open codes and explore our research questions.

Data analysis progressed in the following steps. The first step involved within-case analysis of project team agility, project management practices, and processes in each of the four projects. The second step involved a cross-case analysis of the project team agility and PM practices and processes.

- Within case analysis: we read all the interview transcripts and generated a list of codes. Next, we coded transcripts to identify project team agility related concepts. To categorize them into different groups, we utilized the existing conceptualization by Conforto et al. [17]: (1) Ability to change quickly, (2) Customer and Team interaction, and (3) Delivery Frequency.
- Cross-case Analysis: we removed redundant items in within-case analysis tables, merged similar ones, and took initial steps away from case-specific descriptions to generate more general findings [38].

3.4 *Assessment of Project Team Agility for Each Case*

While the initial case selection was guided by the objective to obtain variances in project team agility across cases, it was not until after data was collected that this assessment could be confirmed. Using Conforto's [17] conceptualization of project team agility, assessments of the level of project team agility, along with supporting evidence, is presented below for each case. Reflecting Conforto [17], we looked for evidence of the ability to change the project plans quickly, effective customer interaction and feedback leading to adaptation, the ability to sense the need for change (from the customer or other stakeholders), and product delivery frequency.

Project C

The Project C team was able to adjust their project plan quickly because it followed typical Scrum project planning methods with relatively short-term cycles (e.g., 2–4 weeks per iteration), detailed iterative planning, and a high-level overall project plan.

> Project C: So, we run our projects using Agile project management.
>
> Q: Are there any specific practices used as a part of Agile?
>
> Project C: I guess we have our sprints that are supposed to be iterative. Every day, I think I saw, we have our stand up in the morning that goes through what was done the previous day, what were some challenges or blockers, what we are hoping to achieve for today. And

> then maybe some qualified saying where we're going for this sprint, what are our objectives for this sprint. So, we have our stand ups.

Project C team was able to maintain effective interaction among BI&A development team, business customer and other stakeholders. For the duration of the project, stakeholders shared the same office space, had regular short daily meetings, and other interactions when necessary. We observed that the BI&A development team, business customer and other stakeholders were enthusiastic about learning from each other about running BI&A projects, and about sharing expertise from the relevant domain (e.g., BI&A, business, data) with others. The Project C project team, especially the BI&A development team members, had past experience with similar projects. Having relevant skills and experience potentially enables more effective and efficient communication among stakeholders from BI&A development team, business customer team and others.

More specifically, Project C team was able to sense the need for change in plans or the product due to continuous sharing and team learning, previous experience of working on similar projects, and a focus on creating short term business impact. Project C team members were able sense the need for change in project plan or product during data preparation, BI&A development, and testing phases of the project.

Project C team was able to receive frequent feedback throughout the project life. For example, project team had regular demo sessions during BI&A model development. A Project C team member explains the importance of feedback from business team:

> Project C: If I cannot get their attention, I don't have a use case. I don't exist if someone is not using or asking for support, because it's like I am a consultant. Someone has to cut the check, so at the end of the day, I have to charge the business to say, "I've spent x amount of time and resources on your use case." If I don't have your support, I don't have a use case. You will have to be part of the solution. You tell me what to do. If you're not telling me anything, I'm not doing anything.

Project C team was able to frequently deliver partial results for business team to review and test once it processed required data. During data processing period, BI&A were able to issue reports in terms data processing metrics.

> Project C: Before going to the next sprint, we make sure that what we have achieved is acceptable. For data, for example, when we hold it, we need to check and make sure that what we have is correct. We do that check. Once we build the model, we present the model to the leaders, and yeah. We get their buy-in.
>
> When you build the model, you have to make sure that your model is performing well, so that's internal, but then you have the challenge from the leaders because they know the business, so they will challenge your model. We have that during the weekly leadership meeting, and then I think that's give us the feedback whether they are okay with the model, yes or no. If there are things that we need to change, then we change them.

As a result, we assess that the Project C team achieved a relatively high level of project team agility according to Conforto's [17] conceptualization of project team agility.

Project M

The Project M team was able to adjust the project plan quickly because the nature of the product required regular releases. As a result, project planning methods were used that had relatively short-term (e.g., weekly partial product release), relatively detailed iterative planning and high-level overall project plan.

Project M team had relatively effective interaction between BI&A team and business customer team. Project M team members had scheduled and ad-hoc meetings throughout the project. Moreover, Project M team members shared previous experience of similar projects and had developed a shared understanding of product and project, communication processes.

> Project M: Let's say that these weekly sessions are officially held, that this does not limit the fact that if something suddenly comes up between the sessions they can have up to daily contacts or every third day, but at least, it is once a week.

Project M team relied on internal customer (business) team feedback at the beginning of the project. After a few initial partial product releases, Project M team started to receive end user related product performance metrics such as product use. As a result, Project M team members were able to recognize the need for project plan changes and were able to receive frequent feedback from internal customers and end user performance indicators.

As discussed above, the Project M team was able to deliver partial product increments due to the nature of the product.

> Project M: Once we had all the definitions, let's say that the deliveries occurred in time, as the dynamics were established, there were some deliveries that were made weekly and others that were made biweekly, but because that was how it was defined.

As a result, we assess that the Project M team achieved a relatively high level of project team agility according to Conforto's [17] conceptualization of project team agility.

Project P

Project P team was not able to change the project plan quickly due to a lack of resources and support and buy-in from the senior management and business teams. Project P team started the project with project planning similar to Scrum, relatively short-term (e.g., 2–4 weeks per iteration), detailed iterative planning and high-level overall project plan. However, due to lack of resources and resistance from some stakeholders, the approach switched to a linear, waterfall type of planning.

> Project P: Then became agile. The stand ups, retros, planning sessions, that lasted ... We had a very good run on that. I believe that lasted ... four months, roughly. At some point, they started to get diluted. I guess it had to do with the fact ... Number one, people have different responsibilities, of course.

Similarly, to project planning and execution, the interaction between BI&A team and customer team was not as effective in the latter part of the project.

> Project P: Other team members had other priorities, of course. So that slowed it (meetings and communication) down. Then we decided to … not officially, but to … dilute the stand ups (daily meetings). Now it's a little bit more informal. We still have questions every now and then. But clearly, it's not as intense as it used to be.

The Project P team did not receive feedback frequently. Project P started with having regular demo sessions (e.g., every other week), but eventually it transitioned to not having regular feedback from customer team.

> Q: In terms of delivering partial results, showing the demos, how often that happened?
>
> Project P: Every two weeks, so at the end of the sprint (previously).
>
> Q. So that continues now, as well?
>
> Project P: Right now, no it's not happening. … We are not doing demos anymore, right now.

Finally, Project P team were not able to deliver product increments frequently. While project work was conducted in short iterations, customer did not review partial product increment performance and did not provide feedback.

> Q: Would you agree that partial results of the project were frequently presented, discussed, and delegated, with the customer?
>
> Project P: No. Particularly, the validation part.

As a result, we assess that the Project P team achieved a relatively low level of project team agility according to Conforto's [17] conceptualization of project team agility.

Project W

Project W team was not able to change the project plan quickly. Although minor changes were implemented relatively quickly, the main product and project change occurred in the later stage of the project when it went 90% over the planned project time. At that point, project performance was tested and assessed as being insufficient by the customer team.

> Q: Were there any process management methodologies or specific practices chosen or defined for this project?
>
> Project W: No, not really. The only thing that we were doing was kind of regular meetings. So we had an executive meeting every two weeks, and then we had like regular meetings with (BI&A development team) for every three or four days just like very quick, half an hour, or something like that just for they to give us the advances that they had.

Project W team was able to have regular and frequent communication between BI&A team and customer team. Moreover, frequency of meetings increased towards the end of the project timeline.

> Project W: We, analytics, had a conference with all members once a week. When the timeline was getting to the end, we were having kind of everyday calls every like by the end of the day where they were giving us the results. But that was only, that was basically by the end of the timeline, probably one month before deadline.
>
> Project W: With the customer, we had three weekly meetings for the execution, one very short, I believe it was on Tuesday, no on Monday, to set up all the activities for the week.

> We call on Wednesdays on how these activities are being executed and if there's an issue to take care of. And weekly meetings on Friday in order to review the completion of all the activities. And also we have our biweekly executive meeting with all the managers and the sponsors of the project in order to review how we were doing and all the banks and orders we were getting. ... sometimes they were a reunion with all the people inside the room, some by video conferencing, and sometimes it was just a call. It was decided during the week if the team need to get together or it was just a call to report advance or issues. It was depending on how big is the issue and how good was the success of activities on that week.

While BI&A provided regular reports regarding product performance, customer team were not able to directly test the final or partial product early in the project life. The customer team was able to test the product in the later stages of the project and found the performance to be insufficient. Prior to testing, the customer team received regular performance reports. However, the customer team remained skeptical and insisted upon direct performance tests.

> Project W: So I think one of the main issues was when they were saying, "Okay, this is working." Then they would say, "No, it is not because this or that." Then they have also other numbers that's theirs and they are not similar. So yeah, those were the main issues, but they were always very kind to listen and do everything on it and to change and to adapt to this new resource that we were obtaining. But they were not very happy about that.

> Project W: Sometimes we were performing some, like, if they give us some ... One of the things we did when they said that this thing was kind of working, we said, "Okay, let's try it ourselves." So analytics asked them to provide with a direct link to software and we use it ourselves like in a real conversation. We did that like a lot of times to have a good sample of the outputs. ... At the beginning of June, or I think a month and a half, month and a half we did it once and then the results were not very well, and then we asked again and they kind of did it and then we checked it again and it didn't work as we were expecting. At the end we said, "Okay, let's do a final test but in place." So, we went to the (business location) to try it.

As a result, we assess that the Project W team achieved a moderate level of project team agility according to Conforto's [17] conceptualization of project team agility.

In summary, we assessed two of the case projects as having high levels of project team agility (teams C and M), one case with a moderate level of project team agility (team W) and one case with a low level of project team agility (team P).

4 Findings

The analyses from the case study are presented below organized by the three principles of CAS and their sub-dimensions. The patterns observed in project are compared with the assessment of project team agility, presented above.

4.1 Principle 1. Match Coevolutionary Change Rate

4.1.1 Poise at the Edge of Chaos

The planning cycle of Agile methods such as the widely adopted Scrum (high-level project plan vs. detailed iteration plan) allows for project adaptation and the potential emergence of innovative solutions while relying on the structure that is sufficient in terms of how detailed planning is for the given project context. The planning cycle can vary depending on the length of each iteration as the iteration length will influence the amount of detailed planning required for each iteration. The degree to which planning is outlined in detail before each iteration will dictate how much change can be accepted and addressed during the project lifetime and how detailed the current iteration plan will affect the short-term performance of the project team.

Planning in projects that use Agile methods is essential, but it typically is bottom-up, based on user stories rather than top-down at the project level. First, the structure of the planning base allows for more relevant adaptation that should follow the change as it is based on user stories. Second, such adaptation would occur quicker as the change is perceived and captured very close to the source (e.g., customer requirements, external environment change, etc.).

In some of our cases, we found similar patterns in terms of short-term stability elements (e.g., more detailed iteration planning) and long-term flexibility (e.g., high-level planning).

> Project M: We establish releases in phases because we always look for results to be obtained as soon as possible, and so as not to wait to finish the whole project, we generate tasks in parallel in the same diagrams or in the same timelines. And this allows us to have weekly monitoring; there may be traffic lights (performance indicators) that allow us to know who is on time, who can be delayed, and who is delayed so that we can jointly meet the objectives.

In this case above, we can see that the most attention is being paid to the current release and its short-term, in this case, daily, performance rather than the planning and performance across the whole project as the interconnected and evolving nature of the project is taken into consideration (i.e., edge of chaos). Also, it is evident that the project team takes into consideration the uncertainty of the project performance at each stage, depending on the uncertainty that stems from customer feedback, technological uncertainty, market turbulence, etc. In the case above, the team is primarily concerned with customer feedback and technological uncertainty to a lesser degree. Thus, we posit that according to CAS Principle 1, Project M is well equipped to sense and match the co-evolutionary rate of change given the project's context and history. Project team processes are established in a way that allows it to sense the changes (e.g., user requirements, data, technology challenges) and adapt in a timely manner due to the ability to balance on the edge of chaos.

> Project C: There's just one schedule that shows all the sprints and then within the sprints … during the sprint, we would decide what the timing is for all the tasks for that. By that, I mean the high level one is just like we started on this day, and we plan to end on this day. Objective, I mean, in terms of Agile project management, the people on the (area/function)

team, they have an idea of what is being done during each sprint. But I don't think there's anything more granular than that. It's just more like sprint one we do data and then et cetera.

Similarly, in the case above, we observe more detailed planning for the current iteration (sprint), and the details for future iterations (sprints) are developed when necessary. These structures ensure that there is a certain stability in terms of what should be done today (e.g., start the day with daily stand-up) and during the current iteration (i.e., edge of chaos). However, there is more flexibility in where the project would head next as the uncertainty regarding project performance and, hence, potential change in the direction is taken into consideration. Here, similarly to Project M, we posit that Project C is well equipped to sense and match the co-evolutionary rate of change considering project goals and the structure of the project team, resources, and processes. We observe the similarity in the focus on short-term performance and continuous adaptation throughout the project timeline. Detailed planning only for upcoming iteration with the inherent flexibility of overall high-level project planning allows both projects to remain at the "edge of chaos" and, thus, be able to adapt to project changes and deal with uncertainty via a combination of short-term detailed planning and long-term high-level planning, and continuous adaption of the long-term plan.

However, the other two cases in our sample did not demonstrate similar adherence to project planning for short-term stability and long-term flexibility. Project P did not have sufficient resources and leadership support to continue with the original plan. The project team started by adhering to traditional linear PM methodology (waterfall) and switched to Agile PM methodology for about four months. However, the project team struggled with adhering to the Agile PM methodology for the length of the project timeline due to a lack of support from the organization's leadership and acceptance of the Agile PM methodology from some stakeholders. As a result, the project team was not sufficiently equipped to sense and match the co-evolutionary change rate and stay at "the edge of chaos" due to returning to detailed long-term planning associated with traditional linear PM methodologies.

Project P: Well, as I mentioned before, it started traditionally. Like a waterfall, make the requirements through emails. Then, it became agile. The stand-ups, retros, planning sessions, that lasted … We had a very good run on that. I believe that lasted … four months, roughly.

Project P: So, we had those planning sessions; we all write down what kind of tasks we needed to do in the two weeks. And we always create the (Atlassian) Jira tickets; we had a board there. Like, just follow that.

The planning and execution of Project W were based on PMI guidelines for a more traditional linear model (waterfall). The choice was justified by previous experience and the simplicity of the methodology. Such an approach brings long-term stability and requires extensive planning prior to the start of the project. However, that comes at the cost of long-term flexibility due to the necessity to follow the plan. Here, similar to Project P, we posit that Project W was not sufficiently equipped to sense and match the co-evolutionary change rate and stay at "the edge of chaos" (i.e., by not using a combination of detailed short-term planning high-level long-term planning).

Project W: It's a mix of many. We generate project plans; we do all the P&Ls, but we don't follow a specific methodology. We have all the fundamentals of project management, but we ... I mean, I've been doing this for a while, but when you try to apply the methodology, it becomes complicated to fulfill all the specifications of that methodology, especially when you have like different projects and different products that get in the way of the methodology.

4.1.2 Co-evolution

Volberda and Lewin [28] theorize that a complex adaptive system needs to match or exceed the rate of external change (e.g., institutional configuration, social movement, etc.). They posit that to be able to do so, organizations should develop routines, capabilities, and measures to monitor relevant rates of change, learn, and make changes. In the case of Agile projects, another source of change comes from the uncertainty and complexity of the requirements. Therefore, depending on the nature of the project, it may potentially face high levels of complexity, uncertainty, and external change.

Learning can be based on customer feedback and other sources. For instance, the learning can occur among the customer and development team and take the form of various "interactions and knowledge sharing, continuous gathering/refining requirements, frequent working software presentations/delivery, and feedback" [33]. Project C and Project M had enough feedback from their customers and stakeholders regarding both the performance of the product and the way processes were established and how all the feedback was incorporated into project plans and tasks later. This enables the sufficient rate of co-evolution as a result of interactions with agents of complex adaptive systems and the environment e.g., Kauffman [39].

Project C: Then, also, in the meantime, we're learning from different stakeholders to understand the way they're working and, also, apply some stuff to our lab, too. So, after a couple of iterations from different labs, I think right now, it's more, I think it's a lot better than what we started. It works more efficiently because everyone knows their responsibility; everyone knows at what stage, what they need to do.

Project M: (Feedback and communications are) good here; every week, there is feedback. There is internal testing in terms of how it performs.

In contrast, in case W, while the customer and developer team had regular communication, the customer team was able to test the product and provide feedback only two times closer to the end of the project. The main challenge here is not in the frequency of communications and the amount of feedback from customers and other stakeholders, it is whether it was sufficient to enable a sufficient rate of co-evolution given the rate of change and uncertainty of the project goals and the nature of the product. According to CAS principle of co-evolution, a project team should be able to match the rate of external or internal change. In the context of case W, communication between development team and customer team for both sides to sense the need for change, but it was insufficient in case of Project W because customer team was not able to test and provide feedback regarding project performance. Thus, we posit that from the perspective of CAS theory Project P and Project W did not match the necessary rate of co-evolutionary change rate.

> Project W: I think a month and a half, we did it once, and then the results were not very well, and then we asked again, and they kind of did it, and then we checked it again, and it didn't work as we were expecting. In the end, we said, "Okay, let's do a final test, but in place."
>
> Project W: So, yeah. I mean, also and, clearly, this isn't working in four or five months. I mean, it was going to be a waste of time, and we were trying to say, "Okay, let's not work this anymore."

In Project P, the development team intended to conduct product demonstrations and reviews after each iteration. However, due to a lack of leadership support and sufficient resources, the feedback from product demonstrations was not regularly collected. Consequently, the development team was not able to receive and implement customer feedback. As a result, the development team was not able to match co-evolutionary change rate. Although the situation is somewhat different in that the development team wanted to conduct product demonstrations and reviews but experienced a lack of interest from the customer side, the result was similar. Project P was not equipped to match the co-evolutionary change rate because it lacked the input required to sense the specific needs for change.

> Project P: Demos? We didn't really do demos. We didn't do retrospectives that much. We didn't want to include other people, because they were getting a little bit annoyed with all these ceremonies. That's the reason. And for demos, it was very hard to get people together. Because in a demo, we should include our VP, their VP, it was a bunch of people, and get them at the same place at the same time; it's kind of hard.
>
> Project P: We tried to get feedback as much as we could.

There is a varying degree of regularity of customer feedback across all four cases. Having more frequent and regular feedback sessions does not necessarily result in a product that creates higher business value for the customer. However, according to the CAS principle of matching co-evolutionary change rate, the frequency of the feedback sessions should be related to the changes to the product requirements, business value views, and customer priorities. Thus, if the product requirements often change based on product performance and customer feedback, such pace of change should be either matched or exceeded by the pace of collecting requirements, delivering iterative business value, and integrating customer feedback. In all four cases that we studied we observed uncertainty related to data management, BI&A model and technology development and related need for change.

To summarize, the case data and analyses of the sub-dimensions show that projects C and M had good abilities to match the coevolutionary change rate. Projects P and W were not able to do this as well. This pattern matches the assessment of project team agility in the projects, suggesting that projects that are able to follow CAS principle 1 will have higher project team agility.

4.2 Principle 2. Optimizing Self-organization

4.2.1 Collective Mindfulness

Vidgen and Wang [13] and Kautz [33] hypothesize that self-organization is another important principle of CAS. More specifically, they posit that in a loosely controlled and interconnected development environment, autonomous stakeholders facilitate flexibility and adaptability. Here is an example from one of the cases where stakeholders are provided with the freedom to collaborate and learn from other stakeholders and project context.

> Project C: For us, it's more so how do we provide an environment for collaboration, development of customer-first type solutions, and then third, developing talent. That's what agile means to us, being able to do all these three things but learning fast. I know people say fail fast, but for us, it's "learn fast and pivot."

Certainly, this is not the case for all four of our cases. For instance, Vidgen and Wang [13] posit that such practices as regular stand-up meetings and pair programming emerged as a practice reflecting self-organization. However, we did not find universal adherence to such practices in all four projects (e.g., more so for M and C and less so for P and W). The same can be said regarding the interconnectedness and autonomy of stakeholders (e.g., physical presence, closeness/commitment to the project). Such close proximity (e.g., shared physical office space for all main stakeholders) and the ability to interact on the spot build a sense of shared responsibility and allow for discipline based on peer and self-observation [13]. Similarly, it enables higher levels of collective mindfulness due to improved awareness to multiple perspectives, sensitivity to multiple perspectives.

> Project C:
>
> Q: Are stakeholders usually in place for the duration of the project?
>
> A: Yes, in fact, we are all in the place, in different offices, in different buildings, because we work there, but yes, they are usually there, if there are meetings or involvements through video conferences or audio conferences, they are in their offices. And, in some cases, we meet in an intermediate office to have face-to-face meetings. After a moment, at the initiation of the project, and then in the middle of it, we had the support of the provider of members of the [organization name] team who were advising us over time; they came from other countries and were physically (here) during a time in the place.

It should be noted that the colocation of the project team in the same spot does come with some drawbacks. One of the participants noted that constant communication among team members could be a distraction from work. Such a negative impact can be reduced by having mixed workspace with separate zones for meetings and solitary, focused work.

> Project C: I think there are downsides, really. Sometimes you want just to be in your own space and working on something. Like when the space is full, there's I'm not sure how many people, at least like 16 plus people in there. So, there's a lot of stuff going on. There are other projects that are going on, so other people have discussions for their project, and people are just moving around. You can get distracted pretty easily.

Project C:

Q: Could you tell us a little bit about the process or activities that were done as part of planning for the project?

Answer: Sizing of the impact, of the business deliverables. Basically, that's to confirm whether it's worth it to do this project or not. Understanding the resource requirement, including the skill set of the people and is there any additional technology or what not required or do we need additional consultant. As well as socializing in general.

Project M presents an interesting case where we can see that project team members and stakeholders would have regular face-to-face meetings at the same location for a period that was deemed necessary. However, once the period of such necessity passed, the frequency and medium of those meetings changed (i.e., from face-to-face meetings to phone meetings. However, developer team remained at the same location. We can see an example of how frequency and medium of communication can be adjusted once sufficient level of collective mindfulness was reached.

Project M:

Q: You mentioned previously you mentioned a bunch of stakeholder projects from the stakeholders; how many of them were usually on site during the project?

A: When we move to weekly meetings, we had one day ... one weekday, we had on site meetings for like three months. Two or three months probably, then we moved them to phone meetings.

In Project P, due to some resistance towards Agile methods and lack of resources, similar settings and conditions were not met. It was also harder for Project P stakeholders to establish sufficient level of collective mindfulness over how the project should have been conducted due to turnover and stakeholder resistance towards Agile PM methods. Finally, the development team and business customer team did not have previous experience working on any BI&A projects.

Project W was not able to reach sufficient level of collective mindfulness and autonomy due to physical distance, insufficient communication, and lack of product demonstration with business stakeholders. Furthermore, the development team and business customer team were part of two different organizations and were in separate buildings. Lastly, similar to project P, the development team and business customer team did not have previous experience working on similar type of BI&A projects.

4.2.2 Sharing and Team Learning

Emergent co-evolution is characterized by "joint learning of customer and development staff as well as software evolution through multiple forms of close interactions and knowledge sharing, continuous gathering/refining requirements, frequent working software presentations/delivery, and feedback" in Volberda and Lewin [28].

When stakeholders from two or more different knowledge areas interact in the project setting, it is expected that some will be more proficient in one area and less so in other areas. In this context, mutual learning and knowledge sharing can address those gaps in knowledge among those groups. That is especially often the case for BI&A projects, where stakeholders from data analytics, data science, various business area expertise, and project management work together. If all or many of those stakeholders have previous experience working each or working with other areas of expertise, it might accelerate sharing and learning and contribute towards knowledge sharing and team learning. Kautz [33] posits that mutual learning reinforces emerging structures of collaboration and interaction.

We can see how sharing previous experience facilitates higher levels of knowledge sharing and enables the reliance on previous team learning. In Project C and Project M, we have project team members and stakeholders who have previous experience in knowledge sharing and team learning. This contributes to project team's collective mindfulness.

> Project C: There are two things, so the deliverables also are in terms of resources because of the way how we operate in this kind of project, we do training, and the fact that we all work together … we learn from each other. As I'm working on this project, I'm learning the project management tools. The team also, they are learning that, and the DS (Data Scientists) and the DE (Data Engineers), they work very closely, so when the DS … I mean, they can learn some DE techniques, the DE can learn some DS technique. From that perspective, it's successful for the team because they are learning new tools.

> Project M: Additionally, the administration of the tool, in a unique campaign or isolated, is very different than when you have to combine different campaigns linked together, which are fired from certain triggers. Therefore, somehow, we were learning together how to be able to maximize the functionalities that the tools gave us. That was faced with other types of challenges, we find ourselves on the road, for example, the execution as such of the media.

> In some way, the level of information or skills or knowledge or experience that came up at first was enough to begin to build the foundations with which this project was started, but the reality is that if you compare those skills versus the current ones, they have been increasing depending on how you are learning from the other things you can do and the different challenges that are presented along the way.

In Project P and Project W, team members and stakeholders did not have any previous experience of working with each other or stakeholders from other areas of expertise on similar projects. As a result, building knowledge sharing and team learning required relatively more time compared to Project C and M. That negatively influences team's level of self-organization and autonomy.

> Project P: That was one thing is that I had to learn: this is the first time I worked with the [topic] … in that sort of way. … Understanding how it would work in practicality at the time of the judication that took me a long time up until I did come back many times and say, "Okay, pretend I don't know anything about this model. Explain it to me". And I had a hard time grasping that information from the level of knowledge that I had.

> Project W: They (the developer team) have learned a lot on how these processes work and all the mistakes or things that, the mistakes they are making, or things that can be improved. Just going through the implementation of this pilot, they learned a lot, so for that, they are very excited. They agreed to keep going with this thing because they understand the value they're getting just from going through this process.

According to the self-organization principle, knowledge sharing and team learning are necessary to facilitate self-organization within a project team. As we can see from projects C and M, knowledge sharing and team learning occurs within project team with continuous communication and colocation as it provide opportunities to learn from each other.

> Project C: …the department around us, they know you guys have those kinds of analytics teams. They are also interested in what you guys do. So, one of my tasks, we do that kind of training (events). So, we have that kind of training, training for all the employees. We're going to train that (many) people, we just want to let them know what we do, in terms of what kind of methods we're using, like kind of Agile process we're doing, and how we're doing it. Then our audience is much bigger. So, for the training, (we talk about) what kind of tool we're using, what kind of standard processes we're doing (for) the project.

4.2.3 Individual Skills

Such a process of mutual learning is enhanced when stakeholders with overlapping knowledge in multiple domains of knowledge are present. For instance, in our cases, the people with background and experience in both business and analytics were invaluable. It usually starts from the Project Manager/Scrum Master of a project, but we observe that with experience, other stakeholders gradually learn from the other areas of expertise (e.g., Data Analysts learn more about the business domain for which they develop analytics models/products).

> Project M: Well, I have 18 years of experience in the financial sector, of which 10 years have been very punctually in the business of (financial services), in the administration of the product, placement channel, sales management, administration of risks team, anyway, all that entails the business strategies to place the different products through the different channels. And in the (financial institution), I have a bit more than three years (of experience), and my activity here has been purely on different campaigns or strategies that we have a contact within the CRM analytics. So, my experience might not be too technical, it is rather commercial or of strategy, but in the end, I am in charge of much more technical teams that are in charge of the execution of the strategy.

> Project W: When I got into technology, I start as an engineer. I was an architect. Then I don't know when I moved to the sales part and then to the sales manager, and then to own a company, and now in business development for the past five years.

While only half of the cases had someone with knowledge of two or more knowledge areas, the absence of such team members can be compensated by those with experience of working in a similar project and mutual learning and knowledge sharing. Although all four project teams had a mix of skills in terms of relevant business area experts and experts in BI&A, Projects C and M had previous and shared experience of working on the similar projects, and Projects P and W did not have shared experience or an experience of working on similar projects or products. Jain and Meso [34] posit that developer skills play an important role in agile methodology. They cite emphasis on the talent and skills of team members, such as being responsive, competent, and collaborative [40]. Having those skills and talents can benefit a project team when project team members learn from each other [41]. Further, such qualities as being responsive and collaborative are crucial when the project team communicates and collaborates over the project's lifetime [41]. Therefore, we can conclude Projects C and M were better prepared in terms of shared and individual skills and knowledge.

Project P:

Q: So, how would you evaluate your team members' levels of skills, both in people skills, technical expertise, or business knowledge?

P: My colleagues' business knowledge? No, not that great. I think we are really missing that part, and that's why we had troubles. People skills, great and not so great. I think they have really nice people, have people skills, but sometimes in business, you have to have some, little bit different people skills. You know, so missing that part a little bit. From the technical, we have really skilled people, on both the engineering and ASI sides. It's like, really high technical knowledge there. Yeah, but they are getting blocked.

Project M:

Q: How would you evaluate the skills of team members? Regarding their technical knowledge and business knowledge, do you think it is enough, or not?

M: As for technical knowledge, yes, they are good. In terms of knowledge in business, let's say they do not know all the products and all the services of the (financial institution).

Project W:

Q: How would you evaluate their technical expertise?

W: High, very high.

Q: How about their business knowledge?

W: Medium-low.

To summarize, the case data and analyses of the sub-dimensions show that project C had the strongest abilities to optimize their self-organization. Project M also had fairly good abilities. Projects P and W were found to be considerably weaker. This pattern again matches the assessment of project team agility in the projects, suggesting that projects that are able to follow CAS principle 2 will have higher project team agility.

4.3 Principle 3. Synchronize Exploration and Exploitation

4.3.1 The Edge of Time

Complex adaptive systems adapt by focusing on the present (e.g., in Agile projects: yesterday and today during daily stand-up, upcoming iteration when the iteration plan is created). Such focus allows for further sustained efficiency of the project as it allows for a project team to focus on the activities and tasks that add to business value in near future or further improve existing processes.

Project C and Project M demonstrated sufficient fit between developed routines, capabilities, and measures to monitor relevant rates of change. These project teams adhered to Agile practices and relied on developed experience and routines with a focus on current iteration. Balancing on the "edge of time" with a focus on the near future (e.g., this iteration) enabled teams to produce, test and demonstrate tangible incremental outcomes of the product periodically.

> Project C: There's just one schedule that shows all the sprints and then within the sprints … during the sprint, we would decide what the timing is for all the tasks for that. By that I mean the high level one is just like we started on this day, and we plan to end on this day. Objective, I mean in terms of Agile project management, the people on the (area/function) team, they have an idea of what is being done during each sprint. But I don't think there's anything more granular than that. It's just more like sprint one we do data and then et cetera.

However, Project P and Project W did not manage to focus on the near future or be at the "edge of time". Project P started with a linear project management methodology (waterfall). Later, Agile methods were adopted for a period of a few months. Consequently, later they returned to more waterfall methodology due to pressures from stakeholders who found Agile methods overly demanding. Here we have an example where suitable methodology, processes, and measures were selected (Agile), but due to lack of support, experience, stakeholder buy-in the project team failed to continue to adhere to an Agile PM methodology.

> Project P: (Developer) Well, as I mentioned before, it started traditionally. Like a waterfall, make the requirements through emails. Then, it became agile. The stand-ups, retros, planning sessions, that lasted … We had a very good run on that. I believe that lasted … four months, roughly. We had those planning sessions; we all write down what kind of tasks we need to do in the two weeks. And we always create the (Atlassian) Jira tickets; we had a board there. Like, just follow that.

In the case of Project W, the project team selected a linear, waterfall methodology for the duration of the project. Here, we can argue that a suitable methodology was not selected because this case represents an attempt to develop and implement new product without much history of success in the given context, both technology and product. Linear PM methodology combined with very few demonstrations or customer product tests did not enable the focus on the near future as the main focus on this project was the end of the project when the product would be developed. However, the business customer team was able to learn that the final product did not meet the requirements, and the only viable alternative at that point of the project was to pivot the project towards the solution with much lower requirements and, hence, potentially much lower business value. According to the CAS edge of time principle, focus on short term, near future impact via frequent customer tests, feedback, and incremental implementation and product delivery would enable project teams to benefit from matching exploration (i.e., learning and innovation) and exploitation (i.e., improvements in productivity, process efficiency, and product extensions and enhancements).

> Project W: It's a mix of many. We generate project plans; we do all the P&Ls, but we don't follow a specific methodology. We have all the fundamentals of project management, but we … I mean, I've been doing this for a while, but when you try to apply the methodology, it becomes complicated to fulfill all the specifications of that methodology, especially when you have like different projects and different products that get in the way of the methodology.

4.3.2 The Modularity of Product and Process

Modularity of design allows further exploration and exploitation of the outcomes of a project. Benbya and Shelvey [30] discuss the importance of modular design. Modularity may also be important beyond simple code reuse. For instance, some of the processes or structures that emerged during the project can be later applied to another project with necessary adaptations. Three of the four cases studied demonstrated the intention to extend the use of knowledge regarding processes, developed technology elements, or products beyond the current project. Two cases aimed to reuse the product and processes result from a successful project.

Project C redeployed processes and experience from previous projects of a similar nature. Furthermore, the Project C team planned to further improve, learn, and incorporate that knowledge and experience in the following projects adapting it for the new set of requirements and project goals. Thus, Project C is well positioned to leverage product and process modularity.

> Project C: For us, it's more so how do we provide an environment for collaboration, development of customer first type solutions, and then third, developing talent. That's what agile means to us, being able to do all these three things but learning fast. I know people say fail fast, but for us, it's "learn fast and pivot." ...: I'm happy if, after this project, they can do their own project. That's a good way to see that they've picked up the learnings. It's not just analytics of the data science; it's also the project management, the project delivery. The nuts and bolts about analytics, but also the tracking, the strategy, the communication. There's a lot that goes into the delivery of a project, so it's easy to see how mature on that curve the team is because they'll come to you asking for specific resources.

Similarly, Project M redeployed processes and experience from a previous project that was very similar but at a smaller scale. Thus, Project M is well-positioned to leverage product and process modularity in the future as per modularity principle of CAS.

> Project M:
>
> Q: Is there additional work to be done or learning from this that will go into the new project?
>
> M: Yes, we plan for many other [type of] products, [example of product] or [example of product].

In the case of Project W, stakeholders found a way to adapt a technology part that failed to meet the original project scope and requirements to be used in a different context. Thus, it is a case of product adaptation and improvement [30] that can be achieved regardless of the project performance. This demonstrates how product and process modularity enables relatively quick adaptation and project benefit maximization according to Principle 3 of CAS which emphasizes the importance of balancing exploration and exploitation [30]. However, the continuity of modularity in process and product in the case of Project C and Project M is predetermined in nature, while Project W product modularity is somewhat situational. As a result, we can posit that project teams that plan for future product and process modularity can leverage and benefit from such modularity to a relatively higher degree than product or process modularity that has ad-hoc nature.

Project W: So, yeah. I mean, also and clearly this isn't working in four or five months, I mean, it was going to be a waste of time, and we were trying to say, "Okay, let's not work this anymore." But in the end, we decided to keep working on this but take a different approach and with a different business model, so we don't waste everything that's already implemented.

Project P is a case of a project where the project team and business stakeholders struggled with product and process modularity. Such modularity was not a part of the original design, as the emerging data was intended to contest the existing solution and replace it only if certain goals were achieved. In the end, the contender was not able to outperform the existing solution and was not used by the organization.

Project P: Oh, that's hard to … we are, even now, we are having trouble figuring out that (product performance sufficient to replace the existing solution). It's not like … Well, there are some certain measures that you need to meet with your model, that they can consider that model. So, I think we reached those. So, I think we met that requirement. But still, that doesn't guarantee that your model will be used, or … So, we are still making the business impact analysis even now because it's not clear yet, how to use. We came out with a … some way of how to use the model, and they're currently evaluating that.

To summarize, the case data and analyses of the sub-dimensions show that projects C and M were effective at synchronizing exploitation and exploration activities. Project M also had fairly good abilities. Projects W and P were weaker than C and M in their fit with this CAS principle. This pattern again matches the assessment of project team agility in the projects, suggesting that projects that are able to follow CAS principle 3 will have higher project team agility.

Table 1 summarizes the findings of each project for the 3 CAS principles and their sub-dimensions. Discussion and implications for these findings follows.

5 Discussion and Implications for Research and Practice

Observed patterns in our case studies found consistent overlap between the principles of CAS theory and the achieved levels of project team agility. This suggests that using the CAS principles to guide project practices could lead to stronger project team agility and enhance business impact and value. Below, we review how BI&A projects that hope to achieve higher levels of BI&A project team agility could benefit from implementing PM methodologies, processes, and resources according to the CAS principles and sub-dimensions. The findings also illustrate that CAS theory is a valuable lens to study agile projects and predict performance. Our work helps address the existing gap in theoretical perspectives for agile projects and the lack of theoretical empirical studies. Each of the three CAS principles are discussed below, along with implications, and suggested recommendations for practice.

Project teams that can match the co-evolutionary change rate should be able to achieve a higher level of BI&A project team agility. This includes staying at the "edge of chaos" via adhering to a high-level long-term plan and detailed short-term plan and maintaining a sufficient rate of co-evolution, continuous feedback and adaptation

Table 1 CAS in BI&A project team agility

CAS principle	CAS dimension	Project C	Project M	Project P	Project W
Principle 1. Match coevolutionary change rate	Poise at the edge of chaos via a balance of high-level long-term plan and detailed short-term plan	• Well-equipped to sense and match co-evolutionary change rate • Managed to stay at the edge of chaos via adhering to high-level long-term plan and detailed short-term plan	• Well-equipped to sense and match co-evolutionary change rate • Managed to stay at the edge of chaos via high-level long-term plan and detailed short-term plan	• Not equipped sufficiently to sense and match co-evolutionary change rate • Not able to stay at the edge of chaos because of lack of leadership support and resources	• Not equipped sufficiently to sense and match co-evolutionary change rate • Not able to stay at the edge of chaos because of lack of communication, transparency, stakeholder buy-in
	Co-evolution via continuous feedback and adaptation	• High rate of co-evolution via continuous feedback and adaptation	• High rate of co-evolution continuous feedback and adaptation	• Insufficient rate of co-evolution due to lack feedback and adaptation	• Insufficient rate of co-evolution due to lack of access to test product prototypes
Principle 2. Optimizing self-organization, focus on people and skills	Collective mindfulness via autonomy, colocation, sufficient communication	• High level of collective mindfulness via colocation, sufficient communication, and project team autonomy	• High level of collective mindfulness via colocation, sufficient communication, and project team autonomy	• Insufficient level of collective mindfulness due to project team turnover, lack of buy-in from some stakeholders and overall lack of support from senior leadership,	• Insufficient level of collective mindfulness due to lack of communication, physical distance between stakeholders

(continued)

Table 1 (continued)

CAS principle	CAS dimension	Project C	Project M	Project P	Project W
	Sharing and team learning	• High level of experience, and knowledge within the organization and each project team of developing similar products and working using Agile PM methods	• Sufficient prior experience, knowledge and know-how, some customer base already experienced how previous versions of the product operate and perform	• Insufficient previous experience, and knowledge (especially on the business stakeholder side), lack of human resources	• Insufficient previous experience in similar projects
	Individual skills	• Sufficient mix of relevant knowledge • Sufficient level of previous experience in similar projects	• Sufficient mix of relevant knowledge • Sufficient level of previous experience in similar projects	• Sufficient mix of relevant knowledge • Insufficient level of previous experience in similar projects	• Sufficient mix of relevant knowledge • Insufficient level of previous experience in similar projects

(continued)

Table 1 (continued)

CAS principle	CAS dimension	Project C	Project M	Project P	Project W
Principle 3. Exploration and exploitation	Edge of time via focus on short term business impact	• High level of exploration via continuous search for ways to improve existing products and methods	• High level of exploration, looking for ways to improve existing products	• Sufficient level of exploration as the development team learned a lot about challenges and explored various ways to address them to reach project goals, sharing with the business side did not occur often enough	• High: the exploration development team learned a lot about challenges and ways to address them to reach project goals, insufficient support and product performance
	Modularity of product and process	• High level of exploitation via leveraging knowledge and experience from previous projects is used in the current one, feedback from previous iterations is also integrated, the product backlog is continuously updated	• High level of exploitation as an implementation of ideas discovered in previous product iterations or even previous projects was achieved	• Sufficient level of exploitation as the project team eventually decided to pivot the product to be applied to other goals	• Insufficient level of exploitation, not much continuance in terms of staff, PM methods, and levels of support from different stakeholders;

among project team, customers, and other stakeholders. The majority of projects using Agile methodologies use either Scrum or its variations (ScrumBan and Scrum/XP Hybrid) [42]. Most Agile projects with these methodologies use time-boxing in the form of 2–3-week iterations as an internal mechanism that allows projects to change and adapt between iterations, but the scope of each iteration remains fixed for its duration. We observed similar planning and execution in short iterations (2–4 weeks) in 3 out of 4 projects. That is also related to the co-evolution principle of CAS. As the length of iteration relates to the flexibility and speed of change implementation (e.g., in the case of a detailed one-week iteration, such planning structure allows for the change to be implemented after one week, and it would require a three-week delay in implementing a change in case of iteration length being that of 3 weeks). According to CAS, the intervals for planning, execution, and change should fit the nature of the project [33].

Our case studies found that some projects may change iteration length throughout a project lifetime. This is done to change the amount of adjustment that can be addressed within one iteration: either speed up development or slow it down. This is similar to Benbya and McKevley [30] theorization of the need for "speeding up the rate of gap reduction between IS effectiveness and changing requirements of users via quickening learning action loops and real-time action on their feedback." However, both Vidgen and Wang [13] and Benbya and McKevley [30] focus on the iteration length for the whole project, as suggested by Schwaber and Sutherland [43], rather than an option to adjust the iteration length throughout the project. While such an option has the potential to introduce further flexibility to the project, it may come at the cost of negotiating terms with the stakeholders to implement such adjustments. As a result, we recommend adjusting iteration length as one of the ways to adjust BI&A project team agility. Shorter iterations may facilitate higher rate of change, when necessary.

Project teams that can optimize self-organization should be able to achieve a higher level of BI&A project team agility. This entails achieving three conditions: (1) achieving high levels of collective mindfulness via team member autonomy, project team colocation, and sufficient communication, (2) a high level of knowledge sharing and team learning via accumulation of relevant knowledge and experience within a project team, and (3) investment into individual skills of each project team member via training and providing opportunities to learn in different projects. Consequently, we would like to emphasize the importance of communication and knowledge sharing in BI&A projects to create shared understanding of the project direction, and the need to invest in individual skills as part of the project team creation. Agents (units) of a complex adaptive system tend to change their behaviors as a result of interactions with other agents and the environment [39]. Existing agile methods have mechanisms and practices that teams should consider enhancing self-organization, For example, Scrum has four formal events for inspection and adaptation that ensure process adjustment in case of deviation outside acceptable limits (e.g., deviation from customer requirements): sprint planning, daily scrum, sprint review, and sprint retrospective. All four formal events in Scrum facilitate multiple opportunities for feedback from customers, project team and other stakeholders. Furthermore, typical Scrum artifacts such as product backlog and sprint backlog are designed to "provide transparency and opportunities for inspection and adaptation" [44, p. 14]. Another example of customer input during the project visioning session that precedes more detailed release planning is an important mechanism of co-evolution from a customer point of view. Finally, Scrum requires a shared understanding of what each completed deliverable represents. Also, in our sample, some projects invited third party agile coaches and consultants to help improve project processes and build individual knowledge. Organizations can commit resources to further disseminate BI&A project practices and successes across the organization by conducting regular organization wide meetings informing potential stakeholders about different projects and success cases.

Similar to findings by Vidgen and Wang [13], Kautz [45], and theorizations of Benbya and McKelvey [30], we found that communication and coordination structures and practices allow establishing collective mindfulness, sharing and team learning, and self-organization. Further, we found examples of the potential cost of those practices, as was pointed in Vidgen and Wang [13], who cautioned against "over-communication between team members." We also found that the time commitment and availability of some team members can vary throughout the project. While the highest levels of agility probably require a full-time commitment from all project stakeholders, it is challenging to replicate for most projects. Nonetheless, if we consider it to be the highest level, we would find that reductions in commitment may negatively influence project team agility. However, the functional roles of each stakeholder might have different influences on the project team agility, depending on how their role changes throughout the project. For instance, some developers and customers can be required to participate in product development and provide relevant feedback more in the initial stages of the project and less so in the later stages. As such, we would advise taking a balanced approach when establishing communication and coordination structures, rules and routines being cognizant of potential cost of "over-communication".

Agile methodologies aim to release useable results as soon as possible (e.g., within an iteration). That means it aims to create an incremental impact on each iteration, typically within 2–4 weeks. That indicates a fundamental focus on delivering business value to the customers with maximum alignment between customer requirements and project output. That process allows for reducing the amount of wasted effort and resources spent on the deliverables that are outside of customer requirements. This reflects another co-evolutionary principle—synchronization of exploitation and exploration [13]. Vidgen and Wang [13, p. 373] define exploration as "concurrent innovation and knowledge creation" and exploitation as "improvement in productivity, in processes, and product extensions and enhancements."

Project teams that can balance sufficient levels of exploration and exploitation should be able to achieve a higher level of BI&A project team agility. This entails focusing on being at "the edge of time" via a focus on the generation of short-term business impact, continuously searching for ways to learn and improve upon existing products, and striving for modularity of products and processes via findings ways to apply previous knowledge, experience and adapt products and processes to current demands and contexts. We found that the releasability and usability of iteration outcomes also vary depending on the nature of the product. One of the potential explanations is the extent to which preliminary results can be evaluated by the customer. In the case of analytics projects, we can see that in the initial stages of data definition, collection, and cleaning, it is challenging to provide potentially releasable or usable" results.

Developing multiple competing analytical models' stages requires a certain level of quantitative methods knowledge from the customer to judge the progress and provide relevant feedback. We saw this in one of the project cases where the customer was the analytics team of the organization. In this case, they had the capability to test preliminary results and provide feedback. Thus, we can see that releasability does not solely depend on the nature of the project but also on the capabilities of the customer to provide relevant feedback. As such, we recommend that practitioners consider both nature of the project (e.g., to what extent can the product be separated into several modules and each can be developed either in parallel or separately?) and knowledge required to provide feedback on product performance. In this case, project team may rely on stakeholders' experience, knowledge, judgement, and critical thinking. Finally, in some cases, a hybrid approach with less agility during data processing part can be combined with higher levels of agility during the model development and testing phase may be advisable.

We found that successful projects are capable of balancing exploration and exploitation or, as Vidgen and Wang [13] describe it as "product innovation" and "process adaptation and improvement." The balance between those two elements depends on the nature of the product, available resources, and adaptation of processes to the project and organizational context. As was theorized and explored before, various processes and structures can and should be changed according to the internal and external rate of change. At the same time, we found that within the confines of a single project commitment, iteration length, customer feedback, and other processes and structures can be adjusted as well. Nonetheless, Vidgen and Wang [13] advise against overreliance on "common sense" when deciding on process adaptation. One of the ways for practitioners to use these concepts is to evaluate product modularity, and plans for process adaptation and optimization. In cases when a situation is deemed to require more process adaptation and optimization, relevant resources should be allocated in order to ensure that implementation of such plans.

6 Future Directions and Limitations

As with any research study, there are limitations to note. This study aims to understand different levels of project team agility and how different factors enable project team agility using CAS as a theoretical framework. Hence, future research could investigate how different theoretical frameworks would help to explain project team agility and the factors that enable it. Other theories and concepts [46, 47], such as ambidexterity [48] and project emergence [49] possibly have potential as explanatory lens.

The exploratory nature of our case sample is limited to BI&A projects in the financial sector. This does not allow us to generalize our findings to a variety of other projects that are different in nature, business sector, etc. Thus, future researchers should replicate our findings in other contexts, types of projects to enhance generalizability of the findings.

We did not quantitatively study project outcomes and performance. Generally, improved performance is not a guarantee, according to Vidgen and Wang [13, p. 364] due to the inherent unpredictability of some projects. It is also important to note that a CAS approach does not guarantee success or survival and that the project outcome is inherently unpredictable. Our results do allow us to argue that complex adaptive systems (CAS) are better positioned to adapt to given situations compared to more rigid types of systems. A complex adaptive system (CAS) should adapt to changes at an appropriate pace. Otherwise, such a system will either waste resources in case of an excessive change of pace or fall behind in terms of change of pace in case of an insufficient change of pace. Future research to study projects that have varying rates of change and required adaptation would be useful.

Vidgen and Wang [13] identify enablers and inhibitors of agility and group them into emergent capabilities of an agile team. Interestingly, in one of the cases, SysCheck, Vidgen, and Wang [13] observe a number of practices that contribute to agility without adhering to any Agile melthodology. Thus, the authors subscribe to the agility view similar to Vial and Rivard [50] in that project agility is not necessary achieved via the adoption of Agile methodology (e.g., Scrum, Kanban, XP). On the contrary, project agility is described as "the ability to create and respond to change" [40, p. 120] and explained in terms of different factors such as structures, processes, and routines. Consequently, future studies may assess project agility in terms of emergent capabilities theorized by Vidgen and Wang [13] supplemented by CAS literature on ISD projects e.g., [30, 33, 34] and findings of this study.

The recent interest in AI in general and Generative AI (Gen AI) in particular leads to growing interest in how AI can benefit different aspects of managing BI&A projects, and we may conjecture how AI can influence BI&A project team agility and future research on this topic. Application of AI and Gen AI in BI&A projects is relatively new with the availability of AI tools such as ChatGPT by OpenAI, Copilot by Microsoft, and Gemini by Google. From the nascent research related to project team agility in BI&A context e.g., [51] and practitioner overviews e.g., [52], we can observe that current and near future applications of AI in BI&A projects appear to be focused on optimization, support, automation, and consequent acceleration of existing project management processes and practices. As such, we can conjecture that project team members will have more time to focus on tasks that require critical thinking, problem solving, and creativity.

At the same time, there are several types of challenges that are faced by BI&A projects that adopt Gen AI in project processes. For instance, de Campos et al. [51] in online survey of 45 practitioners who used Gen AI in projects, identified three main categories of challenges: human and cultural (e.g., lack of trust in AI, corporate culture), technical challenges (e.g., information security, result accuracy), and management and strategy (e.g., alignment between the need for AI with organizational structure, financial costs, and information security). Thus, in the short to medium term, future research should probably focus on theorizing and studying ways to overcome existing challenges of implementing AI in BI&A projects. In the longer term, we posit that with further development of Gen AI, future research could shift focus towards how Gen AI may influence BI&A project team agility or look beyond at BI&A project team agility in terms of Gen AI agility and human project team agility. That would require more complex ways to conceptualize and measure agility from machine and human perspectives and the combined performance.

7 Conclusion

Our study has illustrated the value and fit of Complex Adaptive Systems theory for studying agile projects and project team agility. The CAS principles can provide insights into effective practices and why the practices are valuable. We believe our study adds to the theoretical literature on agile projects and provides advice to practitioners. Overall, there are many agile techniques and practices that organizations should adopt to make their complex projects into successful adaptive systems.

References

1. VersionOne.com. 17th Annual State of Agile Report (2023). https://www.stateofagile.com/?utm_campaign=soa&utm_source=pressrelease&utm_medium=social#ufh-i-338498988-10th-annual-state-of-agile-report/473508. Accessed 26 July 2024
2. Beck, K., et al.: Manifesto for agile software development. Retrieved Feb 9, 2011 (2001)
3. Mcavoy, J., Butler, T.: The role of project management in ineffective decision making within Agile software development projects. Eur. J. Inf. Syst. **18**(4), 372–383 (2009). https://doi.org/10.1057/ejis.2009.22
4. Lee, G., Xia, W.: Toward agile: an integrated analysis of quantitative and qualitative field data on software development agility. Mis Q. **34**(1), 87–114 (2010)
5. Dingsøyr, T., Nerur, S., Balijepally, V., Moe, N.B.: A decade of agile methodologies: towards explaining agile software development. J. Syst. Softw. **85**(6), 1213–1221 (2012). https://doi.org/10.1016/j.jss.2012.02.033
6. Baham, C., Hirschheim, R.: Issues, challenges, and a proposed theoretical core of agile software development research. Inf. Syst. J. **32**(1), 103–129 (2022). https://doi.org/10.1111/isj.12336
7. Han, H., Ma, F., Liu, X.: Transformational leadership and project success: the serial meditating roles of team flexibility and team agility. Front. Built Environ. **9**, 1334413 (2024). https://doi.org/10.3389/fbuil.2023.1334413
8. Krüger, F.: Impact of team agility on team effectiveness: the role of shared mental models, team empowerment, and team reflexivity. J. Manage. Sci. 123–147 (2023). Seiten. https://doi.org/10.5282/JUMS/V8I1PP123-147
9. McConnon, P.J., Chowdhury, S.: Agility in the age of cloud computing. In: AMCIS 2020 Proceedings: Association Information Systems, Atlanta (2020). https://www-webofscience-com.proxy.queensu.ca/wos/woscc/summary/a951de96-5955-4e7a-9fcd-08263101ccba-ff12bea4/relevance/1. Accessed 30 Jul 2024

10. Radhakrishnan, A., Zaveri, J., David, D., Davis, J.S.: The impact of project team characteristics and client collaboration on project agility and project success: an empirical study. Eur. Manag. J. **40**(5), 758–777 (2022). https://doi.org/10.1016/j.emj.2021.09.011
11. Rathor, S., Xia, W., Batra, D.: Achieving software development agility: different roles of team, methodological and process factors. Inf. Technol. People (2023). https://doi.org/10.1108/ITP-10-2021-0832
12. Werder, K., Maedche, A.: Explaining the emergence of team agility: a complex adaptive systems perspective. Inf. Technol. People **31**(3), 819–844 (2018). https://doi.org/10.1108/ITP-04-2017-0125
13. Vidgen, R., Wang, X.: Coevolving systems and the organization of agile software development. Inf. Syst. Res. **20**(3), 355–376 (2009)
14. Ueland, J.S., Hinds, T.L., Floyd, N.D.: Equity at the edge of chaos: applying complex adaptive systems theory to higher education. New Dir. Inst. Res. **2021**(189–192), 121–138 (2021). https://doi.org/10.1002/ir.20356
15. Marchand, D.A., Peppard, J.: Why IT fumbles analytics. Harv. Bus. Rev. **91**(1), 104–112 (2013)
16. Oesterreich, T.D., Anton, E., Teuteberg, F.: What translates big data into business value? A meta-analysis of the impacts of business analytics on firm performance. Inf. Manage. **59**(6), 103685 (2022). https://doi.org/10.1016/j.im.2022.103685
17. Conforto, E.C., Amaral, D.C., da Silva, S.L., Di Felippo, A., Kamikawachi, D.S.L.: The agility construct on project management theory. Int. J. Proj. Manag. **34**(4), 660–674 (2016). https://doi.org/10.1016/j.ijproman.2016.01.007
18. Eckerson, W.: Secrets of analytical leaders: insights from information insiders. Technics Publications, Westfield, NJ (2012)
19. Eisenhardt, K.M., Brown, S.L.: Competing on the Edge Strategy as Structured Chaos, vol. 31, p. 4 (1998)
20. Anderson, P.: Complexity Theory and Organization Science. Organ. Sci. **10**(3), 216–232 (1999)
21. Stacey, R.D.: Complexity and Group Processes: A Radically Social Understanding of Individuals. Routledge (2003)
22. Mitleton-Kelly, E.: Organisation as co-evolving complex adaptive systems (1997)
23. Haeckel, S.H.: Adaptive Enterprise: Creating and Leading Sense-and-Respond Organizations. Harvard Business Press (1999)
24. Highsmith, J.: Extreme Programming (2000)
25. Schwaber, K.: Controlled chaos: living on the edge. Am. Program. **9**, 10–16 (1996)
26. Highsmith, J.A.: Agile Software Development Ecosystems, vol. 13. Addison-Wesley Professional (2002)
27. Choi, T.Y., Dooley, K.J., Rungtusanatham, M.: Supply networks and complex adaptive systems: control versus emergence. J. Oper. Manag. **19**(3), 351–366 (2001). https://doi.org/10.1016/S0272-6963(00)00068-1
28. Volberda, H.W., Lewin, A.Y.: Co-evolutionary dynamics within and between firms: from evolution to co-evolution. J. Manag. Stud. **40**(8), 2111–2136 (2003)
29. McKelvey, M., Alm, H., Riccaboni, M.: Does location matter? Knowledge collaboration in the Swedish biotechnology-pharmaceutical sector. In: Working paper ESSY, Chalmers University Gothenburg (2004)
30. Benbya, H., McKelvey, B.: Toward a complexity theory of information systems development. Inf. Technol. People **19**(1), 12–34 (2006)
31. Parnas, D.L.: On the criteria to be used in decomposing systems into modules. Commun. ACM **15**(12), 1053–1058 (1972). https://doi.org/10.1145/361598.361623
32. Sun, H., Ha, W., Teh, P.-L., Huang, J.: A case study on implementing modularity in software development. J. Comput. Inf. Syst. **57**(2), 130–138 (2017). https://doi.org/10.1080/08874417.2016.1183430
33. Kautz, K.: Information systems development projects as complex adaptive systems, p. 13 (2012)
34. Jain, R., Meso, P.: Theory of Complex Adaptive Systems and Agile Software Development, N.Y., p. 9 (2004)

35. Benbasat, I., Goldstein, D.K., Mead, M.: The case research strategy in studies of information systems. MIS Q. 369–386 (1987)
36. Yin, R.K.: Case Study Research: Design and Methods (2002)
37. Cross, B.L., Brohman, M.K.: Project Leadership: Creating Value with an Adaptive Project Organization. CRC Press (2014)
38. Lee, A.S., Baskerville, R.L.: Generalizing generalizability in information systems research. Inf. Syst. Res. **14**(3), 221–243 (2003). https://doi.org/10.1287/isre.14.3.221.16560
39. Kauffman, S.A.: The Origins of Order: Self-organization and Selection in Evolution. Oxford University Press, USA (1993). https://books.google.ca/books?hl=en&lr=&id=lZcSpRJz0dgC&oi=fnd&pg=PR13&dq=kauffman+1993+complexity&ots=90CKjV8PRo&sig=dHiAiQBJgoPZIfrWkso-R-uuoVw. Accessed 3 Aug 2024
40. Cockburn, A., Highsmith, J.: Agile software development: the people factor. Computer **11**, 131–133 (2001)
41. Chow, T., Cao, D.-B.: A survey study of critical success factors in agile software projects. J. Syst. Softw. **81**(6), 961–971 (2008). https://doi.org/10.1016/j.jss.2007.08.020
42. VersionOne.com.: 12th Annual State of Agile Report (2018)
43. Schwaber and Sutherland.: The Scrum Guide (2017). https://www.scrumguides.org/docs/scrumguide/v2017/2017-Scrum-Guide-US.pdf#zoom=100. Accessed 27 March 2019
44. Schlauderer, S., Overhage, S.: Exploring the customer perspective of agile development: acceptance factors and on-site customer perceptions in scrum projects (2013)
45. Kautz, K.: Beyond simple classifications: contemporary information systems development projects as complex adaptive systems, p. 22 (2012)
46. March, J.G.: Exploration and exploitation in organizational learning. Organ. Sci. **2**(1), 71–87 (1991). https://doi.org/10.1287/orsc.2.1.71
47. Geraldi, J., Maylor, H., Williams, T.: Now, let's make it really complex (complicated): a systematic review of the complexities of projects. Int. J. Oper. Prod. Manag. **31**(9), 966–990 (2011). https://doi.org/10.1108/01443571111165848
48. Gibson, C.B., Birkinshaw, J.: The antecedents, consequences, and mediating role of organizational ambidexterity. Acad. Manage. J. **47**(2), 209–226 (2004). https://doi.org/10.2307/20159573
49. Cooke-Davies, T., Cicmil, S., Crawford, L., Richardson, K.: We're not in Kansas anymore, Toto: mapping the strange landscape of complexity theory, and its relationship to project management. Proj. Manag. J. **38**(2), 50–61 (2007)
50. Vial, G., Rivard, S.: Understanding agility in ISD projects. In: ICIS 2015 Proceedings (2015). http://aisel.aisnet.org/icis2015/proceedings/ManagingIS/5
51. de Campos, A., Melegati, J., Nascimento, N., Chanin, R., Sales, A., Wiese, I.: Some things never change: how far generative AI can really change software engineering practice (2024). http://arxiv.org/abs/2406.09725. Accessed 27 July 2024.
52. Sfondrini, N.: Revolutionizing Agile: How AI and GenAI are Transforming Project Management. Forbes. https://www.forbes.com/sites/forbestechcouncil/2024/07/05/revolutionizing-agile-how-ai-and-genai-are-transforming-project-management/. Accessed 27 July 2024

Fintech and Financial Inclusion—A Review of Risk Management Strategies

A. Shruti and Sreekumar

Abstract Fintech innovations have revolutionarily transformed the financial landscape, especially enhancing economic incorporation by providing usage to fiscal services for underprivileged populations. This chapter presents a comprehensive assessment of risk strategies within the framework of fintech-propelled fiscal incorporation projects. This chapter scrutinizes the multifarious hazards linked with fintech adoption in endorsing fiscal incorporation. This chapter investigates cybersecurity dangers, records privacy problems, administrative observance obstacles, and functional hazards inherent in fintech remedies. This study conducts a comprehensive assessment of risk strategies within fintech-propelled financial inclusion projects, leveraging a 10-year dataset of secondary data. While fintech innovations aim to expand access to financial services, risk must be judiciously managed to safeguard consumers and markets. This analysis underscores crafting nuanced oversight that supports progress yet protects against destabilizing volatility.

Keywords Fintech innovations · Financial inclusion · Risk strategies · Fintech adoption · Cybersecurity risks · Data privacy · Regulatory compliance · Operational risks · Financial services for underprivileged populations · Market stability

A. Shruti
Research Scholar (Management), Biju Patnaik University of Technology, Rourkela, Odisha, India
e-mail: ashruti70@gmail.com

Sreekumar (✉)
Professor, Rourkela Institute of Management Studies, Rourkela, Odisha, India
e-mail: sreekumar42003@yahoo.com

K. D. Strang and N. R. Vajjhala (eds.), *International Program and Project Management—Best Practices in Selected Industries*, Information Systems Engineering and Management 31, https://doi.org/10.1007/978-3-031-80275-1_9

1 Introduction

Financial institutes have had major effect over the financial services market for centuries. However, in this current era financial institute has undergone significant transformations due to rapid technological advancements. This shift not only highlights the security of financial transactions but also improves the overall experience for the people. The emergence of Financial Technology, or Fin-Tech, marks a new era in the financial services industry in India. The Indian economy is improving day by day with an aim to achieve sustainable development, there is an urgent need to ensure the inclusion of people from every societal group. The lack of awareness and financial illiteracy of the rural population hinders economic development because a significant portion of them lack reliable credit-giving options. This is a significant hindrance to the nation's economic progress. The financial institutes have adjusted to these hindrances by introducing technical advances including debit and credit card technologies, online banking, machine payments (ATMs), etc. Although, these financial innovations have changed urban culture, a majority section of the rural population still lacking for the knowledge of these advancements and is not able to use official financial activity [1]. The term "Fintech" has become popular since 2015. Fintech, which is frequently misunderstood as a term reserved for the heavily technical and tech-savvy, refers to the integration of technology, innovations, and improvements into financial services provided to various people. More than the simple need for these services, fintech growth has been encouraged by rapid technological advancement and changes in the finance industry. The integration of technology and the financial sector necessitates the creation of several platforms and applications to make financial services more user-friendly [2]. An industry that includes businesses that use technology to deliver financial services smoothly and efficiently is referred to as "Fintech," which is the abbreviated form of the phrase "financial technology." In the twenty-first century, this kind of service is still developing. In the financial sector, technology is being applied for mobile payments, loans, money transfers, and asset management. Both old and new start-up companies are attempting to replace these ineffective methods with the new traditional transaction system.

Additional instances of technology being utilized in financial transactions include digital wallets, blockchain, mobile banking, peer-to-peer lending, and peer-to-peer payment technology. These aim in bringing further advantage, achieving high efficiency for the financial transactions and most importantly lowering the cost for consumers [3]. A robust and strong financial system is essential to achieve economic power, self-reliance and sustainable development. To bring an improvement in the growth of an economy the financial system must establish friendly circumstances for people, families and private institutions. Every nation requires an inclusive financial system for several reasons, the main one being that it facilitates the effective distribution of productive resources. In addition, financial inclusion will greatly enhance the ability to obtain financial services, which will enhance managing one's finances. Third, an all-inclusive financial system will contribute to the decrease of unofficial sources of funding like moneylenders who attempt to take advantage of people in rural areas. Therefore, by facilitating a wide range of financial services and offering opportunities for safe and secure saving practices, an inclusive financial system will improve efficiency and wellbeing [4]. Besides everything that is financial inclusion and fintech, technological advancements also introduce fresh risks, cybersecurity threats and misuse or privacy of data and finally regulatory challenges. Effective risk management in fintech involves using technology where sophisticated algorithm is been used for real time monitoring, implementing strong measures of cybersecurity and keeping on updating with the regulatory compliance.

Fintech companies can support sustainable growth in the ever-changing financial landscape, handle uncertainty, and establish trust with stakeholders through the integration of risk management methods into their technical solutions. The complexity of financial services and the reliance on technology to deliver them make risk an inherent part of the fintech industry. Regulatory risk is a major concern, as fintech companies must navigate a maze of regulations that differ by jurisdiction, ensuring compliance with laws governing data protection, anti-money laundering, and consumer protection. Given the sensitive financial and personal data that fintech firms handle, cyberattacks are a major concern. Operational risk, which includes technological failures and human errors, can disrupt services and harm reputation. Market risk also exists, as fintech firms compete with traditional financial institutions and other startups, in addition to the volatility of financial markets. Reputational risk is significant, as any negative publicity or customer dissatisfaction can erode trust and hinder growth. Regardless of these risks, effective risk management strategies, such as comprehensive risk assessments, strong internal controls, regulatory compliance, cybersecurity measures, and continuous monitoring, are required to successfully navigate the fintech landscape. Any factor that may pose a challenge, obstruct, or impediment to accomplishing certain goals is considered a risk. The reason for this could stem from either external or internal factors, depending on the kind of risk present in the specific scenario.

When we talk about internal and external factors, we mean those that have an impact on the business from within and outside of it, respectively. Situational gravity may increase if that risk is present. Implementing preventative measures to detect potential risks that may lead to unanticipated events is a more effective approach to handling such a scenario. One could argue that anticipating and managing risks is preferable to waiting for them to happen. An approach to recognizing, evaluating, and mitigating a specific risk is called risk management. It is an ongoing process that is useful when making decisions. Risk management is used to increase the likelihood of positive outcomes occurring in addition to minimizing the probability of unforeseen circumstances, according to the Higher Education Funding Council for England (HEFCE). According to a model known as "Prospect Theory," an individual is more inclined to take a chance than to aim for a certain failure [5]. Risk and the fintech sector are deeply interconnected, as fintech companies operate in a dynamic environment where innovation intersects with traditional financial services. The rapid evolution of technology introduces opportunities for efficiency and accessibility, but it also exposes firms to numerous risks.

These risks span regulatory compliance, cybersecurity threats, operational disruptions, market volatility, credit and liquidity challenges, reputational damage, partnership dependencies, fraud vulnerabilities, and economic uncertainties. However, effective risk management strategies, including rigorous compliance measures, robust cybersecurity protocols, comprehensive internal controls, and continuous monitoring, are imperative for fintech firms to navigate these challenges successfully. By prioritizing risk management, fintech companies can safeguard their operations, protect customer interests, foster trust, and sustain long-term growth in the ever-evolving financial landscape.

Financial technology companies (Fin-Techs) came into existence during the period of global financial crisis of 2007–2008 and have since transformed the financial industry by bringing new technologies especially commercial banks. The rise of fintech has quickened monetary intermediation, pushing more customers to move from old traditional commercial banks to fintech companies. This move is driven by advancement in innovations like big data and cloud computing, which have altogether affected the credit capacities of old traditional banks. Fintech convenience and customization financing services go beyond traditional money management, which reduces the amount of credit that commercial banks offer. Commercial banks are struggling with their outdated administration culture and lack of experience, making them slower to adjust to fintech advancements. Fintech has overcome geographical and worldly constraints empowering commercial banks to draw in a broader client base. Online administrations and manufactured insights presented by money related innovation have supplanted conventional offline labor, driving to take a toll decreases. Fintech cultivates inventive consideration inside commercial banks by prioritizing inventive ability and persistently making unused items and administrations. Commercial banks are effectively collaborating with fintech firms to use innovations like big data for more exact customer distinguishing proof and upgraded chance estimating capabilities. Whereas fintech presents openings for old traditional commercial banks, it moreover presents risks that seem to weaken their solid and steady improvement. Despite revolutionizing money related intermediation, fintech holds conventional budgetary risks and presents unpredicted loss in future. These risks envelop ethical risks and specialized vulnerabilities characteristic in internet-based computer programs and equipment. Existing scholastic investigation transcendently looks at the execution and operational productivity of commercial banks beneath fintech impact, with restricted center on crucial risks fintech postures to these banks [6].

The following section discusses the risk management strategies, risk management process and financial risk management.

2 Risk Management Strategies

When we discuss risk management strategies, broadly there are two types of risk viz. financial risk and non-financial risk as shown in Fig. 1.

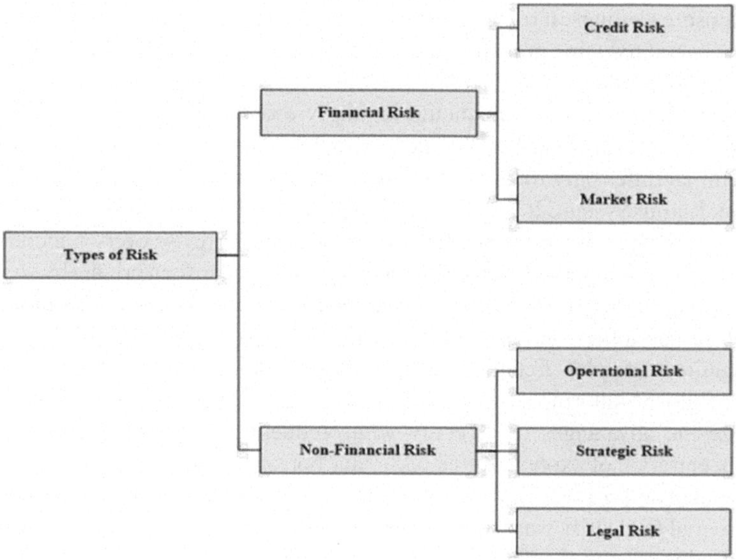

Fig. 1 Types of risk

2.1 *Financial Risk*

Financial risk implies any risk that involves any sort of funding. This is a risk class when there is a possibility of losing either all or most of the money. It relates to the bankruptcy risk that comes from an organization's possible inability to make its debt repayments on schedule. A company's financial risk increases with its debt-to-equity ratio. Many financial transactions, such as buying and selling, investing and borrowing, and other commercial ventures, give rise to financial risk [7].

Credit Risk

A counterparty's or bank borrower's potential failure to uphold their end of the bargain is known as credit risk. For any number of reasons, there's always a chance that the borrower will violate his agreements, which raises the credit risk to the bank. This damage may manifest as an outright default, or they may instead be losses resulting from adjustments to the value of the portfolio due to real or apparent declines in credit quality that stop short of default. There is some credit risk involved in lending money to businesses that are closely associated with market risk factors. Credit risk management aims to lower risk and raise the bank's risk-adjusted rate of return by taking on and controlling credit exposure within permitted limits. Measurement using credit rating or scoring; quantification using an estimate of projected loan losses; scientific pricing; and control using an effective loan review mechanism and portfolio management are all included in credit risk management [8].

Managing Credit Risk involves:

i. Exposure Ceilings: It refers to helping to limit the amount of credit to individual borrowers, reducing potential losses if any default occurs. The limit should not exceed 6–8 times.
ii. Review/Renewal: Lenders should regularly evaluate borrowers' creditworthiness by analyzing financial statements, credit history, and industry trends. Renew credit facilities only for those who meet credit standards and risk tolerance.
iii. Risk Rating System: It helps to develop an in-depth system of six to nine points for rating risks. It is advised clear rating criteria and revise every 6 months.
iv. Scientific Pricing: Predicted loss should be connected to loan pricing by charging high-risk borrowers a premium. Compile past default data, set aside money for unforeseen losses, and price credit facilities based on assessed risk levels. Use quantitative models to determine appropriate pricing.
v. Portfolio Monitoring: The goal of credit portfolio management is to maximize the advantages of diversity while reducing the possible drawbacks of concentration of exposures to a particular borrower, industry, or sector.
vi. Mechanism for Loan Review: A credit audit, regardless of credit operations, is essential to identify warning signs, assess risk, ensure compliance with sanction procedures, and recommend measures to improve credit quality. It is crucial to target significant credit risks and annually review 30–40% of the loan portfolio for LRM exposure.

The credit risk strategy is set forth in the bank's credit risk management policy. Target markets, levels of risk acceptance and avoidance, risk tolerance limitations, preferred degrees of concentration and diversification, and methods for measuring, monitoring, and controlling credit risk are all outlined in these policies. Because of the Bank's continuously improving risk management procedures, it will grow stronger and gain a competitive edge in the marketplace [9].

Market Risk

Being exposed to this type of risk can result from the bank's intentional speculative holdings (own account trading) or from its market creation activities for its clients (dealing). The primary three interdependent but independently controlled transmission channels are how the bank may be impacted by changes in the financial markets. These include variations in the level of interest rate and evolution trend, changes in the foreign exchange rate and, as a result, in the local currency value of the various foreign-currency assets and liabilities expressed, and fluctuations in the financial assets rate that could have an impact on the market value of the portfolio of titles and the value issued by banks [10].

There are 3 types of market risk viz. Interest risk, liquidity risk and country risk

Interest Rate Risk: Interest rate risk is the chance that changes in interest rates will have a detrimental effect on net interest revenue, as it relates to the financial health of an organization. Interest rate fluctuations affect cash flow, earnings, asset value, and liabilities off-balance sheet factors. Part of the earnings perspective involves analyzing how changes in interest rates may affect accrual or reported earnings soon. Monitoring variations in Net Interest revenue (NII), which is the sum of the differences between total interest revenue and total interest expenses, is how this is ascertained [8].

Managing Interest rate risk involves-

i. Liability and assets must be equal.
ii. Banks must determine a precise estimate of the borrower's ability to repay, if there will be late payments, and whether the borrower's credit rating will alter, potentially influencing the interest rate on the loan.
    Based on this, a bank will choose how much of the loan to support
iii. It is important to remember that the demand for loans, the bank's asset balance, and the accuracy of its interest rate will all have a big influence on its profitability forecast. Although banks must comply with regulations regarding loan making, liquidity, and diversification, the profitability of the bank must be taken into consideration.
iv. Diversify maturity: The conventional strategy for mitigating interest rate risk involves distributing investments with stable income over the whole yield curve, from very short-term bonds to very long-term ones.

Liquidity Risk: It means the danger that an organization might not have enough cash on hand to pay its debts when they become due. Since the contractual maturities of bank deposits are usually significantly shorter than those of loans, potential withdrawals of bank deposits must be taken into consideration while managing liquidity. Liquidity is the ability to manage deposits efficiently, lower liabilities, support loan growth, and maybe support off-balance sheet claims. The expected future actions of the assets, liabilities, and off-balance sheet items are used to categorize the cash flows into discrete time buckets. Liquidity risk is made up of call, funding, and timing concerns.

Country Risk: Cross-border transactions, driven by economic liberalization and globalization, have led to a significant increase in country risk. It may not be possible for a country to repay its debts to foreign creditors on time. In contrast to transfer risk, which is connected to possible losses from restrictions on out-of-country remittances, sovereign risk is related to providing finance to a sovereign nation or accepting official guarantees. Political risk occurs when a country's political environment or legislative framework permits the government to take control of a financial entity's assets—as in the case of nationalization—and prevent it from fulfilling previously agreed-upon obligations.

Managing the market risk involves [11]:

i. Experiment with different asset classes.
ii. Go for alternatives of different asset classes.
iii. Within each selected category of asset, try to spread and diversify.
iv. Spread in financial institutions and fund families.
v. Expansion of industry and sectors across horizontal lines.
vi. Expand through fund and portfolio managers.
vii. Expand through different timings and liquidity intensities.

2.2 Non-financial Risk

Risks not related to money are those that affect how a bank markets its products and services and grows its company. Failure in business expansion is one example of these hazards. Poor management, competition, a shortage of products or services, external factors, etc., could all contribute to these risks. In these situations, it is important to carefully examine both operational and strategic risk [5].

Operational Risk
Operational risks are defined as those that result from a failing or subpar internal process, system, people, or occasionally external sources, according to the BCBS (2004). Reputation and strategic risks are not included in the concept; only legal concerns are. Four factors—processes, people, systems, and external factors—are broken down into the explanation of operational hazards in financial institutions by BCBS (2004). The broad field of operational risk studies the hazards posed by the people, systems, and procedures that a business uses to run its operations. It also includes additional risk categories including fraud, legal risk, and environmental or physical risk [12].

Managing Operational Risks involves:

i. Self-assessment exercises regarding the specific operational risk inherent in their operations must be carried out by all bank units.
ii. Banks will create all important risk indicators as requirements which will be alerted in case of increased risk, potential losses, and threats.
iii. Banks should have an insurance policy in case any third party may incur some losses or any mistake.
iv. Training program should be provided to the employees to improve the work internally which would lead to decrease in operational risk.
v. Banks will evaluate, identify, and implement mitigating measures for any operational risk associated with new activities, products, systems, and processes as they generate new product approvals.

Strategic Risk
Strategic Risk: This type of risk results from making poor decisions about resource allocation, poor execution of corporate plans and strategies, or poor response to shifting market conditions. There is other more risk factors that might lead to the manifestation of operations, credit, and market risks. Numerous risk variables are interconnected, meaning that one affects the others. Managing Strategic Risks Involves:

i. Specify the goals and company strategy.
ii. To gauge performance, set up key performance indicators (KPIs).
iii. Determine the risks that may cause performance fluctuation.
iv. Define critical risk tolerance levels and key risk indicators (KRIs).
v. Offer comprehensive risk tracking and reporting.

Legal Risk

It is online financing faces legal risks due to unclear laws, regulatory changes, and contractual conflicts. The global nature of the industry and constant innovation make it susceptible to legal challenges. Compliance with different laws in various countries is challenging, and financial institutions must adapt quickly to avoid legal pitfalls. Privacy issues and contractual disputes can lead to legal disputes and compensation claims, so agencies must prioritize compliance, risk assessment, and user training to manage legal risks effectively.

Let's take the example of banks. Banks increasingly rely on IT platforms for innovative products, enhanced customer experiences, and improved response times. They turn to a wide range of IT services and data from third-party fintech companies to reduce costs and administrative complexity. However, outsourcing these services to FinTechs poses significant security, governance, and management challenges. According to a study by the Bank for International Settlements (BIS), the payments, clearing, and settlement sector is the largest fintech service provider, followed by credit, banking, and financial services. The rise of fintech introduces various risks for banks, as new entrants leverage innovation to offer cheaper services that meet customer expectations. For example, technology platforms that allow customers to switch automatically between different savings accounts or mutual funds to optimize returns. Banks must evaluate their strategic options, deciding whether to partner with fintech companies and retain a portion of the profits or move down the risk chain to take on riskier deals to achieve financial goals. They must also address outsourcing risks, as the increased participation in financial product and service provision complicates the monitoring of operations and risk management activities conducted by third parties.

To protect themselves and their customers, banks need robust processes for due diligence, contract management, continuous control assurance, and monitoring of outsourced operations. The rise of operational risk is significant, as the increased technological interdependencies between banks, fin-techs, and market infrastructures can escalate an IT risk event into a systemic crisis, especially when services are concentrated in a few dominant firms. New goods and services further complicate financial services delivery, making operational risk management more challenging. Legacy IT platforms may struggle to meet the demands of new products and services. "Cloud sourcing" is a cost-effective way for banks to exchange critical resources like software packages and analytics, but it heightens data security, privacy, money laundering, cybercrime, and customer protection risks. Advanced technologies and business models can increase cyber risk if controls do not keep pace. The reliance on APIs, cloud computing, and other technologies that enhance interconnectivity with fintech firms, which may not be subject to the same regulatory standards, could make the banking system more vulnerable to cyber threats and data breaches. The growth of big data also raises compliance risks with data privacy laws, particularly when data is repatriated due to fintech engagements. Regulatory variations across regions add to the complexity, increasing costs for banks. Managing these risks requires banks to develop workforce strategies to ensure employees have the knowledge and capabilities to address fintech challenges. The fintech revolution is transforming the banking sector, and while collaboration with fin-techs offers significant benefits, banks must carefully plan for the associated risks as they adapt to more tech-savvy operations.

The above explained are different types of risk which is been discovered in fintech firms. To cut down this type of risk, there is a process through which it can be minimized which is discussed below.

3 Risk Management Process (RMP)

The goal of risk management is to reduce a financial institution's risk to a manageable level by assessment, oversight, and monitoring that aligns with strategic goals. In addition to focusing on a single risk type, risk management also addresses many risk types, such as those arising from natural catastrophes, fires, accidents, fatalities, lawsuits, financial instruments, or market situations. A methodical approach to analyzing hazards and determining how best to handle them is called risk management [13].

3.1 Set the Context

This is the very first and foremost step of the risk management process. In this step we develop a team for the management of risk, create a plan, define roles and responsibilities, make decisions on budget, define risk tolerance.

3.2 Risk Identification

This is the second foremost step after setting up the context. Knowing the history of the various risks involved, the elements that make a scenario risky, and the causes of that rise in risk are all part of the risk identification process [5]. Focus groups, interviews, surveys, and other methods can be utilized to identify risks. Discussions, historical information and SWOT analysis [13]. Fintech companies needs to figure out various risks such as regulatory compliance risk, operational risk, market volatility etc. This involves evaluating possible potential risk in their technology platforms, processes and business model.

3.3 Risk Analysis

The third step is risk analysis, or we can say risk assessment. The institution has identified a risk in this second stage, and it can now be measured using a variety of methods. Some of the risks in this risk analysis are difficult to quantify, but this one can be. The probability of probability events is padded by this difficulty races. Both qualitative and quantitative analysis can be used for some approach techniques [13]. Assessment and the measurement of the risk impact is the main basic work here. Fintech companies analyze the risk after identification process because to develop strategies to minimize the risk.

3.4 Risk Control

The fourth stage of the risk management process is risk control. The techniques of risk control include insurance and hedging, exposure ceiling fixing, reconstruction and securitization, company diversification, and timely transfer of risk to another party [5]. Five steps will get you through this risk management stage:

(1) reduction of the possibility of an event happening, which means the organization should surely make plans for potential losses.
(2) A risk pass on decision that shifts the risk to a third party, like insurance, is known as a risk transfer.
(3) Minimizing the risk associated with this choice by following a strategy that predicts potential risks before they appear.
(4) The decision-making process will involve setting aside a specific amount of money as a reserve with the intention of foreseeing risks [13].
(5) Fintech companies formulate and implement strategies to minimize risk such as following strict regulatory guidelines, diversifying investing revenue in multiple platforms, deploying robust cybersecurity measures etc.

3.5 Monitoring and Reviewing

This is the last stage of the risk management process. This includes supervising the risks, progress report, matching with the regulations and follow-up. Continuous monitoring and reviewing are essential in fintech to ensure that risk measures taken are effective and to identify new risks as soon as they merge. Regular maintenance of risk management processes allows companies to adapt to changes in the regulatory landscape, to cope with technology advancements and market conditions. There are many examples of this, Stripe, a prominent online payment processing company, has significantly reduced fraud rates by investing in data security and real-time transaction monitoring with artificial intelligence algorithms. This has resulted in safe payment processing for millions of businesses worldwide. Grab Financial, a Grab Holdings subsidiary, has exploited its user base to deliver secure financial services in Southeast Asia, earning millions in consumers' trust through effective risk management. A hedge fund manager controls a collection consisting of various financial assets, such as stocks and bonds, on behalf of a group of investors. The fund is exposed to market, credit, and liquidity risk; thus, the manager must ensure that the fund's risk profile lines up with the level of risk capacity and objectives of the investors. To reduce these risks, the manager enforces a variety of different actions. Diversification is one such method, which means spreading assets across industries, asset classes, and geographic regions to reduce the impact of a single investment. The manager might additionally use financial derivatives such as options and futures to protect against market volatility and control credit risk. Furthermore, the manager may establish loss limits thresholds to limit potential losses in bad market situations, as well as conduct scenario analysis to identify potential risks and analyze the impact of market events on the portfolio. By successfully implementing these strategies, the hedge fund management can reduce risks, increase returns, and give investors greater trust in the fund's performance.

Risk management is a crucial duty for financial institutions, involving identifying and controlling risks to ensure long-term financial stability. It involves assessing potential threats and vulnerabilities, maximizing profit, and optimizing capital funds. Good risk management techniques safeguard resources and preserve a competitive advantage in the marketplace. By implementing risk management protocols, financial institutions can also enhance their reputation and credibility with clients and investors. This proactive approach helps to build trust and confidence in the institution's ability to navigate challenges and uncertainties. The above written process is represented in a diagrammatic view which is shown in Fig. 2.

4 Financial Risk Management (FRM)

Financial risk management refers to unforeseen circumstances due to fluctuations in the financial markets. Financial risks faced by an organization occur when something goes wrong with the people, process and system within an organization for which management strategies are developed, it has been aligned with certain regulations and policies. The identification of financial risks as soon as possible may provide an organization with a competitive advantage.

4.1 Evolution of Financial Risk Management Research

The birth of FRM research has grown to encompass a wide range of risks and their interconnectedness. Initially, research focused primarily on market and credit risk, but the discipline has since diversified to include systemic risk, interconnectedness

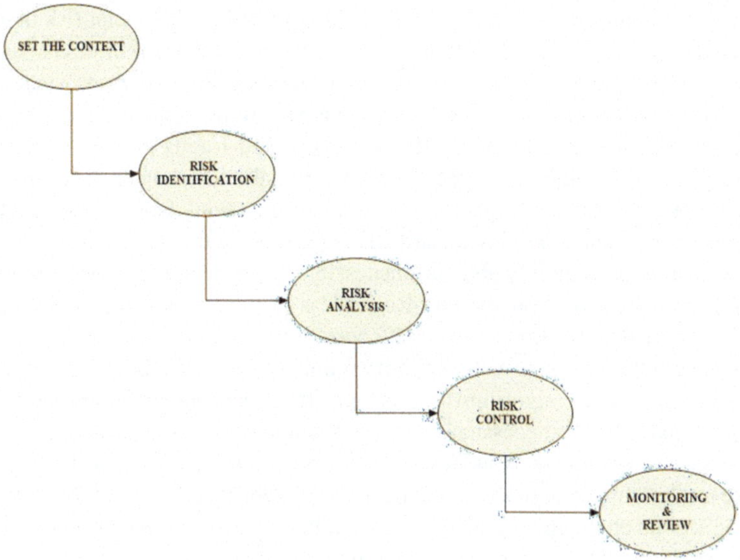

Fig. 2 Risk Management Process Flow

among financial companies, and the impact of macroeconomic variables on financial stability [14]. There was a financial crisis in 2008 which led to the revision, or we can say reexamine the risk management practices and the development of more sophisticated models. Post 2008, research found out that technology has been very important role in the financial risk management that means giving importance to data analytics, artificial intelligence (AI), machine learning etc. which is offering tools for risk assessment and decision making [15]. In addition to this, researchers found out that apart from financial risk, non-financial risk is also there which includes environmental, social, and governance considerations, into traditional risk management frameworks [16].

4.2 Challenges and Criticisms in FRM

Financial risk management research has been questioned and criticized for a variety of reasons. The reliance on quantitative models, particularly during periods of extreme volatility in the markets, has been doubted because they may fail to adequately capture tail risks [17]. The frequent occurrence of "black swan" events, which are distinguished by their low probability and high impact, poses a challenge to traditional risk models [18]. Furthermore, the dynamic nature of financial markets necessitates constant adaptation of risk management strategies [19]. The speed at which technology is developing introduces new dangers, such cybersecurity concerns, that the current structure may not be able to sufficiently handle [20].

One of the challenges is rapid technological innovations. Financial risk management has been drastically revolutionized by technological breakthroughs such as AI, machine learning, blockchain, and big data analytics. Predictive analytics, credit scoring, fraud detection, and market trend analysis are all applications of artificial intelligence and machine learning. Blockchain, noted for its openness and security, is a promising solution for reducing operational and financial risks. Big data analytics examines and evaluates large amounts of financial data, providing insight into market trends and emerging hazards. However, these developments pose issues such as model interpretability and bias, necessitating ongoing adaptation and investment in personnel and infrastructure. Regardless of the hurdles, incorporating this technology is critical for firms to navigate current financial risk management.

While technological improvements carry immense potential, they also pose obstacles. Using complicated algorithms raises concerns regarding model interpretability and bias. Rapid change necessitates ongoing adaptation, talent development, and infrastructural investments. Nonetheless, firms must integrate these technologies to navigate modern financial risk management and remain ahead of developing dangers [21]. The second challenge is globalization. Globalization has transformed the financial landscape, resulting in interconnected markets and novel possibilities as well as challenges for financial risk management. Increased cross-border risk exposure is a serious concern, as economic, political, and regulatory changes in one country can have a global financial impact. The rapid spread of risks, as observed during the 2008 financial crisis, necessitates that risk managers consider market interconnection while developing policies.

Increased global competition can lead to risky behavior, emphasizing the importance of robust risk governance. On the bright side, globalization offers for increased diversification and access to overseas investments, however these come with their own set of dangers and difficulties [22] Cybersecurity concerns offer significant challenges to financial organizations and risk controllers in today's digital financial ecosystem. Data breaches occur when cybercriminals target sensitive financial data, compromising data security and posing major financial and reputational risks. Financial systems are especially vulnerable, as attacks on essential infrastructure disrupt markets. Employee training and awareness programs are critical for mitigating these hazards. Ransomware attacks, in which criminals encrypt crucial data and demand a ransom for its release, can halt financial institutions and result in severe losses.

To remain ahead of evolving threats, financial institutions must implement advanced threat detection and response methods, create incident response plans, and work with cybersecurity specialists and regulators [23]. Regulatory changes have an important role in changing the landscape of financial risk management. Financial institutions have both challenges and opportunities in ensuring compliance and effective risk management strategies in an ever-changing regulatory landscape. Financial institutions have significant challenges in navigating the complexity and volume of regulatory changes. Regulatory agencies often update and revise rules, necessitating significant resources to interpret, execute, and adhere to these changes. Non-compliance with new regulations can result in legal penalties, reputational loss, and operational disruption.

4.3 Integration of Sustainability in Financial Risk Management

Sustainability is increasingly being recognized as a key factor in financial risk management. This approach recognizes that environmental and social factors can significantly impact financial risks. Researchers are incorporating ESG criteria into risk assessment models to align with the financial industry's shift towards sustainable and responsible investing practices. Climate change's potential impact on financial markets has raised awareness, prompting the development of risk assessment tools. These tools aim to understand the relationship between climate-related risks, including physical and transition risks, and their impact on financial stability [24].

5 Relation of Fintech and Financial Inclusion Along with Risk Management

The term "Fintech" is the amalgamation of two words finance and technology. It is a combination of two things. Fintech (Financial Technology) has significantly impacted the landscape of India's financial sector. The amalgamation of fintech has resulted in a wave of remarkable developments, changing the country's economic structure and boosting financial inclusion on a massive scale. This is the emerging type of service in this twenty-first century that is converting from old traditional financial system to new online based financial system. It helps to reduce consumer cost [3]. The authors highlight the key role of FinTech in promoting financial inclusion to support the UN Sustainable Development Goals. They emphasize the various ways in which FinTech can facilitate access to essential financial services and contribute to the overall achievement of the 17 SDGs [25]. Countries have selected financial inclusion to address inequality and achieve Sustainable Development Goals (Goal 1, 2 and 8). The COVID-19 pandemic has encouraged the urgent need for digital financial inclusion, which provides basic financial services to underserved or lower income populations using digital tools. This includes individuals worldwide, such as women, the poor, and rural communities. Financial inclusion offers access to affordable services for meeting various needs, like microenterprise investments or education. By expanding financial opportunities, lower income or underserved income individuals can reduce risks and improve their socio- economic status and living conditions of the people ultimately contributing to poverty reduction and sustainable development [6, 21–23, 26].

Thus, we can as much as there is an improvisation or growth in the fintech companies, they also go through certain unforeseen challenges. Risk plays an important role in fintech companies. Risk management is crucial in the fintech industry due to technological advancements shaping the sector. Looking ahead to 2024, emerging trends will influence risk management practices. While fintech has transformed financial services with innovative solutions, it has also introduced new risks that necessitate effective strategies. Proper risk management is essential, especially in a fast-paced industry like fintech where risks can have significant impacts. Establishing a culture of risk management involves identifying and mitigating risks, as well as training employees to recognize and report potential threats. Regular reviews and updates to risk management strategies are also important for the success and sustainability of fintech companies.

Indian fintech startup Paytm has revolutionized financial services by offering digital wallets and payment systems to unbanked and underbanked communities, particularly in rural areas. This has facilitated small firms and individuals to participate in the digital economy, promoting financial inclusion and economic growth. Paytm uses innovative technology like AI for fraud detection, robust cybersecurity measures, and stringent regulatory compliance to address risks. Regular training on risk recognition and risk management techniques fosters confidence and promotes long-term growth in the fintech sector.

6 Conclusion

Risk is a threat to many financial industries and fintech companies. It is a very negative concept and can hamper many financial activities which is not good for the efficiency and effectiveness of a financial industry. Risk is an inherent component of the financial business, affecting both existing institutions and new fintech startups. The paper discusses how risk affects these company strategy, operations, and overall strength. Understanding and managing risk is essential for navigating the complicated world of finance, which includes credit and market risks, technological and regulatory concerns, operational and strategic risks. Credit risk, or the chance that borrowers would fail to meet their contractual commitments, is an important concern for both traditional financial institutions and fintech enterprises. Banks are exposed to this risk through their lending activities, but fintech companies, which frequently operate in peer-to-peer lending or alternative markets for financing, are also sensitive. Credit risk management is doing a thorough assessment of borrower creditworthiness, using statistical analysis to predict risk, and executing suitable risk mitigation methods. Market risk refers to the fluctuation and unpredictable nature that exists in financial markets, which have an influence on the value of investor portfolios. Traditional banks are subject to market risk due to their trading habits and portfolio of investments, while financial technology companies that provide financial services or trading platforms are also exposed. Managing market risk involves investment diversification, hedged methods, and actual time trend monitoring, particularly within the environment of quickly evolving fintech platforms.

Liquidity risk takes place when institutions lack the ability to satisfy their immediate responsibilities due to an absence of liquid assets. Established financial institutions must keep enough cash on hand to cover deposit withdrawals and operations expenses. Fintech companies, especially those in payment processing or digital banking, encounter liquidity risk in various ways. Managing liquidity risk involves maintaining sufficient reserves, conducting liquidity position stress tests, and creating contingency funding plans. Business operations can be disrupted, and financial losses incurred due to internal failures like technology malfunctions, fraud, or human mistakes. Operational risk impacts both conventional banks and fintech firms, particularly due to their increased dependence on sophisticated computer networks and online infrastructure. Minimizing operational risk requires robust internal controls, security measures, and ongoing advancements in technological resilience. Fintech firms are closely linked to technology, which exposes them to risks related to technology. It consists of cyber risks, data breaches, and mathematical errors. Classical financial institutions are also at risk, particularly as they engage in digital upgrade initiatives to remain competitive. Mitigating technical risk requires allocating resources to cyber security measures, conducting regular risk assessments, and staying up to date on advancements in technology. Risk dominates all aspects of the financial sector, impacting the tactics, functions, and sustainability of both conventional establishments and fintech companies. To handle a complicated risk environment effectively, one must engage in research, creativity, and adjustment to address credit, market, operational, and regulatory risks. Financial institutions can increase their resilience and thrive in a constantly changing environment by adopting robust risk management procedures, taking advantage of technological advancements, and promoting a culture of compliance.

References

1. Berry, D., Ahmed, A.: Fintech symphony: harmonizing risk management in the banking sector. Int. Dev. Plan. Rev. **22**(2), 351–356 (2023)
2. Kandpal, V., Mehrotra, R.: Financial inclusion: the role of fintech and digital financial services in India. Indian J. Econ. Bus. **19**(1), 85–93 (2019)
3. Priya, P.K., Anusha, K.: Fintech issues and challenges in India. Int. J. Recent Technol. Eng. **8**(3), 904–908 (2019)
4. Aggarwal, R.: Financial inclusion in India: challenges and opportunities. Int. J. Res. **1**(4), 557–567 (2014)
5. Jadav, H.N.: Risk management in Indian banks: some issues and challenges. Online J. Multidisc. Subj. **11**(3) (2017)
6. Li, G., Elahi, E., Zhao, L.: Fintech, bank risk-taking, and risk-warning for commercial banks in the era of digital technology. Front. Psychol. **13**, 934053 (2022)
7. Zakariya, N.A., Abdullah, S.N.H.S., Aizat, M.A., Hamzah, N.: Risk management analysis using the enterprise risk management method with FinTech as the risk controller. Proc. Int. Conf. Da'wa Commun. **4**(1), 59–73 (2022)
8. Annapurna, J.: A study on usage of risk management strategies in financial institutions. Int. J. Commer. Bus. Stud. (IJCBS) **5**(2), 7–14 (2023)
9. Singh, A.: Credit risk management in Indian commercial banks. Int. J. Market. Financ. Serv. Manage. Res. **2**(7), 47–51 (2013)
10. Treapat, L.M., Anghel, L.C.: Managing the market risk in banks. In: Strategica International Academic Conference, pp. 147–156 (2016)
11. Shukla, A., Kukreja, M.: A study of risk management in finance sector. Int. J. Concept Manage. (IJCM) **3**(1), 1–8 (2014)

12. Basle Committee on Banking Supervision.: Range of Practice in Banks' Internal Ratings Systems. Basel Committee on Banking Supervision (2000)
13. Asif, M., Hussain, A., Hameed, J.: Dark side whitewashes the benefits of FinTech innovations: a bibliometric overview. Int. J. Bank Mark. **42**(1), 113–132 (2024)
14. Sekhar, S.C., Kumar, A., Reddy, P.R., Rao, B.M.: Risk modelling and prediction of financial management in macro industries using CNN based learning. In: International Conference on Disruptive Technologies (ICDT), pp. 297–300 (2023). https://doi.org/10.1109/ICDT57929.2023.10151085
15. Anh, T.L.H.: Analyzing the behavior of financial risk management and its impact on success of project. Int. J. Res. Vocat. Stud. (IJRVOCAS) **3**(1), 20–22 (2023)
16. Curcio, D., Gianfrancesco, I., Vioto, D.: Climate change and financial systemic risk: evidence from US banks and insurers. J. Financ. Stab. **66**, 101132 (2023)
17. Kandie, S.K., Bogonko, J.B.: Risk management and financial performance of commercial banks listed at the Nairobi securities exchange. Int. J. Finance **8**(2), 40–64 (2023)
18. Khandelwal, P., Sharma, A., Gupta, S., Singhal, S.: Risk management model using Bigan. In: International Conference on Emerging Trends in Engineering & Technology-Signal and Information Processing (ICETET-SIP), pp. 1–5 (2023)
19. Davidescu, A.A., Hapau, R.G., Manta, E.M.: Impact of crises on capital market volatility: a bibliometric analysis. In: The New Digital Era: Other Emerging Risks and Opportunities, pp. 21–53. Emerald Publishing Limited (2022)
20. Kouloukoui, D., Pinheiro, J.Q., Da Rocha, J.L.A., Lima, G.A.S.: Factors influencing the level of environmental disclosures in sustainability reports: case of climate risk disclosure by Brazilian companies. Corp. Soc. Responsib. Environ. Manag. **26**(4), 791–804 (2019)
21. Chen, L., Liu, J.: The impact analysis of financial risk management on ordinary investors' asset allocation. Financ. Eng. Risk Manage. **5**(7) (2022)
22. Xiaoyi, J.: Discussing the construction of a budget management system combining multimedia technology and financial risk management. Adv. Multimed. (2022)
23. Barbara, G., Dorota, K.: Sustainable metrics in project financial risk management. Sustainability **14**(21) (2022)
24. Woodhouse, S., et al.: Climate X: making climate risk data useful and usable for the financial sector. In: EGU General Assembly Conference Abstracts, EGU-8097 (2023)
25. Anakpo, G., Xhate, Z., Mishi, S.: The policies, practices, and challenges of digital financial inclusion for sustainable development: the case of the developing economy. FinTech **2**(2), 327–343 (2023)
26. Odei-Appiah, J., Agyei-Holmes, A., Tackie, N.O.: Fintech use digital divide and financial inclusion. DPRG **24**(5), 435–448 (2021)

Spotlight on Public Administration and Project Management in Nigeria: Discovery of New Frontier Through Emerging Trends and Technology

K. Ibrahim Hassan, Victoria Hunga, Fatile Jacob, and M. Oluwaseyi Adetoba

Abstract Nigerian public administrative system is characterized by bureaucracy, inefficiency, and a need for more transparency. These hamper advancements in project planning, resource allocation, risk management, and communication in project management practices. Also, low technologies affect project lifecycles, streamlining and stifling collaboration, enabling project monitoring and evaluation in project management. The study explored the impact of emerging trends and technology on public administration and project management in Nigeria. It seeks to identify and analyze the new frontiers that have been discovered through the adoption and integration of this technology in Nigerian public service. It highlights the potential for these advancements to revolutionize governance and project execution, ultimately propelling Nigeria towards a new frontier. The research is explorative, and it utilizes an interpretive qualitative paradigm for data collection methods. Secondary data were obtained from relevant literature and reports. The study revealed that emerging trends and technology have significantly transformed Nigeria's public administration and project management practices. Adopting digital tools, data analytics, automation, and artificial intelligence has enhanced efficiency, transparency, and decision-making processes within the public sector.

Keywords Nigerian public administration · Bureaucracy · Project management · Emerging trends · Technology adoption · Digital tools · Data analytics · Automation · Artificial intelligence · Governance transformation

---

K. Ibrahim Hassan (✉) · F. Jacob
Department of Public Administration, Faculty of Management Sciences, Lagos State University, Ojo, Lagos State, Nigeria
e-mail: ibrahim.hassan@lasu.edu.ng

F. Jacob
e-mail: jacob.fatile@lasu.edu.ng

V. Hunga · M. Oluwaseyi Adetoba
Business School, Kazakh British Technical University, Almaty, Kazakhstan
e-mail: victoria.hunga@lasu.edu.ng

K. D. Strang and N. R. Vajjhala (eds.), *International Program and Project Management—Best Practices in Selected Industries*, Information Systems Engineering and Management 31, https://doi.org/10.1007/978-3-031-80275-1_10

1 Introduction

Nigeria, Africa's most populous country, is experiencing considerable changes in its public administration and project management environment, owing to new trends and technologies [1, 2] assert that the emergence of innovative project management practices in public sector agencies has resulted in the development of project management training programs and certifications tailored to the Nigerian public service, which incorporate innovative approaches such as agile methodologies, project management software, citizen engagement, and improving project collaboration and service delivery [3]. It can be argued that emerging trends and technologies have played an important role in this transformation, opening new frontiers and opportunities for advancement by streamlining processes, improving communication, and providing data-driven insights [4].

This book chapter is a construct of new explorations of emerging trends and technology in Nigeria's public administration and project management. Nigerian public service has adopted open government project management initiatives that promote transparency, citizen engagement, and collaboration [5]. Citizens now have access to government information, budgets, and project da-ta using platforms such as open data portals, which have increased public trust, stimulated citizen participation, and fostered a decision-making partnership between the government and citizens [6]. The use of project management tools and practices has become more common in Nigeria. These technologies allow for more successful project planning, monitoring, and assessment. Gantt charts, task management systems, and collaboration platforms are examples of project management software that simplify processes, increase communication, and assure project delivery on schedule [4]. Consequently, it is important to state at this juncture that the efforts of various Nigerian scholars to give a lucid explanation of the meaning of the emerging trends and technology in Nigerian public administration and project management in the twenty-first century created a breeze of change, starting from simple tools and techniques to more complex tools and machinery (hardware and software).

It can be argued that there are growing concerns regarding the relevance of professional project management in the Nigerian public administrative system and the need for trained project managers to lead public sector projects [3]. Hence, Nigeria has used data analytics in public administration and project management because of the abundance of available data. Also, data analytics allow for evidence-based policymaking and project performance evaluation, resulting in more efficient resource allocation and improved outcomes [7]. It is important to state at this juncture that government agencies can make better judgments, discover trends, and forecast consequences by using data to improve efficiency, transparency, and accountability in governance. Adebayo and Omole [2] are of the view that the adoption of some technologies may have unintended environmental consequences, requiring careful consideration and sustainable practices; likewise, integrating sustainability considerations into project management and public administration requires balancing innovation with environmental responsibility [7].

Meanwhile, Nigeria has seen its fair share of successes in project management and public administration, yet numerous challenges continue to hinder significant progress and efficient service delivery. These challenges are complex and multi-faceted, as public funds allocated for projects are susceptible to misappropriation and diversion, leading to project failures and reduced value for money; likewise, projects are often launched without thorough planning, leading to cost overruns, time delays, and scope creep [1]. Feasibility studies may be absent or flawed, resulting in projects that are unsuitable for the public context or lack a clear path to success. This is evident in the just-completed Second Niger Bridge, worth $3.6 billion, built across the Niger River [8]. Also, there is inefficient utilization of scarce public resources as public funds are wasted due to project failures, corruption, and mismanagement, which makes citizens' access to essential services like healthcare, education, and infrastructure inadequate [3].

Poor project management and public administration create bottlenecks and impede economic growth, which is shrouded in inefficiencies and transparency that fuels public cynicism and distrust in government institutions [4]. Furthermore, project management offices lack qualified personnel and standardized procedures, leading to inefficiencies and poor decision-making as public servants and project managers lack the digital skills necessary to operate effectively in an increasingly technology-driven environment [5]. From the above review, it can be argued that implementing new technologies and innovations often requires significant upfront investment, which can be challenging for public entities with limited budgets. Adapting to new technologies and workflows can be met with resistance from some stakeholders who are comfortable with traditional methods [9]; likewise, integrating new technologies with existing legacy systems can be complex and costly, which also leads to a lack of interoperability between different platforms. These solutions can hinder collaboration and data sharing.

It is important to state at this juncture that there is a lack of systematic and sufficient academic or published research works in the available literature on emerging trends in public administration and project management in Nigeria. Evidence from the existing literature and credible scholarship shows that more needs to be done to address this literature gap. Consequently, this chapter examines emerging trends and technology in Nigeria's public administration and project management. This is the gap this chapter tends to fill. Given this, the gaps in the study seek to identify and analyze the new frontiers that have been discovered through the adoption and integration of these technologies in Nigerian public service. It highlights the potential for these advancements to revolutionize governance and project execution in Nigeria.

2 Literature Review

This section of the chapter reviewed previous studies on emerging trends and technology in public administration and project management in Nigeria and the discovery of new frontiers in the adoption and integration of project management technology in

Nigerian public service. A review was done on the revolutionizing governance and project execution space in Nigeria through advancements in project management technology and anchored on institutional theory.

2.1 *Emerging Trends and Technology on Public Administration and Project Management in Nigeria.*

Public administration is inextricably related to the operation of the state. State action has taken various forms for various sociopolitical and economic organizations, and within this diversity, there has been a tendency for today's management of public administrative issues in Nigeria to involve the use of technologies ranging from information and communication, data storage and retrieval, record keeping, office automation, and so on, many of which have global implications [10]. Technology is being used to improve citizen engagement and participation. Online platforms and mobile apps are being used to gather public input, receive feedback, and facilitate public consultations on government projects and policies. For instance, the Edo State Government's implementation of an AI-powered chatbot to streamline citizen inquiries and service requests has led to improved efficiency and responsiveness in public service delivery [11]. In the same vein, the Federal Ministry of Finance has piloted a project to implement a blockchain-based system for tracking and managing government procurement contracts, leading to increased transparency and reduced instances of fraud and corruption [12]. Nigeria has seen emerging trends and technologies in public administration and project management that have the potential to transform the country's governance and project execution. The Nigerian public sector is using a variety of software programs and online platforms to streamline project planning, monitoring, and evaluation processes [6]. This shift towards e-governance promotes transparency, inclusivity, and accountability in decision-making processes. These technologies provide real-time data monitoring, collaborative workspaces, and automated reporting, which improves project productivity and transparency [4]. These developments allow for the acceptance and integration of new frontiers in the Nigerian public sector.

Another emerging trend is the application of artificial intelligence (AI) and machine learning (ML) to project management [5]. AI and ML algorithms analyze large datasets, identify trends, and make predictions, thereby assisting project managers in decision-making and risk assessment. These technologies also support predictive analytics, which helps in proactive problem-solving and resource optimization [3]. For instance, the Lagos State Government's use of IoT-enabled smart traffic management systems to reduce congestion and improve transportation efficiency has been a significant stride, leading to improved project timelines and reduced logistics costs. Furthermore, the adoption of cloud-based project management tools is gaining traction in Nigeria. Cloud technology allows for safe storage, easy access, and seamless collaboration on project-related documents and data [13]. This enables project teams to work remotely and share information in real-time, which improves communication and collaboration.

2.2 Adoption and Integration of Project Management Technology in Nigerian Public Service

Nigeria's public service is on the cusp of a major transformation driven by the adoption and integration of project management technologies [7]. The adoption and integration of project management technology is an emerging trend that holds great potential for improving governance and project execution [13]. The government's commitment to digital transformation, coupled with strategic planning and capacity building, will be key to realizing this vision. Adebayo and Omole [2] identified several factors contributing to this development:

1. Digital Transformation: The Nigerian public service is undergoing a digital transformation, which includes the implementation of various technological solutions. Project management technology plays a crucial role in this transformation by providing tools and systems to streamline project planning, monitoring, and evaluation processes. It is important to state at this juncture that Nigerian public service now adopts digital transformation tools such as cloud-based project management platforms like Trello, Asana, and Microsoft Project Online. Collaboration tools such as Slack, Microsoft Teams, and Zoom, as well as data analytics and reporting tools like Power BI and Tableau. Mobile applications like project management apps provide real-time updates and on-site data collection. All these enhance project management operations.
2. Efficiency and Effectiveness: Project management technology enables government agencies to optimize project management practices, resulting in improved efficiency and effectiveness. Automated workflows, real-time data tracking, and collaborative platforms help streamline processes, reduce administrative burdens, and enhance project outcomes. It is evident to state that this system embraces a centralized data management system that saves time and resources.
3. Resource Optimization: Technology improves resource allocation and utilization. Government agencies now use project management software to better track resource allocation, costs, and schedules. It enables the government to gain significant insights into resource allocation and project planning. This allows the state institution to make data-driven decisions, identify bottlenecks, and efficiently allocate resources, resulting in better project execution and decision-making.
4. Transparency and Accountability: Project management technology pro-motes transparency and accountability in the Nigerian public service. It can be argued that this technology provides real-time project dashboards and progress updates, automated reporting, and centralized data storage as it becomes easier to monitor

and evaluate project performance reports and provide clear visibility for stakeholders. This enhances transparency, facilitates timely decision-making, and reduces the risk of corruption and mismanagement.

5. Stakeholder Collaboration: Technology facilitates better collaboration among stakeholders involved in public service projects. Project management platforms provide a centralized space for online and offline communication, document sharing, and task management. It is evident that technology facilitates cooperation, coordination, information sharing, and knowledge sharing among project teams, government agencies, contractors, and other stakeholders.
6. Data-Driven Decision Making: Project management technology facilitates data collection, analysis, and reporting in the public sector. It helps in the proactive identification and mitigation of potential issues affecting project success rates. Policymakers can access real-time project data, performance indicators, and predictive insights by leveraging sophisticated analytics and reporting. This empowers them to make informed decisions, detect project risks, and respond proactively to challenges.
7. Capacity Building: The adoption of project management technology in the Nigerian public service requires capacity-building efforts. Training programs and workshops can be conducted to equip public servants with the necessary skills to effectively utilize project management tools and systems. This ensures that technology adoption is accompanied by the development of human capabilities.

2.3 Challenges in Adopting and Integrating Project Management Technology

Project management technology has had a significant impact on public administration, revolutionizing the way projects are planned, executed, and monitored [4]. It has enhanced the efficiency, transparency, and accountability of public sector projects. However, the adoption and integration of project management technology in public administration also comes with several challenges. Oladapo and Fagbemi [6] explore some of these challenges:

1. Resistance to Change: Public servants and stakeholders in the public sector often face resistance to the adoption of new project management technologies. It is a prevalent belief that public servants are accustomed to manual methods and are reluctant to embrace new tools and processes [7, 14]. The Ogun state government faced resistance from project managers when introducing an AI-powered project monitoring system. The government in response implemented a change management strategy, including training, phased rollout, support team, and feedback mechanisms. This method, which is also supported by Chung et al. [15] led to widespread acceptance of the system, improving project visibility and decision-making.

2. Budget Constraints: Public sector organizations frequently face budgetary constraints, making it difficult to allocate funds for implementing project management technologies. The costs of procuring software, hardware, and training the staff can be substantial [5, 16]. The Anambra state government faced budget constraints when implementing a blockchain-based procurement and was able to overcome this by partnering with private sector companies, prioritizing cost-saving projects, reallocating funds from outdated systems, and implementing a phased deployment of the blockchain-based procurement system. Nevertheless, organizations can also minimize financial barriers and improve implementation effectiveness by leveraging existing research and embracing best practices.
3. Interoperability and Integration: Public administration consists of several departments and agencies, each with its own set of systems and processes. Integrating project management technologies across these varied domains may be challenging [1]. Interoperability and seamless data exchange across multiple platforms are critical for successful project management. It may require interoperability protocols, data-sharing agreements, and technical skills. When implementing an IoT-based asset management system in Kaduna State, the government faced challenges with ensuring seamless integration across various departmental systems. Farghaly et al. [17] gave insight on the importance of over-coming interoperability such as the creation of an asset management platform and the development of custom APIs and data exchange protocols that enabled the IoT system to communicate with legacy systems suggested. Also, the establishment of a centralized data governance framework to standardize data formats and management protocols across the different departments could be a useful tool.
4. Data Security and Privacy: Public-sector projects frequently involve sensitive and secret information. Implementing project management technology requires robust data security measures to prevent unauthorized access, data breaches, and cyber threats. Compliance with appropriate data protection regulations can pose additional challenges.
5. Skills and Capacity Building: Adopting project management technology demands the development of a skilled workforce capable of utilizing the tools effectively. Nigerian public service faces a significant shortage of personnel with the requisite project management and technology skills [6]. Investing in training and capacity-building programs becomes vital to bridge the skills gap and facilitate effective adoption.
6. Organizational Culture: Public sector organizations frequently have a hierarchical structure and a bureaucratic culture, which might impede the adoption of project management technologies [3]. The technology may challenge existing power dynamics, decision-making processes, and communication routes. Creating a culture of collaboration, knowledge sharing, and innovation is critical to facilitating technology integration.
7. Stakeholder Engagement: Public administration encompasses a wide range of stakeholders, including citizens, elected officials, interest groups, and contractors. Engaging these diverse stakeholders in the project management technology adoption process is critical to success [13]. Getting their buy-in, addressing

concerns, and soliciting feedback throughout the implementation can help overcome resistance and foster a sense of ownership.

To address these challenges, public administration should undertake comprehensive planning, establish clear goals, and develop a phased approach to technology adoption. Engaging external expertise, leveraging best practices from successful implementations, and fostering a supportive organizational environment can assist in the smooth integration of project management technology within the public sector. Consequently, the adoption and integration of project management technology in the Nigerian public service have the potential to transform governance and project execution. By leveraging digital tools, automation, and data-driven decision-making, the Nigerian government can enhance efficiency, transparency, and accountability in project management processes, leading to improved service delivery and socio-economic development.

3 Revolutionizing Governance and Project Execution Space in Nigeria

Revolutionizing governance and project execution in Nigeria requires a multifaceted approach that addresses both systemic challenges and leverages the power of technology and innovative solutions. The Nigerian public sector has enormous potential for positive transformation by prioritizing transparency, accountability, efficiency, community engagement, and strategic infrastructure development [5]. The country can create a more conducive environment for effective project implementation, therefore improving the lives of its citizens. Adebayo and Omole [2] identified potential areas of focus for revolutionizing governance and project execution in Nigeria.

1. Technology and Digital Transformation: Embracing e-governance and other emerging technologies significantly improves governance and project execution. Implementing these solutions can streamline administrative efficiency, increase transparency, and reduce corruption [13]. Digital platforms facilitate citizen engagement, allowing individuals to participate in decision-making and provide feedback on projects. Additionally, using data analytics and artificial intelligence would enhance project monitoring, evaluation, and resource allocation.
2. Strengthening Institutional Capacity: Strengthening institutional capability is critical for good governance and project execution. This involves investing in the training and development of public officials, supporting merit-based recruitment, and ensuring adequate resources are allocated to institutions responsible for project implementation [7]. The World Bank's support for institutional capacity building in Rwanda through the Rwanda Governance Board (RGB) has been successful in enhancing governance structures and processes, leading to improved service delivery and accountability. Understanding the success factors for collaboration across sectors and adopting a new managerial approach has strengthened institutional capacity for AI implementation. Strategies include

partnering with a local university for AI training, establishing an AI Center of Excellence, pairing government workers with AI experts, and implementing a mentorship program [18].

3. Public–Private Partnerships (PPPs): Fostering collaboration between the government and private sector may bring expertise, resources, and efficiency to project execution. Well-structured PPPs can attract private investment, foster innovation, and ensure the timely completion of projects [8, 12]. However, PPP partnerships must include transparency, accountability, and appropriate risk-sharing mechanisms. Various examples of PPPs in Nigeria demonstrate the diverse applications and benefits of such collaborations ranging from municipal household solid waste collection in Lagos, clean energy infrastructural development, and social service delivery amongst others [19–21].
4. Stakeholder Engagement and Participation: Involving citizens, civil society groups, and relevant stakeholders in decision-making processes is critical for good governance and project execution. Establishing mechanisms for public participation, feedback, and grievance resolution can promote inclusivity, accountability, and legitimacy. The United Kingdom's National Health Service (NHS) has implemented robust stakeholder engagement strategies in healthcare delivery, involving patients, healthcare providers, and community organizations to ensure effective service provision and patient-centered care [22]. Local communities should be included in the planning, implementation, and monitoring of projects that directly impact them.
5. Anti-Corruption Measures: Combating corruption is critical for improving governance and project delivery. Implementing robust anti-corruption measures, such as effective legislation, independent oversight bodies (EFCC, ICPC, and CCB), and transparent procurement processes, can help curb corruption and guarantee that funds are used efficiently and effectively [23, 24]. Singapore's Corrupt Practices Investigation Bureau (CPIB) has been instrumental in combating corruption through rigorous enforcement, public education campaigns, and preventive measures, establishing a culture of integrity and accountability in governance [25].
6. Capacity Building and Information Sharing: Investing in capacity-building programs and knowledge-sharing platforms may help government officials, project managers, and stakeholders improve their skills and understanding of governance and project execution. This can involve training in project management, financial management, and best governance practices. The United Nations Development Program (UNDP) has supported capacity-building initiatives in Malawi, focusing on training government officials, enhancing data collection and analysis, and promoting knowledge sharing to strengthen governance and development outcomes [26].
7. Monitoring and Evaluation: Building strong monitoring and evaluation processes is critical for measuring project progress, impact, and effectiveness. Regular assessment assists in identifying bureaucratic bottlenecks, measuring outcomes, and making data-driven decisions for change [3]. Independent evaluation

committees can give objective assessments and suggestions to improve governance and project execution. The African Development Bank's monitoring and evaluation framework for infrastructure projects in Kenya has improved project oversight, performance tracking, and impact assessment, ensuring efficient resource allocation and project delivery [27]. It is critical to note at this point that revolutionizing governance and project execution necessitates a long-term commitment from the government, active participation from citizens, and a supportive policy environment. Hence, a successful public sector would require consistent work, adaptability to changing circumstances, and a willingness to learn from both successes and failures.

3.1 Advancements in Project Management Technology

This section of the book chapter discusses breakthroughs in project management technology. This technology has advanced significantly in recent years, providing numerous benefits to businesses across a variety of sectors, including governance. Adebayo and Omole [2] alluded to some key advancements, as explained below:

1. Cloud-Based Project Management Tools: Cloud technology has transformed project management by offering centralized and real-time access to project data, enabling team collaboration and communication from anywhere. Cloud-based project management solutions include task management, document sharing, time tracking, and progress monitoring to improve efficiency and coordination.
2. Artificial Intelligence (AI) and Automation: AI-powered project management solutions can automate repetitive tasks, optimise resource allocation, and perform predictive analysis. Machine learning algorithms can analyze historical project data to identify patterns, forecast risks, and recommend the best project schedules. This eliminates manual labor, improves decision-making, and enhances project outcomes.
3. Virtual Collaboration Platforms: These platforms enable teams to work seamlessly, regardless of their geographic location. These platforms provide features like video conferencing, instant messaging, file sharing, and virtual whiteboards, which promote successful communication and teamwork. They are particularly useful in governance contexts where teams may be deployed across different locations.
4. Mobile Project Management Apps: Mobile apps have improved project accessibility and convenience. Smartphones and tablets allow team members to access project information, update tasks, and communicate while on the move. Mobile apps also support real-time notifications and alerts, ensuring timely responses and reducing delays.
5. Data Analytics and Reporting: Project management software now incorporates powerful data analytics features. By collecting and analyzing project data, public sector organizations can gain valuable insights into performance metrics, discover trends, and make data-driven decisions. Advanced reporting tools enable the

generation of comprehensive project reports, dashboards, and visualizations for stakeholders.

3.2 *Benefits of Utilizing Project Management Technology in the Nigerian Governance System*

This section of the book chapter explores a plethora of benefits linked with the use of project management technology in governance. However, integrating project management technology into governance practices significantly improves the efficiency, transparency, and overall effectiveness of government initiatives, ultimately benefiting citizens and strengthening public trust in the delivery of public services and projects. Oladapo and Fagbemi [6] revealed how project management technology can transform the governance system in Nigerian public service.

(1) Enhanced Efficiency: Project management technology simplifies administrative procedures, automates repetitive work, and improves resource allocation. This leads to enhanced project efficiency, reduced delays, and improved productivity. In Nigeria, this can be scaled by addressing significant infrastructural gaps such as especially in areas like reliable electricity, high-speed internet connectivity, and accessible transportation networks, which can hinder the effective implementation and scaling of project management technologies.
(2) Improved Collaboration and Communication: Technology-enabled collaboration platforms facilitate seamless communication and collaboration among team members, stakeholders, and citizens. This improves coordination, transparency, and stakeholder engagement, resulting in better governance outcomes. The sustainability of collaboration can be achieved by leveraging public–private partnerships to pool resources and expertise, accelerating infrastructure development cost-effectively and sustainably, and enabling consistent access to project management tools and platforms.
(3) Timely Decision-Making: Real-time access to project data, analytics, and reporting enables timely decision-making. Project managers and policymakers can follow project progress, identify potential risks, and take remedial action promptly, ensuring that projects stay on track and achieve the desired results.
(4) Transparency and Accountability: Project management technology fosters transparency by providing stakeholders with access to project information, progress updates, and documentation. There should be the prioritization of inclusive and equitable access, ensuring that no organisation or project team is left behind in the digital transformation of project management. This fosters accountability among project teams and ensures that governance processes are transparent and accountable to citizens.
(5) Risk Management: Advanced project management technologies can identify and evaluate project risks, allowing for proactive risk management. By analyzing historical project data, AI algorithms may identify potential risks and suggest

mitigation techniques, reducing the impact of unforeseen occurrences on project results.

(6) Stakeholder Engagement: Project management technology can facilitate meaningful stakeholder engagement in governance processes. Virtual collaboration platforms, smartphone applications, and online portals allow individuals to engage in decision-making, provide feedback, and track project progress, thereby promoting inclusivity and democratic governance.

4 Theoretical Framework

Institutional theory emerged in the 1950s and 1960s with the work of scholars like Talcott Parsons, Philip Selznick, and Alvin Gouldner [28]. The theory has now evolved, notably in public administration, into what is known as Neo-institutionalism. Neo-institutionalism highlights the impact of cultural norms and expectations on public and project management practices [29]. Institutional theory provides a theoretical framework for understanding the role of institutions and their impact on public administration and project management in a governance context. This theory provides a valuable lens for understanding how public administration and project management function within a larger social and cultural context [30]. It emphasizes the importance of considering established rules, shared values, and the impact of other organizations on decision-making and behavior. The focus of the theory is on how institutions influence the behavior and functioning of organizations within them. It is important to state at this juncture that institutional theory suggests that governance practices and project execution are shaped by institutional pressures. Scott [31] categorized these pressures into three types:

(1) **Coercive Pressures:** Coercive pressures arise from formal regulations, laws, and policies imposed by external entities such as government bodies, regulatory agencies, or international organizations. These external pressures influence project management practices, governance structures, and decision-making processes.
(2) **Mimetic Pressures:** Mimetic pressures result from organizations imitating the practices of other successful organizations or benchmarking against sectorial standards. In the context of governance, organizations may adopt project management technologies and practices that are perceived as more effective or efficient based on the experiences of other organizations.
(3) **Normative Pressures:** Normative pressures stem from social norms, professional standards, and cultural values. In public administration and project management, normative pressures influence the adoption of specific project management methodologies, ethical practices, and standards of accountability and transparency.

Consequently, the anchored institutional theory that the study is situated on provides a deeper understanding of how these emerging trends and technologies are transforming the Nigerian public administration and project management landscape. The interplay between institutional theory in public administration and project management in the Nigerian public service explores how the formal structures and hierarchies within public administration organizations influence project management processes [5] In Nigeria, the public service is typically organized into various ministries, departments, and agencies (MDAs). These structures often have clearly defined roles, responsibilities, and reporting lines that impact how projects are managed within them. Formal rules and regulations play a significant role in shaping organizational behavior. Project management in the Nigerian public sector is governed by specific rules and regulations, such as procurement laws, budgetary processes, and reporting requirements. These institutionalized rules provide a framework for project planning, implementation, and monitoring [1]. Based on the foregoing, it can be argued that limited financial and human resources pose significant challenges to effective project management in the Nigerian public sector. Institutional factors, such as budgetary processes and bureaucratic procedures, can further complicate resource allocation and utilization for projects [5]. In addition, informal behaviors such as clientelism, corruption, and nepotism affect project management in the public service. These informal institutions undermine the effectiveness of project management practices and hinder project success.

Furthermore, the public service may be under pressure to adopt project management approaches that are in line with global standards or international donor requirements. This can lead to the adoption of project management frameworks and methodologies that align with these external pressures without considering our ecological dimension [3]. Understanding the interplay between institutional theory and project management in the Nigerian public service helps policymakers, administrators, and project managers navigate the complexities of project implementation within the institutional context [4]. By considering the formal and informal institutions that shape project management practices, stakeholders may develop strategies and interventions that promote effective project implementation and improve public service delivery.

Moreover, this theory has been limited and criticized based on its overemphasis on conformity, which downplays individual agency and the potential for innovation within institutions [32]. The criticism of institutional theory's emphasis on stability is supported by [33], who argue that the theory often overlooks institutional change. Future research on institutional responses to external shocks is suggested by Greenwood and Hinings [34], who emphasizes the need for a more dynamic perspective on institutional processes. Institutional theory has also been criticized for its inability to fully account for the context-specific nature of institutional dynamics, which can vary across different cultural, political, and socio-economic settings. Also, institutional theory has been predominantly focused on macro-level, organizational-level, or field-level dynamics, neglecting the micro-level processes that shape individual behaviors and decision-making within institutions [35]. Also, informal factors such as norms and values are difficult to define and assess, making it difficult to properly capture their effect [5]. Policymakers and project managers may use institutional theory to understand the institutional framework in which they operate, identify the institutional pressures influencing governance practices and project management, and adjust their strategies appropriately. This framework helps in understanding the role of institutional factors in shaping governance outcomes and project success.

5 Conclusion

This book chapter revealed that emerging trends and technology have significantly transformed Nigeria's public administration and project management practices. The adoption of digital technologies, data analytics, automation, and artificial intelligence has enhanced efficiency, transparency, and decision-making processes within the public service. Also, the study identifies new frontiers that have emerged because of these advancements. These include the utilization of blockchain technology for secure and transparent transactions, the integration of Internet of Things (IoT) devices for smart governance, and the implementation of predictive analytics for proactive project management.

The study concludes that continual adaptation and innovation are required to harness the full potential of emerging trends and technology in the Nigerian public service. It seeks to spark a discussion on how emerging trends and technology might propel Nigerian public administration and project management toward a brighter future. It spotlights the transformative influence of emerging trends and technology in the Nigerian context and provides insights into the new frontiers that have been discovered because of their adoption. In summary, effective project management and public administration are critical for Nigeria's development. The growing emphasis on professionalism, in-novation, and capacity building suggests that the country is on the right track to improve project delivery and achieve better outcomes for its citizens. Therefore, adopting project management technology in Nigeria can significantly improve the efficiency, effectiveness, and transparency of public administration and project execution. The Nigerian government may optimize resource allocation, develop a digital transformation strategy, foster digital literacy, enhance service delivery, establish regulatory and policy framework, establish monitoring and evaluation mechanisms, and increase public engagement in decision-making processes by leveraging digital technologies, artificial intelligence, cloud computing, and e-governance platforms.

References

1. Anya, O.: Reforming public administration in Nigeria: lessons from international best practices. J. Public Policy Adm. Res. **8**(4), 5–17 (2023)
2. Adebayo, A., Omole, D.: Strengthening project management in Nigeria: a roadmap for success. Proj. Manag. J. **1**(8), 106–123 (2022)
3. Olowogbon, O., Afolabi, M.: Rethinking project management practices for enhancing project success in the Nigerian public sector. Constr. Manag. Econ. **37**(3), 256–274 (2019)

4. Ojo, O.: The challenges of public administration in Nigeria: a critical analysis. Int. J. Public Adm. **2**(3), 5–15 (2022)
5. Edegbue, O.E.: Public bureaucracy and project management in Nigeria: a critical review. Int. J. Project Manage. **38**(8), 1051–1062 (2020)
6. Oladapo, O., Fagbemi, T.: Project management failures in Nigeria: A case study approach. Int. J. Project Manage. **3**(8), 34–49 (2021)
7. Onuoha, C.: Why Nigeria's public projects fail: a look at the underlying causes. Conversation **1**(3), 10–23 (2020)
8. World Bank.: Nigeria: Public Service Review Report (2022)
9. Uchegbulam, I.M., Okolocha, F.C.: Challenges and prospects of project management in Nigerian public sector. Int. J. Public Sect. Manag. **31**(8), 1006–1023 (2018)
10. Hassan, K.I., Fatile, O.J., Ashade, O.A.: Assessment of artificial intelligence in public administration: implications for service delivery in Lagos state public service. BVIMSR's J. Manage. Res. **15**(2), 120–138 (2023)
11. Abdullahi, B., Ibrahim, Y.M., Ibrahim, A., Bala, K.: Development of e-tendering evaluation system for Nigerian public sector. J. Eng. Des. Technol. **18**(1), 122–149 (2020)
12. Verma, S., Sheel, A.: Blockchain for government organisations: past, present and future. J. Glob. Oper. Strateg. Sour. **15**(3), 406–430 (2022)
13. Ajide, O.T., Oyedele, L.O.: Public procurement and project management: a framework for good governance in Nigeria. Public Adm. Dev. **37**(4), 346–360 (2017)
14. Eghe-Ikhurhe, G.O., Bonsu-Assibey, M.O.: The effects of blockchain technology on corporate governance: evidence from emerging economy. Manage. Dyn. Knowl. Econ. **10**(3), 239–250 (2022)
15. Chung, H.Y., Lee, G.G., Kuo, R.Z.: Determinants of public servants' intention to adopt e-government learning. Rev. Public Pers. Adm. **36**(4), 396–411 (2016)
16. Shuaib, M., Alam, S., Daud, S., Ahmad, S.: Blockchain-based initiatives in social security sector. In: Proceedings of the 2nd International Conference on ICT for Digital, Smart, and Sustainable Development, ICIDSSD 2020, 27–28 February 2020, Jamia Hamdard, New Delhi, India (2021)
17. Farghaly, K., Abanda, F.H., Vidalakis, C., Wood, G.: Taxonomy for BIM and asset management semantic interoperability. J. Manag. Eng. **34**(4), 04018012 (2018)
18. Jankin, S., Esteve, M., Campion, A.: Artificial intelligence for the public sector: opportunities and challenges of cross-sector collaboration: mathematical, physical and engineering sciences. Philos. Trans. R. Soc. A **376**(2128) (2018)
19. Ugwu, M.C., Adewusi, A.O., Nwokolo, N.E.: The role of public-private partnerships in building clean energy infrastructure in the United States and Nigeria. Int. J. Manage. Entrepreneurship Res. **6**(4), 1049–1068 (2024)
20. Aliu, I.R., Adeyemi, O.E., Adebayo, A.: Municipal household solid waste collection strategies in an African megacity: analysis of public-private partnership performance in Lagos. Waste Manage. Res. **32**(9_suppl), 67–78 (2014)
21. Iloh, J.O., Bahir, M.: Public-private partnership (PPP) and social service reform in Nigeria: 1999–2007. J. Educ. Soc. Res. **3**(10), 101–108 (2013)
22. Kohler, G., Sampalli, T., Ryer, A., Porter, J., Wood, L., Bedford, L., Rivoire, E., et al.: Bringing value-based perspectives to care including patient and family members in decision-making processes. Int. J. Health Policy Manag. **6**(11), 661 (2017)
23. Manu, P., Mahamadu, A.M., Booth, C., Olomolaiye, P., Ibrahim, A.D., Coker, A.: Assessment of procurement capacity challenges inhibiting public infrastructure procurement: a Nigerian inquiry. Built Environ. Proj. Asset Manage. **8**(4), 386–402 (2018)
24. Pooe, D.R., Mafini, C., Makhubele, D.T.: Investigating municipal procurement challenges in South Africa: a qualitative study. Int. Bus. Econ. Res. J. (Online) **14**(1), 67 (2015)
25. Van der Wal, Z.: Singapore's corrupt practices investigations bureau: guardian of public integrity. In: Guardians of Public Value: How Public Organisations Become and Remain Institutions, pp. 63–86 (2021)

26. Callaghan-Koru, J.A., Gilroy, K., Hyder, A.A., George, A., Nsona, H., Mtimuni, A., Bryce, J., et al.: Health systems support for community case management of childhood illness: lessons from an assessment of early implementation in Malawi. BMC Health Serv. Res. **13**, 1–13 (2013)
27. Chagunda, C.: Development aid, democracy and sustainable development in Malawi–1964 to date. Cadernos de Estudos Africanos **41**, 145–164 (2021)
28. Meyer, J.W., Rowan, B., Scott, W.R.: Institutionalised organisations: formal structure, rationalization, and myth. Am. J. Sociol. **83**(2), 340–360 (1977)
29. Zwass, A.: The institutionalization of project management. Int. J. Project Manage. **8**(2), 83–88 (1990)
30. Sydow, C., Windeler, A.: Institutional theory and project management research: a critical review. Int. J. Project Manage. **26**(5), 503–515 (2008)
31. Scott, W.R.: Institutions and Organisations: Foundations of Organisational Theory. SAGE Publications, Thousand Oaks, CA (2013)
32. Chávez-Rivera, M.E., Ruíz-Jiménez, J.M., Fuentes-Fuentes, M.D.M.: The effects of context and characteristics of women entrepreneurs on innovation performance. BRQ Bus. Res. Q. **27**(1), 73–90 (2024)
33. Greenwood, R., Hinings, C.R.: Understanding radical organisational change: bringing together the old and the new institutionalism. Acad. Manag. Rev. **21**(4), 1022–1054 (1996)
34. Tina Dacin, M., Goodstein, J., Richard Scott, W.: Institutional theory and institutional change: introduction to the special research forum. Acad. Manag. J. **45**(1), 45–56 (2002)
35. Thornton, P.H., Ocasio, W., Lounsbury, M.: The Institutional Logics Perspective: A New Approach to Culture, Structure, and Process. OUP Oxford (2012)

Project Risk Management (PRM) in Situations of High Complexity and Deep Uncertainty

Yuri G. Raydugin

Abstract Two limitations of the traditional PRM are outlined and two compelling paradigm shifts are introduced. First, the traditional PRM assumes that risks are standalone, noninteracting events. This results in overlooking a substantial part of project risk exposure stemming from risk interactions, especially in complex projects. If project complexity is ignored, project schedule and cost contingencies yielded by the Monte Carlo methodology could be too low, resulting in project failures. The nonlinear Monte Carlo modeling technique, which counts risk interactions, is introduced in this chapter. This epitomizes shifting from linear (traditional) to nonlinear PRM in complex projects. Second, in the traditional PRM, the project alternative selection is based on subjective assumptions regarding the external project environment. This overlooks ongoing dynamics and deep uncertainties in the external environment. This assumes that the initially predicted demand for produced products will keep meeting stakeholders' expectations until the project is decommissioned. What if these assumptions are wrong from the get-go? Or if the demand dwindles in operations due to dynamics in the external environment? The developed project might resemble a "bridge to nowhere" in both cases. Dynamic adaptive methodology (DAM) that handles deep uncertainties in the external environment is introduced and supported by a shift from the traditional predict-then-act to monitor-and-adapt decision-making. This represents the second paradigm shift. The power of systems dynamics supported by implications of game theory is called for to facilitate these two paradigm shifts. Two practical, although highly simplistic, modeling examples on nonlinear Monte Carlo and external environment modeling are provided to demonstrate the introduced paradigm shifts.

Keywords Project risk management · Risk interactions · Monte Carlo methodology · Nonlinear PRM · Project complexity · Dynamic adaptive methodology · External environment uncertainty · Systems dynamics · Game theory · Paradigm shifts

---

Y. G. Raydugin (✉)
Risk Services & Solutions Inc., Calgary, AB, Canada
e-mail: yuri.raydugin@risk-service.com

K. D. Strang and N. R. Vajjhala (eds.), *International Program and Project Management—Best Practices in Selected Industries*, Information Systems Engineering and Management 31, https://doi.org/10.1007/978-3-031-80275-1_11

1 Introduction

Any concept underlying any practical methodology in any area of knowledge (from medicine and engineering to business, to politics, to physics) has underlying postulates and assumptions that define conditions of methodology's applicability. This means that every methodology has a pertaining often implicit "envelope of validity" reflecting on situations when the underlying assumptions are valid. Although, it becomes ineffective, inefficient, irrelevant or harmful outside of that envelope. The borders might be delineated either qualitatively as written statements on conditions when the postulates and assumptions lose validity or quantitatively as relevant mathematical cut-off criteria and parameters.

Sometimes the underlying assumptions and, hence, the validity envelopes are either not fully recognized or they might just get forgotten or ignored. Applying a methodology outside of its validity envelope either knowingly or unknowingly is a proven recipe for failures at best and disasters at worst. In some unfastidious cases it might be also compared with placebo effect in medicine. This chapter analyses underlying assumptions of the widely used traditional project risk management (PRM). Reasons for its chronic and repetitive failures are analyzed and explained. Cases when the traditional PRM gets thrown out of its validity envelope are examined. This analysis is focused on complex projects featured by risk interactions as well as deep uncertainties in their external environments.

This chapter is a high-level recap of a recent book on the topic [1]. As such it has a much narrower scope and provides a helicopter-view narrative. It cannot keep all even quite important topics of [1] related to high complexity and deep uncertainty in external project environment. Inevitably, included topics are discussed in sketchy way. Although the importance of paying proper attention to project complexity and deep uncertainties are accentuated. Scenarios and implications of ignoring these two cornerstone concepts are outlined. As opposed to [1] where each reviewed PRM methodology is introduced through its three-component system (context and requirements, a process and tools), this chapter zeroes in on the contextual component mostly. This allows get zeroed in on underlying concepts, postulates and assumptions and their validity envelopes. Methodologies replacing the traditional PRM are shaped and justified.

To a certain degree this chapter is biased, being aimed mostly at assessments of risks with impacts on quantifiable project objectives such as durations and capital budgets which risks are a subject to Monte Carlo modelling. Taking into accounts project complexity and, hence, risk interactions in complex projects allow the introduction of nonlinear Monte Carlo methodology as an upgrade of the traditional (linear) Monte Carlo modelling. This is the first fundamental paradigm shift outlined in this chapter. The second paradigm shift zeroes in on thoughtful predicting and monitoring of the external project environment that defines and judges on ultimate value of any project. Most radical situations when a project is being developed either on wrong predictions from the get-go or when external environment slowly but surely changes in operations quite drastically are discussed. In such cases the project loses its value completely, becoming "a bridge to nowhere". A dynamic adaptive methodology (DAM) is introduced as a remedy against scenarios like these and justifies introduction of the new monitor-and-adapt decision-making paradigm.

After review of the traditional PRM's limitations in Sect. 2, the pointed-out paradigm shifts are initially introduced separately in Sects. 3 and 4. Section 5 argues that high complexity and deep uncertainties would rather go hand in hand. Besides [1], a few other primary sources are referred to in this chapter. The traditional PRM and some deviations from it are fully described in [2]. Project complexity aspects and modern risk quantification methodologies are fully reviewed in [3]. Discussion on deep uncertainties is inspired by Marchau et al. [4]. Systems dynamics methodology so important when handling high complexity and deep uncertainties is taught in [5]. Thoughtful selection of project decision-making criteria is based on game theory fundamentals of [6]. Some valuable project complex ideas are borrowed from [7].

2 Traditional PRM's Validity Envelope: Three Hidden Underlying Assumptions

2.1 Are Projects that Simple?

Regarding risk assessments, the traditional PRM methodology has two ingredients: qualitative (scoring) and probabilistic (Monte Carlo) methods [2]. The first is aimed at rough assessing and addressing of identified project risks. A risk assessment matrix (RAM) measuring probabilities and impacts of risks as scores is employed. As RAM's accuracy is notoriously low, it cannot be used for development of adequate project contingencies. To develop project contingencies more adequately, the traditional PRM utilizes Monte Carlo models based on uncertainty variations—general uncertainties (aka "ranges") and uncertain events (aka "risks")—around project schedule's activities and estimate's cost accounts [2, 3]. Integrated schedule and cost risk analysis (SCRA) is considered a sort of standard by proponents of the traditional PRM.

Apparently, before risks get assessed or addressed, they should be identified to comprehensively represent project risk exposure. Traditionally, a simplistic bowtie diagram linking causes of a risk, a risk event itself and its impacts on selected project objectives is utilized as a primary risk identification as well as addressing tool [2]. Putting aside any possible risk identification flaws stemming from and described by various realizations of psychological and organizational bias, a question may be risen if the outlined risk identification method represents project risk exposure adequately in its entirety. What is an actual validity envelope of such approach?

Intuitively, one might get curious why several identified project risks could not interact in certain circumstances. Indeed, the classical engineering fault-tree analysis (FTA) methodology [1] to assess reliability of technical systems (from equipment to produce chemicals from oil to modern weaponry) presumes multiple possible interacting events (aka risks) responsible for failure modes. Snowballing and compounding of them is presumed by the FTA methodology. Indeed, whereas the most of standalone risk events could be predicted and mitigated relatively easily, their unexpected and sometimes quite bizarre sequences and combinations may result in terrible catastrophes. (The Challenger and Chernobyl disasters illustrate this point well.)

In the traditional PRM in place of quite sophisticated fault trees presuming risk interactions, used is its highly-pruned version—a bowtie diagram. A risk event, which is called in FTA analysis a top (or resulting) event, "feels" only few immediate predecessors—its causes. Any events before these causes including possible interactions are cut off and ignored. A main explanation of bowtie diagram popularity is its simplicity and convenience to use by non-technical project team members. Although this simplicity vs. informativity trade-off goes with a cost: any project risks must be considered standalone by default. (An analogy from physics could be drawn to illustrate this "stand-aloneness": molecules of rarified gas that do not interact.)

Hence, this is a first major assumption underlying the traditional PRM and defining its validity envelope: all project risks are non-interacting standing-alone events. An immediate question comes up: when the traditional PRM is adequate? In the light of the defined validity envelope the answer to this question is banal and trivial: the traditional PRM is adequate when interactions among project risks can be legitimately ignored. In short, the answer is—when projects are simple enough [1, 3]. Indeed, in terms of the FTA concept this hints that a project eligible for the traditional PRM should be simple enough to deprive risks of possibility to interact there should not be any fault tree but just its "stubs" in a form of bowtie diagrams for noninteracting risks. Immediately this leads to cognizance that the project as a system shall be made up of few rather independent and non-interacting parts. What if there are quite a few interacting project parts that do allow pertaining risks to propagate and interact?

Introducing of a "simple project" notion attached to the traditional PRM is done to ignite discussion on project complexity undertaken in Sect. 3. Besides "simple", introduced are three other project categories: complicated, complex and chaotic [1, 3, 7]. These three do not obey the traditional PRM rules and call for a PRM methodology upgrade.

2.2 Is External Project Environment Matter?

Presumably, a project selected for development promises acceptable economic value to its stakeholders which is largely controlled by future features and states of its external environment. For instance, a sustainable demand for produced products should be ensured until decommissioning in several decades. This presumes that a particular project gets a "frozen" scope (including nameplate capacity and a slate of produced products), economic goals, budget, duration, etc. These goals become "written in stone" or baselined during project's final investment decision (FID) and the project is being developed in post-FID execution phases. A vague doubt sneaks in regarding the described selection of a pursued project alternative. This doubt stems from two slightly different scenarios.

First, presuming that a selected project alternative is somehow selected a right way and upon project completion the demand for offered products is pretty high making stakeholders economically happy. This happiness might hold on for few years until the demand becomes dwindling as external project environment conditions (combinations of market, economic, regulatory, competition, technological, social, geopolitical, etc. aspects) become evolving. In other words, a new state of the external environment may differ from the one used for selecting the initial project alternative. Second, what if a wrong project alternative gets selected from the get-go because the state of the external environment has been wrongly understood? Such situation happens not too often being a limiting case of the previous scenario. Although when it does happen, the project deserves to be labelled as "a bridge to nowhere".

A comparable scenario materialized in the case of Channel Tunnel project [1, 4]. Namely, a predicted state of its external environment (economic and market aspects) used for selecting of project alternatives was based on overoptimistic and careless linear projections of demand for travelling between Britain and France. The actual demand did follow that linear projection but just for few first months. After this competition got kicked in resulting in dwindling Eurotunnel revenue. Such systems behavior is called "boom-and-bust" in systems dynamics modelling [5]. This "tunnel-to-nowhere" situation, which corresponded to the "bust" new state of the external environment, held for more than two decades making the project stakeholders extremely unhappy.

This "tunnel-to-nowhere" example hints on a need for careful investigation of project external environment before it is used for selecting a right project alternative. (Some valuable insights from system thinking and systems dynamics modelling can be used to predict external environment futures including linear projection forecasting fallacy [1, 5].) This in turn presumes that uncertainties handled by the traditional PRM—general uncertainties and uncertain events—do not represent the entire uncertainty universe clouding a considered project. Indeed, multiple interrelated uncertainties in the external project environment are being brewed perpetually during entire project lifespan. If that corpulent chank of uncertainty is overlooked, "successful" managing of trivial general uncertainties and uncertain events attached to a wrongly selected project alternative becomes futile.

Hence, this is a second major assumption underlying the traditional PRM and defining its validity envelope:

- any uncertainties in external project environment are fully and rightly understood at the beginning of a project;
- a predicted future state of the environment used for selection of a pursued project alternative is rightly forecast;
- that future state of the external environment must stay unchanged until project decommissioning to keep selected alternative fully relevant all along.

Again, a question arises when the traditional PRM is adequate? In the light of the defined validity envelope the answer to this question is again banal and trivial:

the traditional PRM is adequate when external project environment can be ignored believing that selected project alternative must stay well-grounded for several decades until project decommissioning. In short, the answer is—rather never.

2.3 *Is Adequate Decision-Making Used by the Traditional PRM?*

The categorial answer in the last line of the previous section should be a reason for some catastrophizing: pinned down incapacity of predicting the future states used for project alternative's selection smells like a grave traditional PRM's nescience. (By the way, the word "nesceince" sounds in Russian exactly as "not science". In this regard, a true science is required to handle external project environment properly which calls for a cardinal change of a decision-making.). As a matter of fact [1, 4], the traditional PRM employes so-called "predict-then-act" decision-making possibly even without realizing this which is supported by a traditional project work process. Indeed, a sort of external environment future gets selected (a "predict" part) somehow first; based on this a preferable project alternative gets selected (an "act" part) somehow; pertaining risks are identified and successfully managed (also the "act" part) until project completion. That is when the traditional project work process stops short. The issue with this is that uncertainties associated with dynamics in external project environment snick back in during operations or sometimes even earlier (Sect. 4).

To handle such occurrences thoughtfully, a new "monitor-and-adapt" decision-making paradigm is introduced and promoted in [1, 4]. In simple terms, predicting highly complex processes and trends in the external environment for several decades ahead until project decommissioning is quite an arrogant intent and it is certainly out of rich. Monitor-and-adapt approach is a compelling band-aid solution for this. (The discussion on the monitor-and-adapt paradigm is provided in Sect. 4). As a matter of fact, Sect. 2.2 fully defines conditions when monitor-and-adapt decision-making shall replace the predict-then-act: when dynamics in the external project environment is really matter which is further discussed in Sect. 4.

To finalize this section, it should be stated that using the traditional decision-making paradigm often presumes quite subjective and vague decision-making criteria when defining external environment futures or selecting project alternatives under the traditional predict-then-act decision-making. Employed decision-making criteria might not be even clearly formulated opening the door to bias and subjectivism. The new monitor-and-adapt paradigm inevitably brings forward quite sophisticated and robust decision-making criteria that are borrowed from mathematical game theory [6] (Sect. 4). Concepts discussed in this and the previous sections are formalized as a dynamic adaptive methodology (DAM) (Sect. 4).

3 Project Complexity Implications

When applying traditional PRM methodology to simple projects, any complexity implications are ignored. Simply, there is no a consistent way to incorporate them: a simple project is viewed as a system consisting of just one part (aka itself) or considering its parts is simply irrelevant. Although, what might be applicable, say, to a small repetitive equipment maintenance project at an existing refinery, might not be a fit for purpose when managing risks of a unique and complex mega project devoted to building of a new refinery. This section overviews all required complexity concepts applicable to PRM.

3.1 Project System Complexity Types

First, all projects can be treated as systems which have interacting parts. As soon as this approach is adopted, the validity envelope of the traditional PRM gets abandoned. As argued in [7], to handle project complexities four project system types should be told apart:

- Simple systems: systems of linear functional projects.
- Complicated systems: fully knowable and predictable project systems with high number of structural parts which can be reduced to several simple ones.
- Complex systems: not fully knowable but somewhat predictable project systems with a high number of structural parts which cannot be reduced to several simple ones.
- Chaotic systems: completely unknowable and unpredictable nonlinear project systems.

Simple projects could be understood as ones when involved project disciplines contribute directly to overall project outcome and do not take part in any input–output relations. Any meaningful interactions take place inside of separate project parts only. Complicated project systems are characterized by a high number of simple structural parts. A program that covers quite a few simple projects could be a logical example of a complicated project when it still could be reduced to a number of simple projects. Project's parts are rather independent although input–output (or sequential) relations among them do exist. This type of complexity could be dubbed input–output complexity.

Complex project systems cannot be reduced to several simple or complicated projects. The concept of reducibility is paramount for understanding of system's complexity [7]. Any attempts to reduce complex systems to several simple ones destroy the former in terms of their adequate representation. Various interactions among system's parts are primary factors shaping system's complexity that prevents the reducibility. This type of complexity is sometimes called feedback or interdependence complexity. Chaotic systems usually come from complex systems when interactions among complex system's parts become excessively intensive. (The transition from complex to chaotic state could be understood as a phase transition in physics sense of this term: individual characteristics of parts are melted down due to highly intensive interactions to form a completely different system's state.)

3.2 Types of Project Part's Interactions

Part's interactions referred to in the previous sections have three basic realizations as follows [1, 3, 7]:

1. Pooled: a simplest type when each part contributes a discrete input to a project at large and is not sequence bound.
2. Sequential: part's contribution (output) becomes an input to other parts.
3. Reciprocal: a most complex type consisting of feedback loops and mutual impacts.

Roughly, simple projects are characterized by pooled interactions only; complicated ones should feature both pooled and sequential interactions among their parts; complex and chaotic projects should support all three interaction's types. A well-oiled car engine could serve as an example of a complex system. Based on my knowledge of mechanics, a fuel system could be seen as taking part in pooled interaction directly providing gasoline to the engine. Cylinders send torque to transmission taking part in sequential interaction. A cooling fan kicks off when the engine sends a signal about overheating, both taking part in a feedback-loop reciprocal interaction.

This narrative does not include chaotic systems yet. As a matter of fact, an ideal complex project cannot not end up in chaos if its parts interact as presumed by "part's interaction baselines". In the engine example, the chaotic systems emerges when the engine becomes too jumpy: its parts become working outside of designed envelopes. This immediately presumes deviations from or perturbations to the part's interaction baselines.

3.3 Types of Deviations from the Interaction's Baselines

Apparently, any possible deviations from the ideal interaction baselines use the same interaction channels as the parts themselves [1, 3, 7]. Simply, these deviations give rise to non-ideal part's interactions. Such "contaminated" interactions can be represented by a simple formula:

$$X = X_0 + \Delta X, \tag{1}$$

where X are the resulting non-ideal part's interactions, X_0 represent the ideal part's interactions (the baseline), and ΔX points to the deviations. Moreover, the word "deviations" above (especially, if combined with words "possible" and "from project objectives and baselines" [1–3]) points to project risks in the traditional PRM sense. Now we may start using the traditional word "risk". However, instead of its traditional PRM meaning which presumes standalone non-interacting risks only, it will be used to point to risks taking part in interactions.

Formula (1) allows extract three types of deviations ΔX and move from part's interactions to risk's interactions as follows [1, 3]:

- From pooled part's interactions to internal risk amplifications (ΔX_A).
- From sequential interactions to knock-on risk interactions (ΔX_K).
- From reciprocal interactions to risk's compounding (ΔX_C).

Internal risk amplifications stem from the perturbated pooled interactions. These could be understood as intra-risk interactions (an interaction of a risk with itself) pertaining to standalone although amplified risks. For instance, a project that has issues with its engineering team would certainly have much more severe engineering errors and omissions risk than a similar project having a strong engineering team. Simply, depending on internal inefficiencies in a project system, severity of a given standalone risk might vary from project to project. When internal risk amplifications are fully ignored in the traditional PRM, any occurred risk could be characterized by a generic deviation ΔX_0. Apparently, taking specific internal risk amplifications into account would lead to recalibration of the deviation evaluated as $[\Delta X_A - \Delta X_0]$. In a way, this is a fine-tuning of risk consequences. This is a first modification of the traditional PRM.

Knock-on and compounding risk interactions are perturbations to sequential and reciprocal interactions, correspondingly. For example, engineering errors and omission might be discovered with a delay as late as in project commissioning phase to exacerbate regular commissioning risks; a regular engineering errors and omissions risk and a risk of scope change driven by stakeholders when happening simultaneously would exacerbate each other, etc. Indeed, earlier occurred risks might knock on down-the-line risks making them more severe; if two risks happen simultaneously their combined impacts may be much higher than if they occurred independently.

Roughly, only internal risk amplifications shall occur in simple projects; both amplifications and knock-on interactions shall characterize the complicated ones; complex projects should support all three risk interaction types. Now positioning of the chaotic projects becomes clear: they emerge from complex projects (usually) when pertaining interactions become excessively strong [1, 3]. (An analogy from physics could be drawn again to illustrate interactions: molecules of a gas under high pressure become interacting.). In the theoretical narrative above there are two gaps. First relates to referred project system parts: what are they? Second points to introduced deviations ΔX: their nature shall be further clarified especially as related to project parts and their misalignments.

3.4 Types of Project Subsystems and Parts

Literature on system's complexity contains multiple sophisticated approaches, definitions and taxonomies [7]. Unfortunately, very few of them answer to context of practical PRM. To overcome this difficulty, three project subsystems are introduced [1, 3, 7]:

- External environment subsystem (EES): relevant economic, market, technology, competition, social, regulatory, geopolitical, legal, etc. aspects of the external project environment including external stakeholders.
- Project structure subsystem (PSS): what should be delivered (scope, locations, technologies, facilities, internal stakeholders, interfaces, etc.).
- Project delivery subsystem (PDS): how the project would be delivered (project organization, applicable procedures, consistency of procedure's implementation, skills, leadership, bias, etc.).

External stakeholders apparently belong to the EES as they project various aspects of external project environment on a project. Internal stakeholders including operations that will take the project over upon its completion are treated as a part of the PSS: they shape it. (Telling apart internal and external stakeholders as well as including internal stakeholders to the PSS is done on purpose to tell apart the project team and the rest of the owner organization.)

These three subsystems are main project parts referred to in the previous section at high level. However, these three subsystems can be further broken down to components as shown in [1, 3]. For instance, the PSS could be represented as project's structural parts and interactions among them (e.g., interfaces); the two major parts of a PDS are well represented by all applicable project procedures and consistency of their implementation; the EES parts represent the entire pallet of the relevant external environment aspects. These subsystems are not further specified for simplicity although their next-level components are presumed [1, 3].

3.5 Two Cardinal Subsystem's Misalignments

An ideal and rather utopian project system presumes full EES—PSS—PDS alignment comparable with subsystems of ideally working car engine. Unfortunately, such full alignment represented by X_0 in formula (1) is practically unachievable. By this reason, three types of deviations ΔX from X_0 are introduced above. Now they should be associated with pertaining subsystem's misalignments or mismatches.

As a matter of fact, EES—PSS—PDS conjunction is a hierarchy. The EES is on the top of it ruling the other two subsystems. Indeed, nobody wants a project that is gravely misaligned with its external environment as its products might be of no interest and demand. This accentuate a paramount need for deep understanding and predicting of project external environment not only when it is being initially predicted but, what is more important, for entire project lifespan. The challenge is formidable as the external environment is a moving target due to numerous brewing economic, market, technology, social, geopolitical, etc. trends, tendencies and drifts.

This points to the first possible EES—PSS misalignment and attached uncertainties. Apparently, it might have two typical root causes. First, when the EES is understood and predicted wrongly for the entire project lifespan from the get-go. Second, when the EES is well understood and rightly predicted for the first few years or just months. Presuming that EES is somehow defined and fixed (either wrongly or rightly), the PSS could be shaped to answer to the fixed EES. This could be undertaken when weighing several alternatives regarding project structure (elements of scope, locations, technologies, etc.). When a most preferable project alternative gets somehow selected, any EES—PSS misalignments and associated uncertainties are pronounced fully eliminated.

Finally, a project delivery (aka the PDS) should be shaped based on the given PSS. However, some weaknesses could pertain to various parts of the PDS. These weaknesses contribute to the PSS—PDS misalignment: the PDS might not be up to a challenge to deliver PSS. Certainly, weaknesses like these could reside not only in PDS. Some of them could impede selecting of a right PSS, some are reasons for wrong understanding and predicting of the EES itself. Hence, they could contribute to both EES—PSS and PSS—PDS misalignments. All those weaknesses could be dubbed "project system issues". Moreover, due to their long-term nature they can be called static or chronic.

As opposed to the long-term static or chronic system issues, there could be dynamic ones. Indeed, even if the EES—PSS—PDS conjunction has been ideal initially, some dynamic events happening in one of the subsystems would disturb and misalign the entire project system at least temporary. The events and changes could be quite slow in the EES due to ongoing processes in the external environment. Nevertheless, EES—PSS misalignment would loom. Terms "external environment states" or "future states" are introduced in [1, 4]. This means that the external project environment might have various attributes (combinations of economic, market, technology, social, geopolitical, etc. aspects and characteristics). Some of them are more realistic, some of them are rather exotic, some of them are not even initially known. So that a predicted external future state used for PSS shaping may become eroding and evolving being eventually replaced by another external environment state. Usually, although not always, this is a slow but sure trend. As a result, the initially selected project alternative losses ground and value.

As opposed to mostly slackly evolving future states in the EES, way more dynamic events might happen in the PSS and PDS. These are usually called "uncertain events", "risk events" or just "risks" in the traditional PRM presuming their occurrence probability. (Along with them, there are also "general uncertainties": they are simply "ranges" around baseline values not characterized by probabilities.) Any such risk event when occurs disturbs the PSS—PDS alignment at least temporary. This mechanism is explained in detail in Sect. 3.6.

Although, as opposed to the traditional PRM, internal risk amplification of a standalone risk means its recalibration due to revealed chronic system issues (Sect. 3.3).

Apparently, such recalibration is overlooked by the traditional PRM. Although, a most dramatic contradictions with the traditional PRM are cross-risk interactions involving pairs of risks: knock-on and compounding interactions (Sect. 3.3). The role of chronic system issues in cross-risk interactions is discovered and investigated in Sect. 3.6. Logically, in the light of the two discussed EES—PSS and PSS—PDS misalignments, the narrative regarding the subsystem's misalignments shall be split to two parallel topics as nature and features of these two EES—PSS and PSS—PDS misalignments are very different.

First, the idea that all three types of risk interactions associated with PSS—PDS misalignments depend on chronic system issues requires further investigation. Section 3.6 compares bowtie diagrams for standalone noninteracting risks (the traditional PRM case) with newly developed bowtie diagrams for the three types of risk interactions [1] to demonstrate the role of chronic system issues in these three cases. The rest of Sect. 3 is devoted to shaping of nonlinear Monte Carlo methodology that takes PSS—PDS misalignments and risk interactions into account through evaluation of pertaining chronic system issues.

Second, a completely new terminology and concepts are required when studying future states of the external environment that shape the EES. Simply, EES—PSS misalignments cannot not be adequately described using the approach utilized for the PSS—PDS misalignments which is based on project risks that start interacting. EES—PDS misalignments do stem from quite fundamental, mystifying, vague and convoluted multiple trends in the external environment. Following [1, 4, 6], a new more general taxonomy of uncertainties is introduced in Sect. 4. (Actually, this new much broader taxonomy does cover PSS—PDS misalignments and attached uncertainties as a limiting case too. Although, the EES—PSS misalignments are in the primary focus of this new taxonomy that is fully overlooked by the traditional PRM.). Along with the new uncertainty taxonomy, a new decision-making approach is introduced in Sect. 4—monitor-and-adapt. This is thought as an inevitable upgrade of the traditional predict-then-act decision-making.

Justification for the new decision-making approach is the same as in the case of the new uncertainty taxonomy pointed out above. The EES future states are way more uncertain and fully uncontrollable by a project team. Moreover, sometimes they are simply unknown and unpredictable. Simply, it is not practically possible to predict EES future states for the entire project lifespan that in turn calls for ongoing monitoring, learning and adapting keeping in mind the new uncertainty taxonomy. As further argued in Sect. 4, this nudges to introduce a notion of "deep uncertainties" for EES—PSS misalignments as opposed to (or rather along with) "shallow uncertainties" pertaining to the PSS—PDS misalignments.

3.6 Role of Chronic System Issues in Risk Interactions and PSS—PDS Misalignments

Branching out of the traditional PRM to its upgraded version where risks interact requires proper understanding of risk interaction's mechanisms. First, it makes sense to start with a review of the well-known traditional bowtie diagram depicting standalone noninteracting risks [2]. It would be interesting to fathom how an occurred risk influences the PSS and PDS (EES—PSS misalignments are not considered in this section as pointed out above). Second, as soon as anatomy of a standalone noninteracting risk is better understood in presence of the PSS—PDS misalignment, this should allow venture into investigating of mechanisms of the three types of risk interactions.

Figures 1, 2, 3, 4 and 5 demonstrate bowtie diagrams that represent:

- a standalone noninteracting risk (as in the traditional PRM) [2];
- a modified standalone noninteracting risk ($\sim\Delta X_0$) that does points to the PSS—PDS disturbances [1, 3];

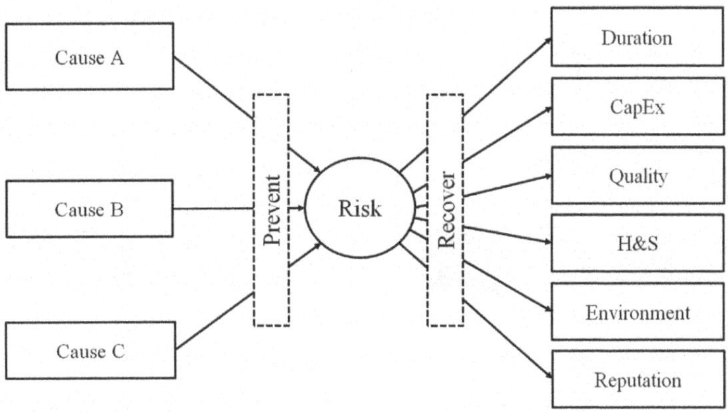

Fig. 1 The traditional bowtie diagram

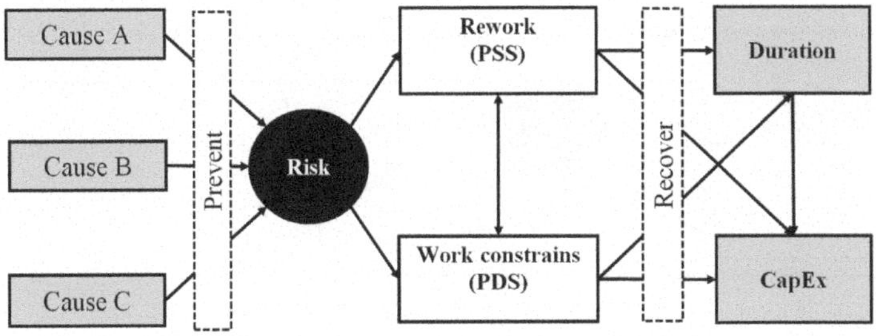

Fig. 2 A modified bowtie diagram

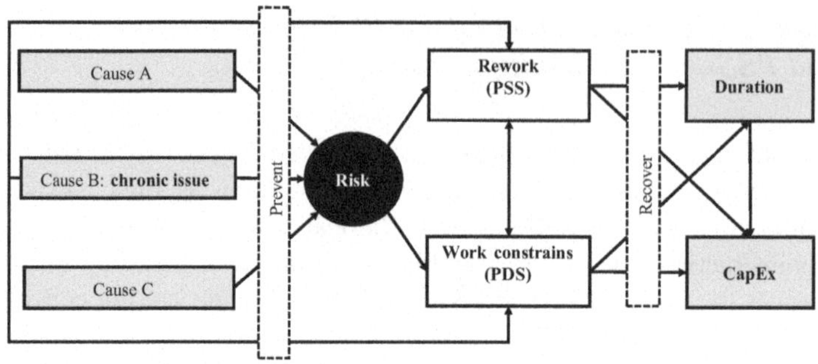

Fig. 3 A bowtie diagram for internal risk amplification

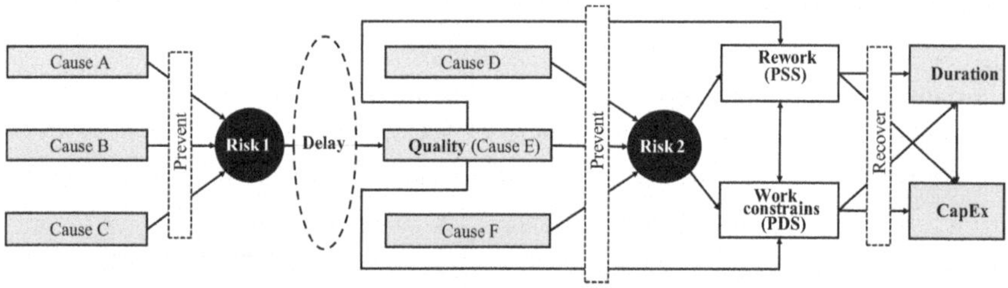

Fig. 4 A bowtie diagram for knock-on risk interaction

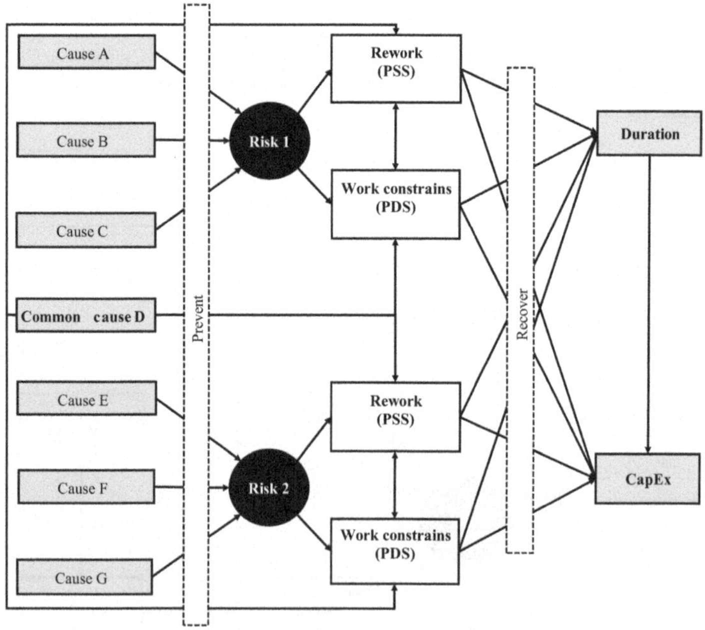

Fig. 5 A bowtie diagram for risk compounding

- a standalone internally amplified risk ($\sim\Delta X_A$) [1, 3];
- a knock-on interaction of two risks ($\sim\Delta X_K$) [1, 3];
- a compounding of two risks ($\sim\Delta X_C$) [1, 3].

Standalone noninteracting risks. Figure 1 is nothing else but a bowtie diagram in a form usually used in the traditional PRM representing causes, a risk event and impacts on project objectives. It does not bring forward any new information being used as a starting point.

Figure 2 represents the same standalone noninteracting risk as Fig. 1 although it reveals a mechanism of risk impacts on objectives through disturbing the PSS and PDS which results in the induced dynamic PSS—PDS misalignment. (The same as the other bowtie diagrams (Figs. 3, 4 and 5) it keeps only two quantifiable objectives of interest: duration and CapEx.). As the modified bowtie diagram of Fig. 2 is used as a building block for the other three risk-interaction diagrams, it is worthwhile to explore it in detail.

Namely, any seasoned project practitioner could argue that an occurred project execution risk results in perturbations to a project system stemming from emerged rework exacerbated by emerged work constraints [1, 3]. (This level of understanding is above the traditional PRM.). First, four sorts of emerged rework disturbing the PSS due to an occurred risk could be told apart:

- additional work;
- adjustment of previously finished work;
- scrapping previously finished work;
- scrapping unfinished work.

Apparently, additional work and adjustment of previously finished work requires extra time and funds. Both kinds of scrapping would require some extra time and expenditures too. Moreover, they certainly imply corresponding sunk costs related to wasted worked man-hours, materials, equipment, etc. Second, the immediate rework as well as continued normal work are impossible right after the risk occurrence as the risk may result in emerged so-called work constrains disturbing the PDS due to an abruptly emerged lack of required

- information (updated drawings, data sheets, interface management registers, etc.),
- labour,
- materials,
- equipment,
- permits.

Project delays and extra costs stemming from work constrains can be explained by a need to eliminate the listed constraint factors first before the actual rework starts. Apparently, both the reworks and the constraints invoke lower work productivity. Their sources are not explicitly shown by Fig. 2 pointing only to disturbed PSS and PDS. (Further discussion on the role of the reworks and the work constraints including out-of-sequence works and work availability concept is provided in [1] where systems thinking approach including pertaining feedback loops is used.)

The same as in the case of the traditional bowtie diagram of Fig. 1, preventive and recovery barriers are shown by Fig. 2. (Similarly, these two barriers are shown by the other three bowtie diagrams for risk interactions (Figs. 3, 4 and 5). Although besides handling logical links belonging to the traditional bowtie diagram, these two barriers are also aimed at suppressing risk interactions.)

Internal Risk Amplification Mechanism. Bowtie diagram of Fig. 3 explicitly shows a role of long-term chronic system issues. Namely, if one of causes turns out to be a chronic system issue, it can distort both the PSS and the PDS in addition to the usual impact of an occurred risk shown by Fig. 2. As a matter of fact, among causes shown by bowtie diagrams of Figs. 1 and 2, there could be givens and facts which could be also called "background causes" [1]. They usually do not lead to risk events themselves but they play an important role in presence of some other regular causes. For instance, existing environmental regulations (a given/a fact/a background cause) do not lead to a regulation-driven scope change risk unless they get combined with some other regular causes. In other words, some regular causes without the environmental regulations in place might not lead to such risk. In addition, a special category of background causes are chronic system issues pointing to given weaknesses in particular project disciplines and functions. They might not lead to particular risks on their own, only in combination with some other regular causes. To be more specific with this example, a regulation-driven scope change risk could be identified using three-part naming [2] as follows.

"Due to (A) existing environmental impact assessment (EIA) permit requirements (a given/a fact/a background cause); (B) low maturity of engineering discipline (a chronic system issue); (C) ambiguity in project scope (a regular cause), EIA-driven project scope change could be mandated (a risk event), resulting in delays and extra costs (impacts on objectives)."

In the traditional bowtie diagram's causes follow FTA's OR-logic only [1] when each cause could lead to the risk event independently. Although, the above-introduced example tells apart three types of causes that should also follow the AND-logic which is fully ignored by the traditional PRM. The causes (A) and (C) separately and without each other don't lead to the discussed risk event. The cause (B) if it is recognized as a chronic system issue becomes a background cause. In this example it cannot result in the risk event on its own. Bowtie diagram for internal risk amplification (Fig. 3) shows a general case when the cause B wears two hats: it plays a role of both a regular cause and a chronic system issue. It leads to a risk event as any other regular cause and in addition it hits the PSS and PDS as a chronic system issue.

In simple terms, a project-specific chronic system issue exacerbates and inflates emerged reworks and work constraints invoked by an occurred risk. A regular risk corresponds to the PSS—PDS misalignment described by the deviation ΔX_0 if the cause B is just one of regular risk causes (Fig. 2). When the cause B is recognized as a project-specific chronic system issue, it describes the PSS—PDS misalignment by deviation ΔX_A when both PSS and PDS are additionally disturbed due to its presence. This explains internal risk amplification of a standalone risk and the calibration difference [$\Delta X_A - \Delta X_0$]. As regular causes and chronic system issue are not told apart in the traditional PRM, this mechanism is overlooked ($\Delta X_A = \Delta X_0$).

One of explanations to such overlooking may be a fact that PRM results reviewed at final investment decision (FID) are derived before EPC contracts are awarded post-FID. Simply, preferred bidders and their strengths and weaknesses are unknown as yet. Any specific chronic system issues pertaining to EPC contractors that could additionally disturb any part of the PSS and PDS cannot be taken into account pre-FID being unknown then ($\Delta X_A = \Delta X_0$). As a result, standalone unamplified placeholder-type risks are considered (Figs. 1 and 2).

Knock-on Risk Interaction Mechanism. Sometimes an impact on quality objective of an occurred risk is overlooked being discovered with a substantial delay. As argued above, engineering errors and omissions might be discovered as late as in commissioning and start-up phase exacerbating regular risks. (By this reason, PRM practitioners call this final project phase "a banquet of consequences" [1, 3].) As a result, poor quality induced by the first risk becomes a chronic system issue for the second risk (cause E of Fig. 4.) At fact, after this the same internal risk amplification mechanism (Fig. 3) kicks in as Fig. 4 shows. Apparently, in this case the chronic system issue stems specifically from a flawed engineering quality assurance/ quality control (QA/ QC) discipline. If this chronic system issue does not exist, which corresponds to robust QA/ QC, these two risks become de-coupled and independent.

Risk Compounding Mechanism. Finally, true project practitioners' nightmare looms when two risks that seemed independent occur simultaneously resulting in much higher consequences in comparison with a case if they occurred independently. Mechanism of combining of two risks is shown by Fig. 5. When the two risks do not occur simultaneously, we get two decoupled modified bowtie diagrams (Fig. 2). To make them happen simultaneously, there should be a sort of common cause (cause D of Fig. 5) that have a capacity to trigger both of them. This common cause might or might not be a regular risk's cause for either of the risks. (Fig. 5 shows a case when the cause D is not a regular cause, just a trigger.) When these two risks do occur simultaneously, they mutually disturb each others' reworks and work constraint thanks to the common cause. This results in additional PSS—PDS disturbances.

Putting aside less likely origins of the compounding common causes, it is believed that they are chronic system issues residing in planning and scheduling. Indeed, a thoughtful amendment of a project schedule may prevent simultaneous risk's occurrences when pertaining risks prone to compounding are separated in time and possibly in space, which corresponds to decapacitating of the compounding common causes D as the compounding trigger (Fig. 5).

In terms of mathematical logic, introduction of the three bowtie diagrams for risk interactions instigated by underlying chronic system issues symbolizes activation of AND-logic pertaining to FTA [1] along with the regular OR-logic. As soon as all elements of AND-logic are removed, the discussed bowtie diagrams get reduced to the traditional ones (Figs. 1 and 2).

Overall Project Risk Exposure Due to Multiple Risk Interactions. The introduced bowtie diagrams for the three types of risk interactions are generic building blocks for overall project risk exposure as one risk may participate in several instances of risk interactions. Namely, each risk might be internally amplified (Fig. 3) before it takes part in several knock-on (Fig. 4) and compounding (Fig. 5) interactions with other risks. This requires development of corresponding risk interaction mapping to navigate the actual risk exposure which includes interactions.

Certainly, drawing a synthesized bowtie diagram that includes all generic bowtie diagrams of all instances of risk interactions (Figs. 3, 4 and 5) could be quite onerous chore although still possible. Creating a table with significant risk interactions would suffice. (Utilizing the risk mapping (risks mapped to particular tasks) used in the traditional Monte Carlo modelling should help out too [1, 3].)

Such table in the case of the traditional PRM is trivial. This is just a regular risk register used for linear Monte Carlo modelling [1–3]. The challenge is not only to develop an upgraded risk interaction table (aka nonlinear Monte Carlo risk register) but to assign particular mathematical expressions to each interaction instance. This can be done based on qualitative knowledge of risk interaction mechanisms and their bowtie diagrams (Figs. 3, 4 and 5) as argued in the next section.

3.7 Abandoning the Traditional PRM's Linear Case: Nonlinear Monte Carlo Methodology

The traditional (linear) Monte Carlo modelling may be considered a substitute to linear regression analysis to generate and analyse overall project risk exposure R. Then the contribution of N identified standalone noninteracting risks R_1, R_2, R_3,…R_N (called predictors [1, 3]) could be represented through their additive linear contributions

$$R = R_1 + R_2 + R_3 + \cdots + R_N, \tag{2}$$

The words "standalone" and "noninteracting" above are reflected in formular (1) by absence of cross-predictor nonlinear contributions such as $R_1 \times R_1$, $R_1 \times R_2$, $R_1 \times R_3$, $R_2 \times R_3$, etc.

As argued in [1, 3] and recognized in statistics, the linear regression formula (2) is a limiting case and simplification of more general nonlinear regression formular that should contain nonlinear contributions proportional to $R_i \times R_j$. This means that using of the simplified linear formular (2) must be justified. Using of the linear formula (2) presumes implicitly that any nonlinearity contributions can be ignored or neglected due to the nature of a studied system. It could be easily figured out that this assumption is true in the case of simple projects even without internal risk amplifications. Meantime, in a general case a nonlinear formula shall be used which reads

$$R = R_1 \times M_1 + R_2 \times M_2 + \cdots + R_N \times M_N \tag{3}$$

where M_i are nonlinearity multipliers coupling a given risk with all the other risks as well as with itself. For example, for risk R_1 it takes a form

$$M_1 = 1 + \beta_{11} \times R_1 + \beta_{12} \times R_2 + \cdots + \beta_{1N} \times R_N \tag{4}$$

The introduced nonlinearity parameters β_{ij} could be called "coupling parameters" in all cases when $i \neq j$ as they couple risks R_i and R_j. However, when $i = j$ and a given risk R_i interacts with itself the term "coupling" cannot be used. For example, when $i = j = 1$ risk R_1 of formula (2) should be replaced as follows.

$$R_1 \rightarrow R_1 \times (1 + \beta_{11} \times R_1). \tag{5}$$

Getting back to types of risk interactions, formula (5) represents nothing else but the internal risk amplification of risk R_1. In terms of formula (1), nonlinear parameter β_{11} reflects on deviations ΔX_A which differ from ΔX_0. As in the traditional PRM, standalone internally amplified risks (contributing to ΔX_A) and standalone nonamplified risks (contributing to ΔX_0) are not told apart, this means $\Delta X_A = \Delta X_0$ and $\beta_{11} = 0$.

If a project system supports only standalone internally amplified risks (a simple project), formula (5) can be used to replace all the risks of the linear formula (2) to its nonlinear substitutes. Accordingly, an existing traditional linear Monte Carlo model could be modified using formula (5). Namely, project risks should be re-calibrated to reflect on $\Delta X_A - \Delta X_0 \neq 0$ as $\beta_{ii} \neq 0$ due to recognizing of chronic system issues (Fig. 3). The nonlinearity parameters $\beta_{ii} \neq 0$ for particular risks R_i shall reflect on power of the pertaining chronic system issues. As a matter of fact, this approach was used initially when developing an early version of the nonlinear Monte Carlo methodology: only internal risk amplifications were considered.

How about the other two types of risk interactions—knock-on and compounding?

As discussed in [1, 3] the same logic can be used to specify nonlinearity parameters as in the case of the internal risk amplifications. Bowtie diagrams for these two interactions (Figs. 4 and 5) shall be used keeping in mind corresponding chronic system issues belonging to QA/QC and planning/scheduling functions, correspondingly. Certainly, in the case of these two cross-risk interactions formula (5) shall be replaced: only cross-risk coupling parameters when $i \neq j$ can be used as opposed to the symmetric case $i = j$ relating to the internal risk amplifications.

As shown in [1, 3], in a case of two risks R_1 and R_2 ($i = 1$ and $j = 2$) that take part in an instance of a knock-on interaction the following replacement shall be carried out:

$$R_2 \rightarrow R_2 \times \left(1 + \beta^{K}{}_{21} \times R_1\right), \tag{6}$$

where an earlier risk R_1 impacts down-the-line risk R_2 which is described by the knock-on nonlinearity parameter $\beta^K{}_{21}$ coupling the two risks. (As pointed out above, this parameter would likely refer to some particular QA/QC chronic system issues.)

As shown in [1, 3], in a case of two compounded risks R_1 and R_2 the following replacement is required

$$R_1 \rightarrow R_1 \times \left(1 + \beta^C{}_{12} \times R_2\right) \tag{7}$$

$$R_2 \rightarrow R_2 \times \left(1 + \beta^C{}_{21} \times R_1\right) \tag{8}$$

where cross-risk compounding nonlinearity parameters $\beta^C{}_{12}$ and $\beta^C{}_{21}$ couple the two risks. It would be reasonable to assume that $\beta^C{}_{12} = \beta^C{}_{21} = ½\, \beta^C$ as the two risks R_1 and R_2 impact each other simultaneously in a reciprocal way. (As argued above, in the case of risk compounding some particular planning/scheduling chronic system issues could stand behind. Those should be assessed to evaluate parameters $\beta^C{}_{..}$)

Similarly to formula (5), formulas (6), (7) and (8) could be used to modify an existing linear Monte Carlo model by replacing standalone noninteracting risks by their interacting analogues in the amended nonlinear Monte Carlo risk register. Formally, this model should become a nonlinear Monte Carlo model that includes all pertaining risk interactions with yet unassigned parameters β_{ij}. Certainly, before this model could be practically used, the involved risk interactions shall be specifically evaluated, both qualitatively and then quantitatively. The evaluations shall be carried out for two cases: before and after addressing both risks and their interactions in accordance with Figs. 3, 4 and 5.

First, an audit of involved project disciplines that may bear relevant chronic system issues standing behind instances of identified interactions should be undertaken. The chronic system issue's list shall be formulated. Second, qualitative assessments of these chronic project system issues should be undertaken. Finally, quantitative assessments of corresponding nonlinearity parameters β_{ii}, $\beta^K{}_{ij}$, $\beta^C{}_{ij}$ attached to corresponding chronic system issues shall be carried out keeping in mind their addressing. The steps of this approach are discussed in further detail below (Sect. 3.8) including a simplistic nonlinear Monte Carlo modelling case (Sect. 3.9).

3.8 Shaping Nonlinear Monte Carlo Methodology

Dynamic Risk Pattern. Figure 6 demonstrates a sample dynamic risk pattern showing risk interaction mapping that introduces cross-risk interactions among six sample project risks R_1–R_6. (A post-FID project schedule consists of just two tasks for demo purposes.) Where applicable these risks are considered already internally amplified. Apparently, the dynamic risk pattern is based on the traditional risk mapping used by linear Monte Carlo modelling [1–3]. Corresponding cross-risk

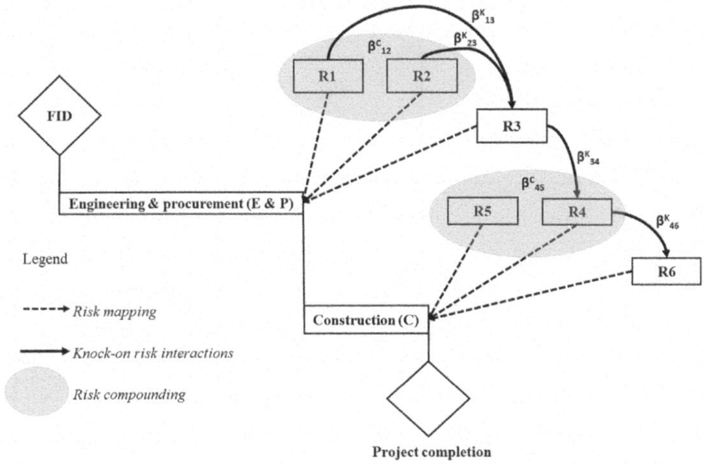

Fig. 6 Example of a dynamic risk pattern for project execution: cross-risk interaction mapping

coupling nonlinearity parameters $\beta^K{}_{ij}$, $\beta^C{}_{ij}$ (formulas (6), (7), (8)) can be assigned to each instance of cross-risk interactions, as Fig. 6 shows.

A system issue assessment matrix (SIAM). Figure 7 introduces system issue assessment matrix (SIAM) [1] to assess pertaining chronic system issues standing behind the involved risk interactions of the three types. The SIAM can be used when assessing chronic system issues either at the discipline level (less specific although still practical) or at individual risk level (more specific although more onerous) through audit and qualitative assessment of applied procedures and consistency of their implementation. (These two factors describe a project PDS as mentioned in Sect. 3.4.) In other words, the SIAM assesses maturity of the PDS at the discipline level or risk level. (Although, as discussed below, when all system issues are roughly considered to belong to the same category characterizing the entire project at large, that would correspond to rough PDS maturity assessment at the project level.)

To a certain degree the SIAM that actually utilizes a scoring method resembles a risk assessment matrix (RAM) used by the traditional PRM. (Certainly, corporate audit procedures shall clearly describe each score category of the 3 × 3 SIAM for typical projects a company works on.) Both matrices—RAM and SIAM—have comparable (i.e., low) accuracies. The purpose of the SIAM is to attach a score-based qualitative issue assessment category—minor, important or critical—to each pertaining chronic system issue standing behind each risk interaction.

After undertaking qualitative assessments using the SIAM, more precise quantitative assessments of involved nonlinearity parameters β_{ii}, $\beta^K{}_{ij}$, $\beta^C{}_{ij}$ shall be carried out that could be used in the nonlinear Monte Carlo model formally developed when using formulas (5)–(8) for each project risk and each instance of risk interactions. However, bridging the gap between the qualitative and quantitative assessments of risk interactions and their chronic system issues, some benchmarking indicators for

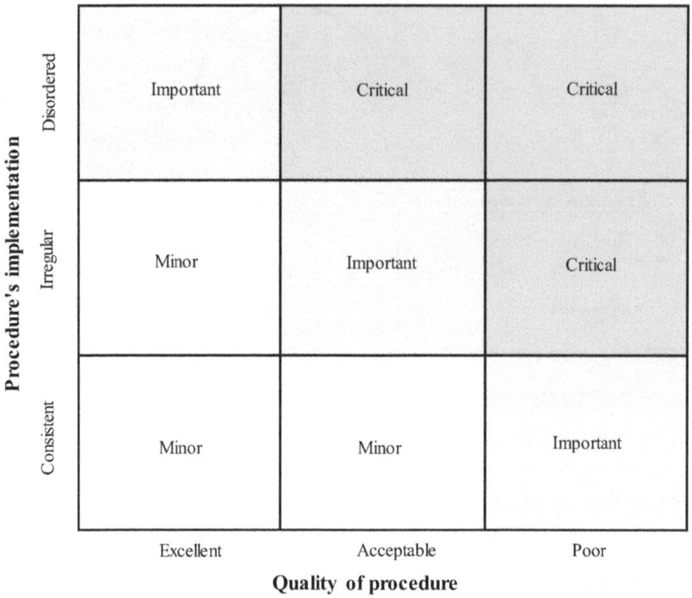

Fig. 7 A systems issue assessment matrix (SIAM)

the entire project shall be introduced keeping in mind results of the linear Monte Carlo modelling.

Project-Level Calibration. What if all identified chronic system issues are critical, how much bigger a project risk contingency due to corresponding all-critical risk interactions is expected to be? If all the issues are important, what range shall risk contingency occupy in comparison to the linear case. Apparently, the linear and nonlinear Monte Carlo models shall produce the same contingencies by default when all interactions are ignored. However, if all the issues are minor, what additions to the linear risk contingency can be expected? The above questions allude to the risk appetite and risk tolerance concepts [1–3]. For instance, a project owner organization may decide that exceeding a linear Monte Carlo contingency for 25% should correspond to all critical issues. The range of 10–25% should correspond to all important issues. Exceeding for up to 10% should be produced by all minor issues. Accordingly, these ranges introduce two 10% and 25% dividers between these three ranges.

This approach does correspond to the project-level chronic system issue assessments when all issues are considered belonging to the same particular category. Despite such rough approach cannot be specific enough to practically assess issues at the discipline or risk level, it can be successfully used for calibration purposes to derive specific values of 10% and 25% dividers in terms of nonlinearity parameters β_{ij}.

Namely, the newly developed nonlinear Monte Carlo model can be obtained from the linear Monte Carlo model when using formulas (5)–(8). It keeps parameters β_{ii},

$\beta^K{}_{ij}$, $\beta^C{}_{ij}$ as placeholders. This nonlinear Monte Carlo model should be run several times as follows. A single project-level calibration parameter $\beta_R = \beta_{ii} = \beta^K{}_{ij} = \beta^C{}_{ij}$ should be used in every run. Of course, parameters β_R shall be different in every undertaken run to try all values from $\beta_R = 0$ (the linear Monte Carlo case) to quite high numbers β_R (all critical issues cases) to scan possible range of the parameter β_R and pin down the two dividers. (Actually, it is presumed that all $\beta_R < 1$, whereas in practice, $\beta_R \ll 1$ [1, 3].)

Comparison of contingencies produced in every run with the linear contingency ($\beta_R = 0$) shall allow pin down $\beta_R(10\%)$ and $\beta_R(25\%)$ dividers. (Certainly, the same confidence level shall be used for the comparison purposes as in the linear Monte Carlo case (e.g., P70)).

Issue Assessments at Discipline or Risk Levels. After this, all individual chronic system issues can be assessed semi-quantitatively (by belonging to one of three parameter's ranges) through assessments of their individual nonlinearity parameters β_{ii}, $\beta^K{}_{ij}$ or $\beta^C{}_{ij}$ at discipline or risk level. Namely, in accordance with the initial qualitative SIAM-based assessments, each individual nonlinearity parameter shall belong to a particular range as follows:

- Any critical issue should get its nonlinear qualitative parameter higher than $\beta_R(25\%)$ value;
- Any important issue should get its nonlinearity parameter from the range $[\beta_R(10\%); \beta_R(25\%)]$;
- Any minor issue should get its nonlinearity parameter from the range $[0; \beta_R(10\%)]$.

Moving from the qualitative assessments using SIAMs to the quantitative assessments of the chronic system issues through their nonlinearity parameters is comparable with transition from qualitative risk assessments using RAMs to assessments required as inputs to linear Monte Carlo models. It could also succumb to similar realizations of bias [1–3].

Finally, chronic system issues belonging to a particular category and their parameters belonging to corresponding ranges shall be compared with each other to come up with more specific β_{ii}, $\beta^K{}_{ij}$, $\beta^C{}_{ij}$ values attached to particular instances of risk interactions. In other words, a sort of hierarchy of system issues and their nonlinearity parameters inside the three ranges shall be mulled over to finalize their quantitative assessments.

Issue Assessments Before and After Addressing. The algorithm of conversion of linear Monte Carlo models to their nonlinear versions presumes two basic cases: before addressing and after addressing. Although, as opposed to the linear Monte Carlo modelling, addressing here means also addressing corresponding chronic system issues. Namely, bowtie diagrams for the three interaction's types (Figs. 3, 4 and 5) point to possibility of preventive addressing aimed at taming of the chronic system issues.

Improving corresponding project procedures and ensuring their more consistent implementation is a general recipe for the system issue's addressing. As discussed above, in the case of knock-on interactions addressing QA/QC system issues shall be in the primary focus; addressing risk compounding may require attention to planning and scheduling, etc. Chronic system issues responsible for internal risk amplifications may belong to any project discipline depending on risk context.

A heads-up warning should be appropriate here that addressing of some of the chronic system issues is not quick and easy, its results shall not be overoptimistic. To be realistic, some audit findings and collected lessons learned might be possibly used for actual improvements only in some future projects. Hence, the after-addressing assessments should not be optimistically biased.

At the same time, addressing chronic system issues has its own merit and it might be undertaken even if the nonlinear Monte Carlo modelling is not planned at all. In terms of project system's complexity, successful addressing of chronic system issues means reducing the project complexity through making risks less interactive. Complex projects might get characteristics of complicated ones and complicated projects might feel more like simple ones after the successful addressing of chronic system issues.

Schedule and Cost Nonlinearity Parameters. Finally, the conversion of the linear Monte Carlo models to its nonlinear analogues described in this section is done for a general case when schedule and cost risks and schedule and cost modules of the Monte Carlo models are not told apart. In practice, integrated schedule and cost risk analysis (SCRA) shall be used as a realization of the Monte Carlo methodology to consider schedule-driven costs. Accordingly, linear L-SCRA and nonlinear N-SCRA models shall be in the focus of these discussions. This immediately results in two separate sets of nonlinearity parameters in place of the generic parameters β_{ij} used in schedule and cost modules of the nonlinear Monte Carlo model.

3.9 Nonlinear Monte Carlo Modelling Example

Figures 8 and 9 demonstrate results of L-SCRA and N-SCRA modelling for a case study project of a deterministic duration 1,310 days and a cost estimate of $1B [1, 3]. The former intakes six sample standalone noninteracting risks R_1–R_6. The latter handles the same although now interacting risks as shown by Fig. 6. All risks are considered internally amplified. Corresponding interactions are formalized through nonlinearity parameters using formulas (5)–(8). The N-SCRA addressing included chronic system issues' addressing along with regular addressing of standalone noninteracting risks from the linear Monte Carlo model.

This case study is fully described in [1, 3]. Provided distributions demonstrate the difference between project contingencies when risk interactions are not considered (the linear Monte Carlo case) and when they are (nonlinear Monte Carlo case). Table 1 summaries and compares results for these two modelling cases. Apparently, the N-SCRA yields significantly higher contingencies. This difference is even more substantial for unshown before-addressing case.

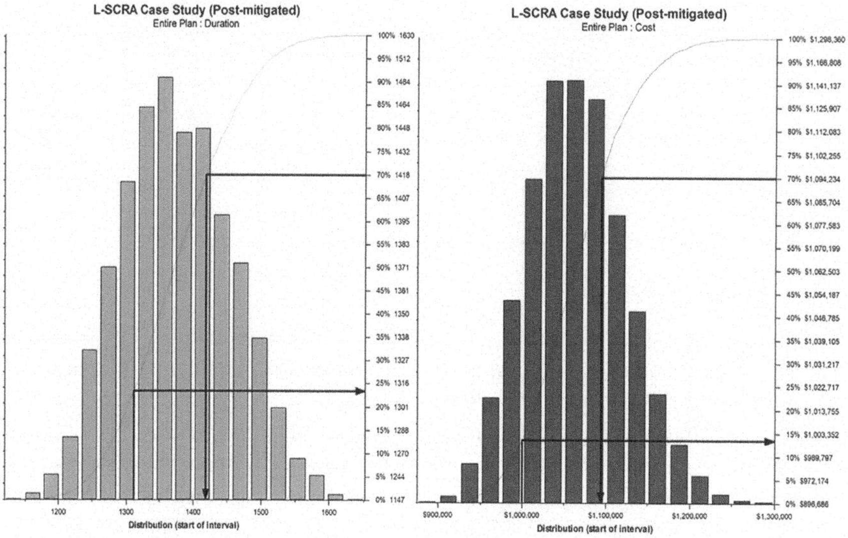

Fig. 8 Results of L-SCRA modelling: **a)** a duration distribution; **b)** a cost distribution

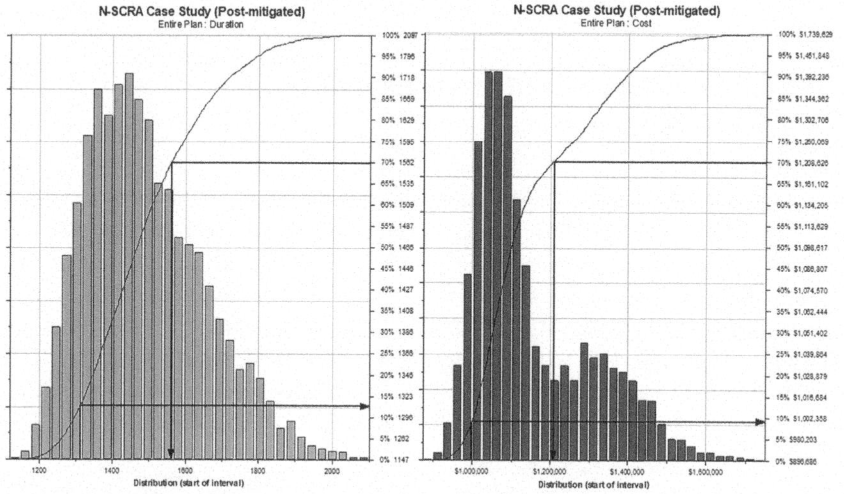

Fig. 9 Results of N-SCRA modelling: **a)** a duration distribution; **b)** a cost distribution

Figure 9b for the cost distribution should be a reason for concern. Namely, appearance of the second peak on the cost distribution is a sign that the project is prone to falling to chaos. Indeed, a single or small group of risks become dominant after taking part in several risk interactions as resulting consequences become overwhelming. As shown in [1, 3], two risks R_4 (construction quality) and R_6 (commissioning and

Table 1 Comparison of L-SCRA and N-SCRA contingencies at P70 confidence level

	Deterministic	Monte Carlo P70 contingency (after addressing)				
		L-SCRA	N-SCRA	L-SCRA, %	N-SCRA, %	N-SCRA above L-SCRA case, %
Duration, d	1310	1418	1562	8.2%	19.2%	10.2%
Cost, $'000	1,000,000	1,094,234	1,208,626	9.4%	20.9%	10.5%

start-up) are responsible for the second peak due to their interactions (Fig. 6): when these two risks are excluded, the second peak disappears. Possible participation of the commissioning and start-up risk R_6 in creating of the second chaotization peak does not contradict to the jargoned name ("a banquet of consequences") of the final project phase. If the second peak gets associated with a confidence level above the adopted criterion (P70 in this case), the project would end up in a chaotic state with probability 70% or higher [1, 3]. (Actually, that was exactly what happened in the before-addressing case that is not considered here.) Proper addressing of pertaining interacting risks and chronic issues standing behind them is a way to reduce the chaotization likelihood.

4 Deep Uncertainties and Monitor-and-Adapt Decision-Making

The traditional PRM handles uncertainties stemming from the PSS—PDS misalignment only. Moreover, the PSS is often considered already preconceived, selected and frozen so that only the PDS is looked at as a loose end. The EES is also considered frozen by then being predicted somehow. This implies that the selected PSS is fully aligned with the EES. Hence, a possible EES—PSS misalignment is either does not exist or rather ignored. This is quite cushy vision of the project world. First, it reflects on the traditional project development work process when in early project phases the EES is fixed (somehow) for the entire project lifespan including operations (Fig. 10). Second, a matching PSS is selected (somehow) with an intent to fully eliminate not only EES uncertainties but possible EES—PSS misalignment too. Finally, a matching PDS is shaped with an intent to ensure flawless delivery of the project and get rid of PSS—PDS misalignment as much as possible. These efforts get baselined during the final investment decision (FID) followed by execution phases (Fig. 10).

In terms of decision-making such project work process is fully aligned with so-called "predict-then-act" decision-making: each consecutive project phase is based on decisions ("predictions") made in a previous phase to be implemented ("act") without looking back. This approach does make some managerial sense as it allows to get a project team focused on future activities and goals not on previous

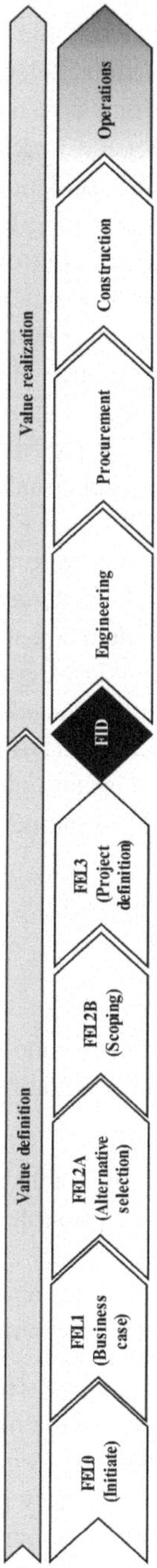

Fig. 10 A traditional project work process for predict-then-act decision-making

ones. Although this otherwise robust focused-activity-focused-result approach drags several pitfalls behind it.

In the traditional PRM, selecting of the workable ESS is undertaken by corporate "oracles" and communicated to project teams; selection of a most optimal project alternatives (i.e., PSS freezing) is not often consistent and it could be based on subjective, untransparent and feeble decision-making criteria. As a result, there are two possible outcomes. First, the EES selection could be made wrong from the get-go. This in turn would result in the wrong PSS inevitably. Second, a more common outcome is that the external project environment becomes evolving upon project completion. Namely, the initial rightly predicted EES and its future state start evolving and disappearing being replaced by a new EES state some time in operations. The first scenario accentuates importance of initial deep understanding and proper handling of the EES. A consistent selection of project alternatives is required as a seamless transition form the EES to PSS: the EES—PSS alignment seems to be a weak point in the traditional PRM. The second one is about monitoring the EES in operations until project decommissioning.

To address these above-listed issues a new more comprehensive uncertainty taxonomy was introduced in [1, 4] (Sect. 4.1). A methodology for consistent selection of project alternatives based on robust decision-making criteria of the mathematical game theory [6] is introduced (Sect. 4.2). A comprehensive new project work process that complies with the new monitor-and-adapt decision-making is shaped (Sect. 4.3). The dynamic adaptive methodology (DAM) to handle deep uncertainties is introduced in Sect. 4.4. A simple business case devoted to modelling aspects of external project environment and the DAM implementation is discussed in Sect. 4.5.

4.1 New Uncertainty Taxonomy

Initially, a taxonomy of uncertainties going beyond the traditional PRM was introduced in 1950s in [6]. Three decision-making situations were told apart in the context of mathematical game theory as follows.

- Certainty: a decision is known to lead invariably to a specific outcome;
- Risk: a decision leads to one of a set of possible outcomes when each outcome can occur with known probability;
- Uncertainty: a decision has as its consequence a set of possible specific outcomes although probabilities of these outcomes are completely unknown.

In context of projects, the certainty category would point to project objectives and baselines; the risk category would be about general uncertainties and uncertain events of the traditional PRM; the uncertainty category seemed to refer to states of the external project environment which was largely ignored by the traditional PRM. Recently this three-part taxonomy was further elaborated in [4]. After minute conditioning in [1], it was finalized for the project management context by introducing seven uncertainty categories or levels L0–L6 [1]. These levels are introduced based on future EES and PSS states as follows.

- Uncertainty level zero (L0): single fully defined (fixed) EES and PSS futures;
- Uncertainty level one (L1): single modified fully defined (fixed) EES and PSS futures;
- Uncertainty level two (L2): single defined (fixed) EES future and a single PSS future with minimal variations;
- Uncertainty level three (L3): a single defined (fixed) EES future and a single PSS future with probabilistic variations;
- Uncertainty level four (L4): a single defined (fixed) EES future and several known PSS futures to choose from;
- Uncertainty level five (L5): multiple known EES and possible PSS futures;
- Uncertainty level six (L6): completely unknown EES and PSS futures.

Levels L0 corresponds to project objectives and baselines in project environment; level L1 also considers some legitimate allowances [1]. Both L0 and L1 levels belong to "full determinism" category (previous certainty category) when both ends of the EES—PSS link are fully defined and fixed. Levels L2 and L3 that previously belonged to risk category now dubbed as "shallow uncertainties" [1] as they are based on a single known EES future and a single known PSS future (i.e., on already selected project alternative) but with variations, the same as in the traditional PRM. The EES side of the EES—PSS link is fixed; the PSS side of the link is uncertain "shallowly". Levels L5 and L6 are referred to as "deep uncertainties" [1, 4] as they presume multiple EES futures that are believed to be either known (L5 uncertainties) or unknown (L6 uncertainties). Apparently, in both cases, both sides of the EES—PSS link are loose and uncertain "deeply".

The remaining L4 uncertainty belongs to in between deep and shallow ones. Indeed, a fixed EES future allows several PSS futures (or rather several project alternatives). These latter are considered known being generated to choose from. By this reason, this L4 uncertainty is called "deep-to-shallow" uncertainty in [1].

All in all, the traditional PRM handles only full determinism (L0 and L1) and shallow uncertainties (L2 and L3). Deep (L5 and L6) and deep-to-shallow (L4) uncertainties are outside of the traditional PRM. Meantime, these missed uncertainties are cardinally important for proper understanding of the external environment future states (the EES), selection of a workable one to get fixed and used for alternatives selection (the PSS) to minimize possible EES—PSS misalignment. Methodologies to handle deep (L5 and L6) and deep-to-shallow (L4) uncertainties that are overlooked by the traditional PRM are introduced in Sects. 4.2 and 4.3. These presume the change of the decision-making to monitor-and-adapt along the way.

4.2 Handling Deep-to-Shallow Uncertainties: Risk-based and Economic-based Alternative's Selection (RBEBAS)

This section describes an attempt to branch out of the traditional PRM that handles full determinism (L0, L1) and shallow L2 and L3 uncertainties to consistently deal with deep-to-shallow L4 uncertainties. In terms of the project work process (Fig. 10), alternatives shall be selected in FEL2A phase right after a business case developed in FEL1 phase gets finalized. Presuming that a workable EES future state is predicted somehow and fixed in FEL1 phase, a next challenge is to select a most optimal project alternative to shape the PSS that answers to the selected EES. Presumably, such selection process shall keep in mind maximum possible economic and other benefits anticipated by project stakeholders. As such this should also presume minimization of possible risks. Risk-based and economic-based alternative's selection (RBEBAS) methodology was introduced in [2] which was developed for testing the waters outside of the traditional PRM. This was undertaken without keeping in mind deep L5 and L6 uncertainties in external project environment that were considered somehow fully resolved by the FEL2A phase.

The RBEBAS methodology that can be considered a bridge from a project business case to actual project development requires thoughtful selecting of an optimal or preferable alternative to work on. This presumes generating of viable project alternatives first as combinations of feasible location's, technology's, produced product's slates', execution strategy's, etc. options. Each shortlisted alternative shall be treated as a standalone potential project and evaluated against previously identified business risks. Each considered alternative must answer to the preselected external environment future to minimize EES—PSS misalignments. This selection implies one-to-many relations between a fixed external environment future and several generated alternatives to choose from.

First, it makes sense to keep only alternatives that ensure acceptable (e.g., low or "green") risks after their viable addressing. In mathematical terms, this shall be a necessary condition. Despite this is important for project stakeholders, they should be even more interested in specific economic and other benefits. Second, by this reason some economic indicators (e.g., internal rate of return, return on investment, return on capital, etc.) should be employed. As most of the usual economic indicators are derivatives of net present value (NPV) of project cash flow (CF), the NPV CF evaluations for each shortlisted alternative survived after consideration of risk exposures can be understood as a sufficient condition for selection of a preferable alternative.

These two selected necessary and sufficient conditions can be reformulated in the language of mathematical game theory [6]:

- Minimax—minimization of maximum possible losses;
- Maximin—maximization of minimal possible benefits.

In simple terms, an alternative with acceptably low risk exposure after addressing and best NPV CF could be selected. Once again, the RBEBAS methodology unconditionally presumes that project business case somehow predicts external environment future underlying any considered alternative (shallow-to-deep uncertainty L4). What if the EES is wrongly predicted from the get-go or it becomes evolving in operations thanks to deep L5 and L6 uncertainties?

4.3 Handling Deep Uncertainties: Monitor-and-Adapt Decision-Making

If the future state of the EES gets wrongly predicted from the get-go, results of the RBEBAS methodology become irrelevant. A selected project alternative would become a bridge to nowhere as it will be out of touch with external environment reality. Similarly, results of the RBEBAS that may be adequate initially could end up in a situation when a selected alternative gets out of touch with reality because the ESS has been evolving significantly enough in operations.

These two scenarios nudge to revise decision-making to ensure that the external environment is kept an eye on during entire project lifespan including operations. The new approach—a shift from the traditional predict-then-act to monitor-and-adapt decision-making—was proposed in [4]. Several methodologies supporting this were also reviewed in [4].

According to the traditional approach, the uncertainty gets reduced to full determinism by the project completion when all project actuals become known. Every previous project phase “predicts” what shall be acted on in the next phase. For instance, a predicted in FEL1 phase state of external project environment which is used for selection of a project alternative in FEL2A phase is never got back to. Meantime, according to the monitor-and-adapt decision-making, overall uncertainty might start growing again because of evolving external project environment. It might get to deep uncertainty levels some time in operations being a moving target. (Sometimes this might happen even before project completion resulting in so-called change orders.) By this reason, monitoring, learning and adapting aspects become compelling during the entire project lifespan.

A best-of-breed dynamic adaptive methodology (DAM) developed in [1] is reviewed in Sect. 4.4 which fully answers PRM context of capital projects in private sector during entire project lifespans. It fully supports the monitor-and-adapt decision-making and integrates various PRM methodologies—from RBEBAS to Monte Carlo (either linear or nonlinear, depending on project complexity)—as fragments in particular project phases. Figure 11 compares the two decision-making approaches in terms of perceived levels of uncertainties L0–L6 in various pre-FID and post-FID phases. Seemingly, these two fully coincide prior to FEL2 phase when the RBEBAS is undertaken. Although, as argued in Sect. 4.4, methodologically these two decision-making approaches do differ even in the early project phases thanks to the featured DAM's process steps as discussed in the next section.

4.4 Handling Deep Uncertainties: Dynamic Adaptive Methodology (DAM)

In early project phases FEL0 and FEL1 (Fig. 10) the DAM allows to meaningfully map future states of the external project environment and make a robust decision on selection of its workable future state. This corresponds to initial elimination of the deep L5 and L6 uncertainties and avoiding of wrong predictions as much as possible. Upon project completion, the DAM allows monitor the external environment a way that helps out to pin down conditions when a project alternative selected in FEL2A and later realized by project completion becomes more and more irrelevant and considerably erodes project value in operations.

In the light of the discussion in Sect. 4.2, the RBEBAS methodology should become a part of the DAM process as soon as the deep uncertainties get resolved initially. Although some additional steps and conditions discussed below shall be applied in the broader DAM context. First, the DAM prescribes employing of advanced modelling techniques—exploratory modelling (EM) and scenario development (SD) [1, 4]—to predict external environment futures in place of the more intuitive approach. This is aimed at reducing the threat of wrong predicting of the external environment futures. Second, along with the minimax and maximin criteria used by the RBEBAS methodology, selection of an adequate EES future states calls for additional game theory criteria. As a matter of fact, the two RBEBAS criteria are called satisficing ones in the game theory [6]: they are used for comparison of decisions in terms of established performance thresholds. They are aimed at resolving deep-to-shallow L4 uncertainty in FEL2A phase.

Third, when dealing with deep L5 and L6 uncertainties in FEL0 and FEL1 phases, assessment of future states requires a totally different approach based on so-called efficiency criteria. Namely, when selecting a most plausible EES future state further used in FEL2A phase for RBEBAS, one should be sure that this state occupies quite broad domain in the overall future state's space. This is a sort of warranty that the selected future state is stable and not too sensitive to change in the external environment. By this reason an applicable efficiency criterion is called domain criterion [6].

A fundamental need for the domain criterion stems from the fact that deep L5 and L6 uncertainties cannot be characterised by probabilities of their realizations [1, 4] as opposed to shallow L2 and L3 uncertainties. This does not allow any ranking of external environment futures based on probabilities of their realizations which are simply unapplicable and irrelevant (Sect. 4.1). Hence, some substitute criteria are required. The classical efficiency (or domain) criterion is aimed at sizing of a domain occupied by a given future state in the overall future state's space.

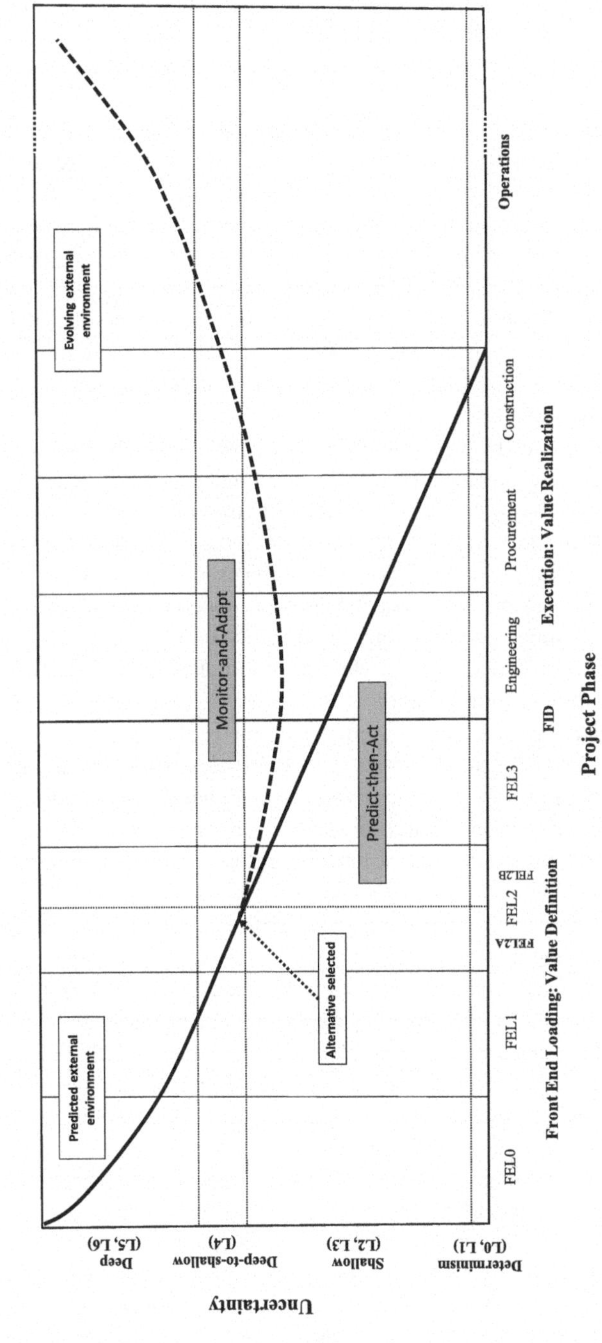

Fig. 11 Perception of project uncertainty for predict-then-act (solid line) and monitor-and-adapt (dotted line) decision-making paradigms

(Both the overall space and domains occupied by states could be defined mathematically through pertaining modelling parameters during EM and SD modelling, which utilizes systems dynamics methodology [1, 5] (Sect. 4.5).) The larger the occupied domain, the more stable a state is regarded in terms of its insensitivity to the external environment changes.

Although, this satisficing domain criterion is a snapshot related to a given moment in time as the occupied space might shrink down and even collapse over time. This means that a future state initially meeting the domain criterion might become too sensitive to upcoming changes. Another state might emerge that would better comply with the domain criterion. In other words, the classical game theory's efficiency (domain) criterion misses important time aspects which nudges to introduce time-domain criterion in addition to the classical space-domain criterion [1]. Indeed, it is important to know if an initially selected future state that has complied with the space-domain criterion is long-lived or it would shrink down and disappear rather shortly after project completion.

According to the generic deep uncertainty theory developed in [4], both satisficing (minimax, maximin) and efficiency (domain) criteria shall be applicable for decision-making simultaneously. However, in the project environment, applicability of the efficiency and satisficing decision-making criteria are separated in time: the efficiency criteria (both space and time domain criteria) shall be used in early FEL0 and FEL1 project phases when working with deep L5 and L6 uncertainties in the ESS; the satisficing criteria (minimax and maximin) shall be used after this in FEL2A phase when using RBEBAS methodology to select an optimal project alternative when handling deep-to-shallow L4 uncertainty.

It is interesting that the satisficing minimax criterion is also used implicitly when handling shallow L2 and L3 uncertainties in post-FEL2A phases as applicable to Monte Carlo modelling of contingencies. Namely, a higher decision-making confidence level for selection of project duration and budget contingencies directly corresponds to more strict satisficing performance threshold: higher confidence-level contingencies (e.g., P70–P90) cover more worse realizations of risks (Sect. 3.9); more aggressive confidence-level criteria (e.g., P30–P50) automatically presume weaker satisficing criterion.

In terms of mathematical logic, efficiency and satisficing criteria should be treated as necessary and sufficient decision-making conditions, correspondingly. Joint applicability of these criteria ensures so-called robust decision making (which is a formal term introduced in [4]). Apparently, application of the domain criteria first (in FEL0 and FEL1 phases), and satisficing criteria second (in FEL2A phase) is separated in time (i.e., belongs in different project phases).

Using EM and SD modelling supported by the efficiency criteria zeroes in on consistent predictions of the EES future states as opposed to usually subjective predictions featuring the traditional predict-then-act decision-making.

Practical realization of the DAM methodology as well as the underlying monitor-and-adapt decision-making could be undertaken by implementation of a DAM process. Its steps are fully described in [1] and not reviewed here. At higher level the DAM algorithm might be illustrated by dynamic adaptive project work process shown in Fig. 12 which is an upgrade from the traditional process (Fig. 10). The former does comply with monitor-and-adapt decision-making; the latter follows the traditional predict-then-act decision-making concept.

The initial part of the process of Fig. 12 resembles the traditional project work process of Fig. 10. Although, the major difference is development and implementation of a monitoring plan. It should have the following features and points.

First, the monitoring plan shall comprise results of EM and SD modelling when selecting an initial workable future state of the external environment which shall be used for RBEBAS in FEL2A phase. Although, along with selection of an initial future state a substitute state shall be identified that would potentially replace the initial state some time in operations due to changes in the external environment. Second, a substituting project alternative shall be selected based on the expected substitute future state too by using a second instance of the RBEBAS. However, in comparison with the classical or standalone RBEBAS methodology, its version that gets integrated in the DAM process comes with an additional feature called flexibility constraint. It imposes certain compatibility conditions on both initial and substituting project alternatives.

It might seem too premature or too farsighted to consider a new future ESS state and matching substitute project alternative way beforehand. However, there is a compelling reason to do this so early. Namely, there is a threat to get locked in the initial project alternative a way that it would be impossible or too costly to switch over to a substituting project alternative when time has come [1, 4]. Hence, the flexibility constraint imposes technical and other applicable conditions on both alternatives—the initial and the substituting. This anti-locking constraint shall be reflected in the monitoring plan.

For example, switching over to a new project alternative some time in operations might require availability of an additional parcel of land. If it is not reserved beforehand developing of the new alternative in a given location would not be viable resulting in unfortunate locking in the initial dying-away project alternative. Although, in some cases strong flexibility constraints are not that compelling as simple or unexpensive measures (e.g., modularization of equipment) could be applicable.

Third, the systems dynamics EM and SD modelling allows to formulate so-called adaptation signals (AS) and adaptation tipping points (ATP) [1, 4] that delineate validity envelops of future states in terms of mathematical parameters used in modelling (see, for example, Sect. 4.5). The former indicate early-warning when approaching to the validity envelope's borders; the latter delineate those borders parametrically. These findings should be a part of the monitoring plan too.

For example, products produced by a project in operations might occupy 50% market share initially; a first AS might be established at the level when market share drops to 45%; a second AS might point to 40%; an ATP could be established at the 35% level when restructuring must become compelling and should be triggered: switching over to a new substituting project alternative gets initiated. (Similarly,

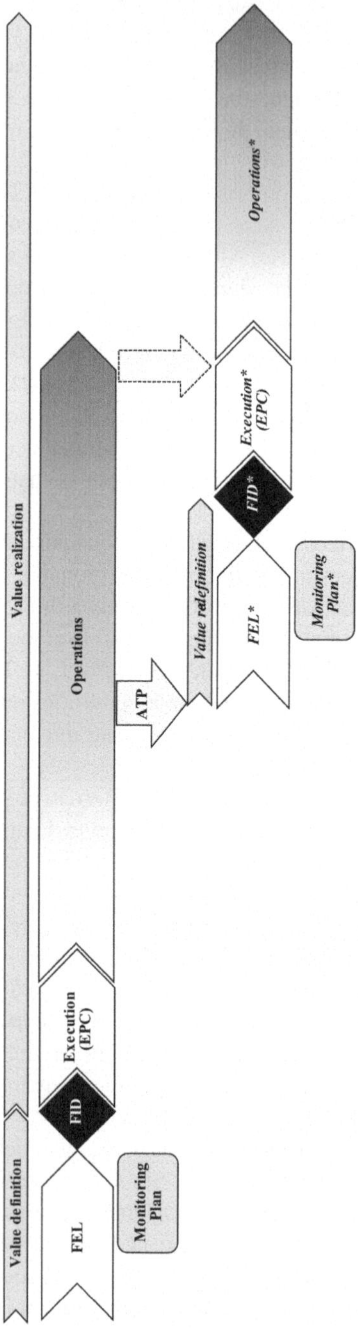

Fig. 12 A dynamic adaptive project work process for monitor-and-adapt decision-making

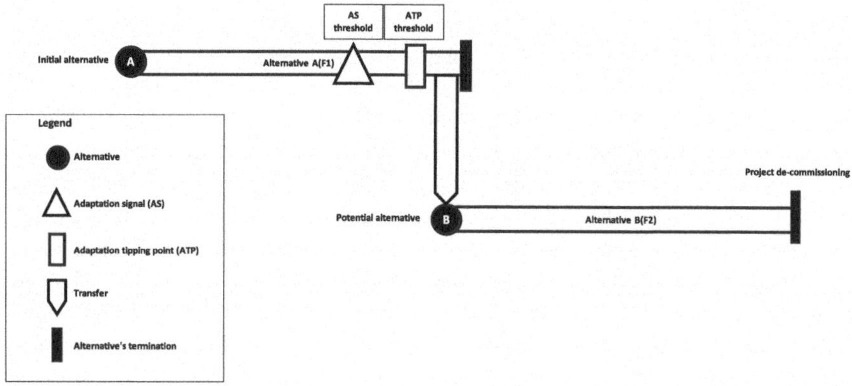

Fig. 13 "Subway map" chart for a case of two project alternatives

upside or favourable thresholds could be established for growing market shares above 50% when project expansion should be contemplated.)

Figure 12 shows just one such ATP for simplicity. After triggering the restructuring, the new instance of the project workflow shall be initiated; the new future state should be formulated based on EM and SD modelling results; a new project alternative shall be selected (or rather brushed up on) using the second instance of the RBEBAS in accordance with the new future state; the updated monitoring plan gets formulated, etc. After execution of the new alternative, operations begin working in a new mode as shown by Fig. 12. Whereas the work processes of Figs. 10 and 12 are applicable to projects in the private sector [1], their analogue in government environment is demonstrated in [4] which is called "a subway map" (Fig. 13). It shows two project alternatives A and B. The first one is based on the initial future state F1; the second answers to a new substituting future state F2. One AS and one ATP thresholds are exhibited.

Finally, the DAM process integrates all other applicable PRM processes—from the RBEBAS, to either linear or nonlinear Monte Carlo methodology (depending on project complexity). These processes get integrated to the DAM process as local fragments used in particular project phases. From this viewpoint, the DAM process could be declared a Great PRM Unifier [1].

4.5 DAM Process Modelling Example

This section demonstrates a simplistic systems dynamics modelling of external economic and marketing environment of a project that shall select a project alternative based on either ongoing dominance of a traditional product A or its new substitute B. (Actually, this case is mostly devoted to initial steps of the DAM process to illustrate

EM and SD modelling.). As the project presumes 30-year operations upon completion, its external environment shall be adequately investigated to select a right future state workable for selection of a right project alternative initially. Conditions for switching over to the other alternative shall be also investigated.

The traditional product A might be understood as internal-combustion-engine cars, power generation using hydrocarbons, pipeline gas supply, etc. In turn, the alternative new product B might be thought as electric vehicles, renewable energy generation, LNG gas supply, etc. Besides the classical supply—demand implications two non-economic and non-market factors are considered referred to as trade wars and new regulations. Presumably, trade wars stemming from geo-political dynamics might impact supply of both products due to tariffs, shortage of required parts, sanctions, etc. The new regulations reflecting on new social trends influenced by governments might suppress demand for the traditional products and encourage demand for the new ones.

A simplistic model developed in one of available systems dynamics modelling software packages is introduced in [1] as Fig. 14 shows.

The EM and SD modelling based on supply–demand intelligence data for both products as well as on multiple scenarios related to timing and intensity of trade wars and on various scenarios of adopted regulations allows find a pattern in market share behaviour. Results of one of modelling runs are shown in Fig. 15.

Zone A (or domain A) points to dominance of product A for the first half of the project lifespan. Zone B (or domain B) represents grows of market share of product B and decline of product A. (It would be also worthwhile to point out that the total market share for both products will be growing over time, which is not further discussed here.)

In terms of future states of external economic, market, geo-political, social, regulatory, etc. environment, zone A epitomizes the initial future state that could justify development of a project devoted to the traditional product A. In accordance with Fig. 15, switchover to the new product B might occur some time around or after the middle of the project's lifespan. From both space-domain and time-domain criteria, a project alternative devoted to the project A is preferable. (Satisficing minimax and maximin criteria used in RBEBAS are not considered here.) All applicable flexibility constraints shall be identified and considered to allow potential future switchover to project B in future. A corresponding monitoring plan should include required adaptation signals (e.g., a product A market share drops below 70%) and an adaptation tipping point (e.g., a product A market share drops below 60%). As systems dynamics modelling (EM and SD) is not supposed and cannot predict the exact timing of switchover from zone A to zone B, the DAM process should be in place during entire project lifespan to comply with the monitor-and-adapt decision making concept.

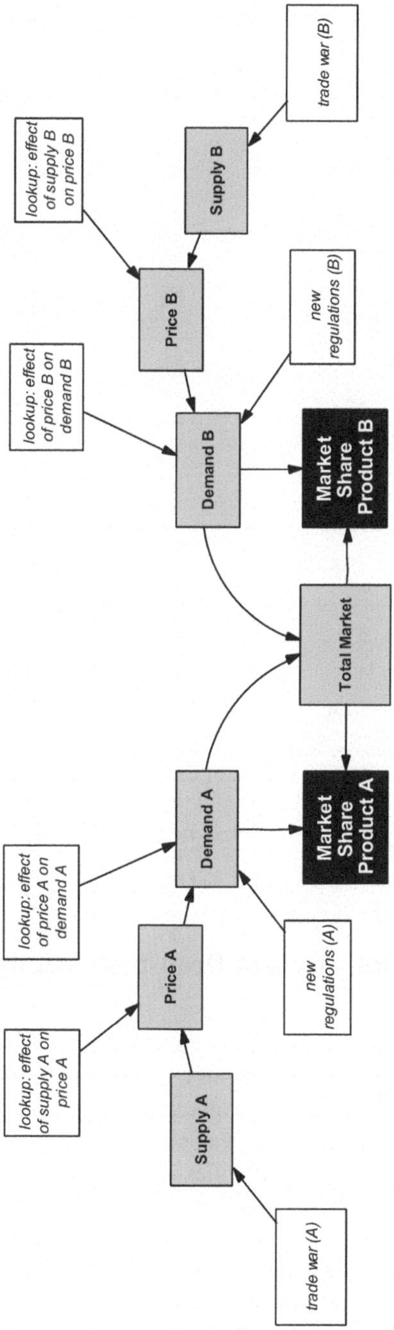

Fig. 14 A systems dynamics model for market shares of products A and B

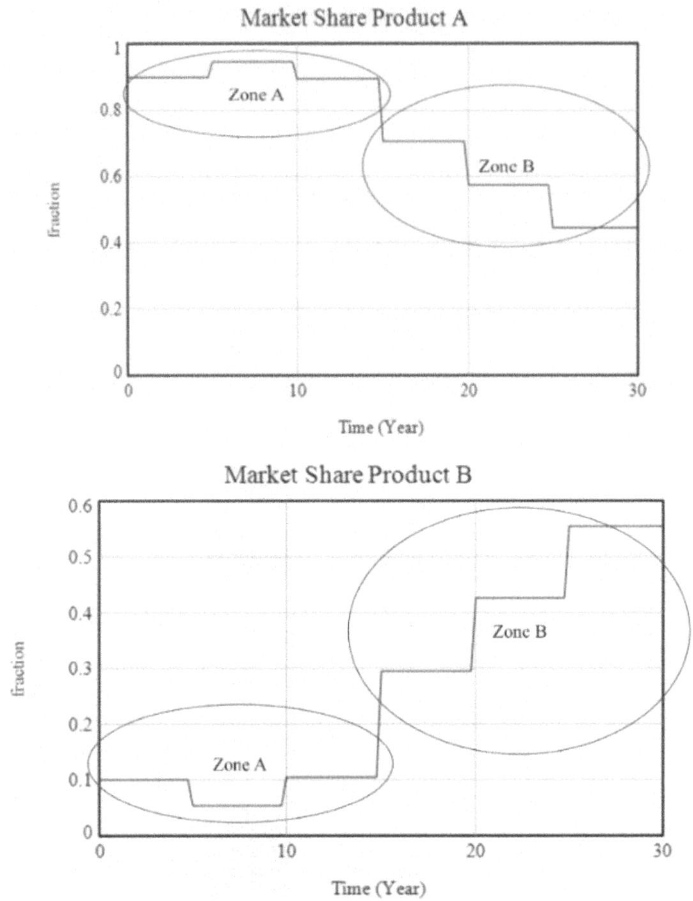

Fig. 15 Expected market shares of the products A and B

5 Uniting High-Complexity and Deep-Uncertainty Paradigms

So far, the discussion of the two new paradigms related to implications of high project complexity and deep uncertainty has been undertaken independently. The complexity concept and the nonlinear Monte Carlo methodology have been introduced (Sect. 3) followed by introduction of deep uncertainties and the dynamic adaptive methodology (DAM) in Sect. 4. However, one might guess that more complex projects must tend to be easier exposed to deep uncertainties. Certainly, situations when a complex project is not much exposed to its external environment or when a simple project is are thinkable. Although such cases might be treated as rather exotic.

Figure 16 provides some additional food for thought by pointing to areas of applicability of nonlinear Monte Carlo and dynamic adaptive methodologies. This

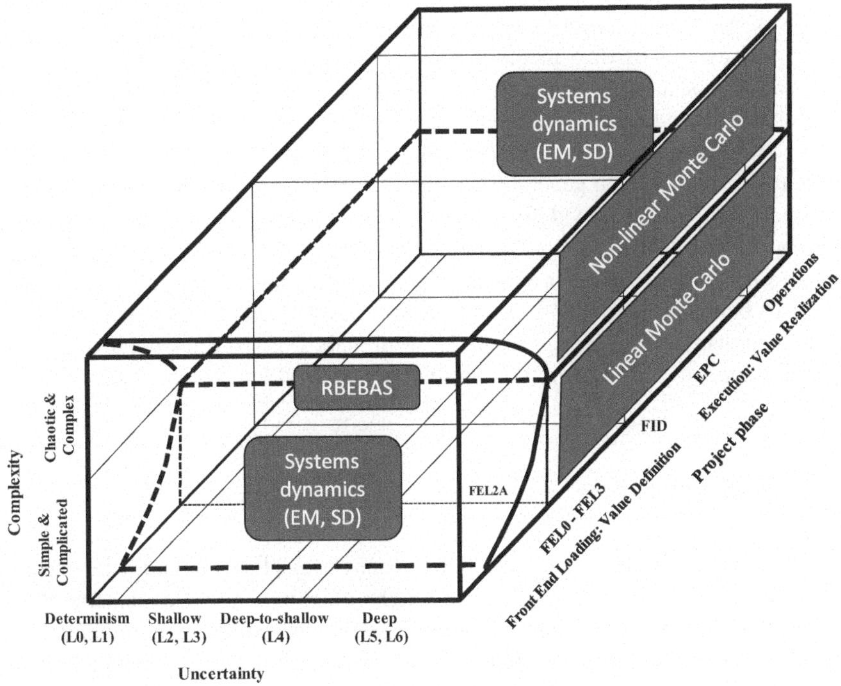

Fig. 16 Methodologies of choice depending of levels of complexity and uncertainty

could be understood as an upgrade of Fig. 11 handling uncertainties in various phases of project development to its 3D version though comprising the complexity concept.

The DAM process is relevant to all phases of project development, there is a short period when the DAM process could be put on a backburner: from FEL2A until some point in execution (earliest) or by the beginning of operation (latest). Before and after that period the DAM is cardinally important calling for systems dynamics methodologies—exploratory modelling (EM) and scenario development (SD)—to handle deep uncertainties in external environment. On the other hand, this short period is exactly when either linear or nonlinear version of the Monte Carlo methodology is applicable depending on project complexity. (This is just a reminder that any instance of Monte Carlo modelling should be associated with a previously selected single project alternative.)

Regarding the two main types of project subsystem's misalignment discussed in Sect. 3.5, Monte Carlo methodology handles only PSS—PDS misalignments and shallow uncertainties, whereas the DAM process is paramount when handling EES—PSS misalignments and deep uncertainties. Although, as argued above, the DAM approach covers all types of uncertainties anyway (Sect. 4.1). This is another reason to take the DAM process as a Great PRM Unifier [1].

From the angle of game theory, efficiency (domain) criteria should be applicable in FEL0 and FEL1 project phases when working with deep L5 and L6 uncertainties in the ESS. The satisficing criteria (minimax and maximin) shall be applicable in FEL2A phase when using RBEBAS methodology to select an optimal project alternative to handle deep-to-shallow L4 uncertainties. As pointed out in Sect. 4.4, the satisficing criterion is also used in post-FEL2A phases when using the Monte Carlo methodology for selecting contingency development confidence levels.

Figure 16 represents a general case when the DAM concept (including systems dynamics) could be adopted for projects with any level of project complexity. Although, it should be mandatory in the case of complex and chaotic projects whereas applying it to simple or complicated projects might be deemed as overshooting. Figure 16 being an upgrade of Fig. 11 does not show alternative's switchover situations demonstrated by Figs. 12 and 13. Although, Fig. 16 does point to system dynamics applications in operations as a part of the DAM process.

6 Conclusion

This chapter being a high-level recap of the recent book [1], leaves out many details and even some important topics related to high complexity and deep uncertainty in project environment. For example, systems dynamics is introduced in this chapter as the EM and SD modelling methodology only when exploring future states of external project environment. Although, it had equally important application in a form of so-called rework cycle modelling [5] that could be a substitute to the Monte Carlo methodology (both linear and especially nonlinear) to evaluate project contingencies [1]. Cost escalation modelling to estimate a special cost-escalation and exchange-rate-volatility project contingencies due to dynamics of highly imbalanced markets as a part of the external project environment is also left out, etc. Even the included topics are discussed in sketchy and rather unelaborated way. At the same time, two major playscripts of project failures are being focused on. First, handling project risks as standalone noninteracting events is a proven way to overlooking of a substantial chunk of overall project risk exposure in complex and chaotic projects. This would certainly result in grossly underestimated duration and budget project contingencies. Simply, contribution of risk interactions to project contingencies in complex and especially chaotic projects could be larger than contributions of standalone risks. This is a recipe for project failure in a form of megascopic delays and overspendings. To meet this challenge, a nonlinear Monte Carlo methodology is introduced that complies with the nonlinear PRM paradigm and takes into account all relevant risk interactions consistently.

Second, ignoring or misunderstanding of deep uncertainties in external project environment would not allow select most optimal project alternative to ensure that expected economic benefits are satisfactory to its stakeholders. In a worst-case scenario, a project might become a sort of a bridge to nowhere being fully disconnected with external environment realities. Furthermore, forgetting about ongoing dynamics in the external environment in operations might result in eroding of the initial project value even if it has been defined right way initially: the initially predicted external environment future becomes a moving and escaping target that might eventually result in destroying of project value.

To meet this challenge, a best-of-breed dynamic adaptive methodology (DAM) that follows an advanced monitor-and-adapt decision-making is proposed based on scanning of the external environment, learning about changes and adapting to them. The DAM process integrates all other relevant PRM processes as local fragments in particular project phases. As such it plays a role of the Great PRM Unifier [1]. As this chapter's main intent has been to push apart boundaries of the traditional PRM, a phrase attributed to a prominent philosopher of science Sir Karl Popper could be cited. He used to say that "our knowledge is finite while our ignorance is infinite." A reader shall decide how successful or futile is this attempt to make our PRM knowledge slightly less finite.

References

1. Raydugin, Yu.G.: Decision-Making in Situations of High Complexity and Deep Uncertainty. Springer Nature, Cham (2024)
2. Raydugin, Yu.G.: Project Risk Management. Essential Methods for Project Teams and Decision Makers. Wiley & Sons, New York (2013)
3. Raydugin, Yu.G.: Modern Risk Quantification in Complex Projects. Nonlinear Monte Carlo and System Dynamics Methodologies. Oxford University Press, Oxford (2020)
4. Marchau, V.A., Walker, E.W, Bloemen, P.J., Popper, S.W.: Decision Making under Deep Uncertainty. From Theory to Practice. Springer Nature, Cham (2019)
5. Sterman, J.D.: Business Dynamics. Systems Thinking and Modeling for a Complex World. McGraw Hill Education, Chennai (2010)
6. Luce, R.D., Raiffa, H.: Game Theory. Dover Publications, Garden City (1989)
7. Gull, S., Williams, T.: Project complexity for the uninitiated. In: Raydugin, Yu.G. (ed.) Leveraging Risk and Uncertainties for Effective Project, pp. 459–469. IGI Global, Hershey (2017)

Incorporating Negative Values into the Simple Additive Weighting (SAW) Under Uncertain Conditions: An Applications in Project Manager Selection Problem

Mohammad Azadfallah

Abstract Negative number/value plays an important role in expressing our preferences (i.e., temperature, economic growth, etc.). However, it has received different attention than positive references in the literature. On the other side, many of our decisions have a percentage of uncertainty and are often expressed as interval data. In this paper, the decision method used is the Simple Additive Weighting (SAW) method. This method was developed to be used when the data are of an interval nature. However, the conventional SAW method with interval data cannot deal with negative data. The critical question is, "What can we do when using the data set containing negative numbers in the SAW method with interval data"? This is the problem we wish to address in this chapter. Further, Salehi and Izadikhah (Decision Science Letters 3:225–236, 2014) extended the SAW technique using interval numbers. We use the Salehi and Izadikhah algorithm to calculate the alternative scores, but we change the procedure of calculating the conventional arithmetic operations to the arithmetic operations proposed by Yamanaka and Oishi (RIMS Kokyuroku Bessatsu B54:71–98, 2015). The results have demonstrated our model to be both robust and efficient.

Keywords Negative numbers · Interval data · Simple additive weighting (SAW) · Decision-making · Uncertainty · Interval numbers · Arithmetic operations · Salehi and Izadikhah algorithm · Yamanaka and Oishi method · Robustness and efficiency

---

M. Azadfallah (✉)
Islamic Azad University, Science and Research Branch, Tehran, Iran
e-mail: m.azadfallah@yahoo.com

K. D. Strang and N. R. Vajjhala (eds.), *International Program and Project Management—Best Practices in Selected Industries*, Information Systems Engineering and Management 31, https://doi.org/10.1007/978-3-031-80275-1_12

1 Introduction

Decision making is an important issue for management to find the optimal alternative to feasible ones. Furthermore, decision making with some evaluation criteria is called Multi-Criteria Decision Making (MCDM) [1]. According to Uzun et al. [2], the concept of Multi-Criteria Decision Analysis (MCDA; also, often called MCDM) has been used since life began on the surface of the earth. It might interest you to know that even animals use instinctive reasoning or judgment to make their own decisions. If animals can make good use of this concept, what can humans, who are far more intelligent than animals, achieve? Although this concept has existed since the origins of life, it was not documented or recorded to assist in making decisions when the need arose. Nevertheless, the title of MCDM was used for the first time in a paper by Zeleny (1975) [3]. MCDM can be defined as, screening, prioritizing, ranking or selecting the alternatives based on human judgment from among a finite set of decision alternatives in terms of multiple usually conflicting criteria [4], and is one of the most widely use decision methodologies in the science, business, and engineering worlds [5]. In general, a decision-making process to tackle MCDM problems involves three stages, namely [6].

1.1 Structuring a Decision Problem

This stage should plan the type of actions that should be taken for each of the following activities:

- Identifying Decision Maker(s) (DMs),
- Defining the goal(s),
- Analyzing the feasible alternatives that will be evaluated,
- Determining the criteria for evaluating the consequences of each alternative.

The activities in this stage are similar among all MCDM methods and provide information that is then used as the basis for determining what type of MCDM method can be applied or to develop a new method to solve the problem.

1.2 Determining and Applying an MCDM Method

To control the applicability of MCDM methods to the problem, an MCDM method should not be chosen before the decision problem has been structured. This stage can cover the following activities:

- Weighting the importance of each criterion with respect to the goal,
- Scoring the preference of each alternative with respect to criteria,
- Calculating the overall weighted scores of alternatives,
- Ranking of all feasible alternatives based on the overall weighted scores.

The differences among MCDM methods lie in this second stage.

1.3 Determining the Final Recommendation

The higher the overall weighted score is the preferable the alternative will be. The obtained results should be examined further by performing a sensitivity analysis that can be used to answer "what-if" questions that DMs might have. Over the past few decades, several interesting tools (i.e., AHP, TOPSIS, VIKOR, etc.) to support selection decision making have been presented [7]. The decision method used in the current research is the Simple Additive Weighting (SAW) method. SAW is the oldest, most widely known and practically used method [8]. Churchman and Ackoff (1954) first utilized the SAW method to cope with a portfolio selection problem [9].

SAW can be considered the most intuition and easy way to deal with MCDM problems, because the linear additive function can represent the preferences of DMs [9]. Difficulty with this method emerges when it is applied to multi-dimensional decision-making problems. Then, in combining different dimensions, and consequently different units, the additive utility assumptions are violated, and the result is equivalent to adding apples and oranges [10]. However, if all the elements of the decision table are normalized, then SAW can be used for any type and any number of attributes [11]. The advantage of this method is that [12]:

- Intuitive method with simple algorithm,
- Able to compensate between variables,
- Suitable for the evaluation of a single alternative.

Also, according to Salehi and Izadikhah [8], the advantage of this method is that it is a proportional linear transformation of the raw data, which means that the relative order of magnitude of the standardized scores remains equal. Liu [13] believe that, in practical decision-making process, the acquisition of information always shows some uncertainty; due to time pressure, lack of knowledge, and the DMs limited expertise related with problem domain [14]. Therefore, these data may be having some structures such as bounded data, ordinal data, interval data, and fuzzy data [15]. The interval numbers are a commonly used mode [16]. Therefore, Salehi and Izadikhah [8] extended SAW technique for using interval data.

On the other side, according to Saaty and Ozdemir [17], the negative of a number compared with that number, is a particular number that when added to the given number equals to zero. The introduction of negative numbers was brought about out of necessity by the development of algebra as science providing general methods for the solution of arithmetic problems, regardless of content or given numerical data.

The need for negative numbers in algebra arose in the solution of problems that reduce to linear equations with one unknown. A possible negative answer in problems of this kind may be interpreted in several ways, for example: oppositely directed segments, motion in a direction opposite to a chosen one, and debt. Historically, although the first set of rules for dealing with negative numbers was stated in the seventh century by the Indian mathematician Brahmagupta, it is surprising that in 1758 the British mathematician Francis Maseres was claiming that negative numbers;

> … darken the very whole doctrines of the equations and make dark of the things which are in their nature excessively obvious and simple.

Maseres and his contemporary, William Friend took the view that negative numbers did not exist. However, other mathematicians around the same time had decided that negative numbers could be used if they had been eliminated during the calculations where they appeared. It was not until the nineteenth century when British mathematicians like De Morgan, Peacock, and others, began to investigate the 'laws of arithmetic' in terms of logical definitions that the problem of negative numbers was finally sorted out [18]. Nevertheless, according to Alinezhad and Khalili [19], attributes (also often called criteria) of C_j, $j = 1, \ldots, n$, with negative desirability from the perspective of DM, which means that their lower amount is more desirable to the DM. negative attributes are usually as loss or cost.

According to Saaty and Ozdemir [17], real life problems can involve subtracting numbers from other numbers (further, comparisons) and the result may be negative numbers. While, the conventional interval SAW method cannot deal with negative value (i.e., temperature, economic growth, etc.). Therefore, in this chapter we addressed the problem of ranking alternatives (containing negative numbers in MCDM problem) with interval data which is a special case of uncertainty in data.

The paper is organized as follows. The second section briefly describes the method considered in this study (i.e., SAW, interval number, MCDM problem with interval data, interval SAW, and interval arithmetic). In third, fourth, fifth, and sixth sections, prior work, research design, propose procedure, and numerical example are discussed, respectively. Discussion is provided in the next section.

1.4 Simple Additive Weighting (SAW) Method

Mathematically, SAW method can be stated as follows [20]:

Suppose the DM assigns a set of importance weights to the attributes, $W = \{w_1, w_2, \ldots, W_n\}$. then the most preferred alternative, A^*, is selected such that (Eq. 1);

$$A^* = \left\{ A_i | \max_i \frac{\sum_{j=1}^{n} w_j x_{ij}}{\sum_{j=1}^{n} w_j} \right\} \tag{1}$$

where x_{ij} is the outcome of the ith alternative about the jth attribute with a numerically comparable scale. Usually the weights are normalized so that $\sum_{j=1}^{n} w_j = 1$.

1.5 Interval Number

In the following, we describe the basic definition and operations of interval numbers.

Definition 1. Let $a = [a^l, a^u] = \{x|0 < a^l \le x \le a^u\}$, then a is called a nonnegative interval number. Especially, a is a nonnegative real number, if $a^l = a^u$.

For convenience, throughout this paper, all the interval arguments are nonnegative interval numbers, and let Ω be the set of all interval numbers, $M = \{1, 2, \ldots, m\}$, $N = \{1, 2, \ldots, n\}$ and $T = \{1, 2, \ldots, t\}$; $i \varepsilon M$, $j \varepsilon N$, and $k \varepsilon T$.

Definition 2. Let $a = [a^l, a^u]$ and the $b = [b^l, b^u] \varepsilon \Omega$, and $\lambda \ge 0$, then:

$$a + b = [a^l, a^u] + [b^l, b^u] = [a^l + b^l, a^u + b^u]; \tag{2}$$

$$a * b = [a^l, a^u] * [b^l, b^u] = [a^l b^l, a^u b^u]; \tag{3}$$

$$\lambda a = \lambda [a^l, a^u] = [\lambda a^l, \lambda a^u]. \text{ Especially, } \lambda a = 0 \text{ if } \lambda = 0 \tag{4}$$

Definition 3. If $W = (W1, W2, \ldots, Wn)$ is a crisp weight vector, such that:

$$W_j \ge 0, \sum_{j=1}^{n} W_j = 1, j \varepsilon N, \tag{5}$$

Then W is called normalized weight vector.

Definition 4. Let $W_j = [W_j^l, W_j^u] \varepsilon \Omega (j \varepsilon N)$ be non-normalized interval weights, then:

$$[\widehat{w}_j^l, \widehat{w}_j^u] = W_j / \sum_{j=1}^{n} W_j = \left[W_j^l / \sum_{j=1}^{n} W_j^u, W_j^u / \sum_{j=1}^{n} W_j^l \right], j \varepsilon N, \tag{6}$$

Are called normalized interval weights of $W_j (j \varepsilon N)$, [14, 21].

1.6 MCDM Problem with Interval Data

In many real-life problems, the data of the decision-making processes cannot be measured precisely and there may be some other types of data, for instance, interval data and fuzzy data. In other words, the DM would prefer to say his/her point of view in these forms rather than a real number because of the uncertainty and the lack of certain data, especially when data are known to lie within bounded variables, or when

Table 1 Decision matrix with interval data

Alternative	Criteria			
	C_1	C_2	...	C_n
A_1	$[x^l_{11}, x^u_{11}]$	$[x^l_{12}, x^u_{12}]$	...	$[x^l_{1n}, x^u_{1n}]$
A_2	$[x^l_{21}, x^u_{21}]$	$[x^l_{22}, x^u_{22}]$	...	$[x^l_{2n}, x^u_{2n}]$
$\vdots$	$\vdots$	$\vdots$	$\vdots$	$\vdots$
A_m	$[x^l_{m1}, x^u_{m1}]$	$[x^l_{m2}, x^u_{m2}]$	...	$[x^l_{mn}, x^u_{mn}]$

facing missing data, judgment data, etc. nevertheless, suppose $A_1, A_2, \ldots, A_m$ are m possible alternatives among which DMs have to choose, $C_1, C_2, \ldots, C_n$ are criteria. With which alternative performance are measured, x_{ij} is the rating of alternative A_i with respect to criterion C_j and is not known exactly and only we know $x_{ij} \in \left[x^l_{ij}, x^u_{ij}\right]$. A MADM (Multiple Attribute Decision Making, also often called MCDM) problem with interval data can be concisely expressed in matrix format as shown in Table 1 [22]:

$w = ([w^l_1, w^u_1]; [w^l_2, w^u_2]; \ldots; [w^l_n, w^u_n])$, where W_j is the weight of criterion C_j.

1.7 SAW Method with Interval Data

This approach may be described as follow [8]:

Step 1. Establish decision-making matrix with interval data $\left[X^l_{ij}, X^u_{ij}\right]$.

Step 2. Determine the laters of criteria by using the interval Entropy method $\left[W^l_j, W^u_j\right]$ (as discussed latter in this chapter-next section).

Step 3. Establishment of normal of decision-making matrix with interval data [ńl$_{ij}$, ńu$_{ij}$].

- For benefit type criteria:

$$\acute{n}^l_{ij} = x^l_{ij} / \sum_{j=1}^{m} \left(x^l_{ij}\right) + \left(x^u_{ij}\right), \; j = 1, \ldots, m, i = 1, \ldots, n, j \varepsilon B, \tag{7}$$

$$\acute{n}^u_{ij} = x^u_{ij} / \sum_{j=1}^{m} \left(x^l_{ij}\right) + \left(x^u_{ij}\right), \; j = 1, \ldots, m, i = 1, \ldots, n, j \varepsilon B, \tag{8}$$

- For cost type criteria:

$$\acute{n}_{ij}^{u} = \left(1/x_{ij}^{u}\right)/\sum_{j=1}^{m}\left(1/x_{ij}^{l}\right) + \left(1/x_{ij}^{u}\right), j = 1, \ldots, m, i = 1, \ldots, n, j \varepsilon C, \quad (9)$$

$$\acute{n}_{ij}^{u} = (1/x_{ij}^{l})/\sum_{j=1}^{m}\left(1/x_{ij}^{l}\right) + \left(1/x_{ij}^{u}\right), j = 1, \ldots, m, i = 1, \ldots, n, j \varepsilon C, \quad (10)$$

Step 4. The calculation of utility function for each index.

$$\begin{aligned} P^{l}(A_i) &= \sum_{j=1}^{n} \acute{n}_{ij}^{l} w_j^{l} / \sum_{j=1}^{n}\left(w_j^{u} + w_j^{l}\right), \\ P^{u}(A_i) &= \sum_{j=1}^{n} \acute{n}_{ij}^{u} w_j^{u} / \sum_{j=1}^{n}\left(w_j^{u} + w_j^{l}\right), \text{ and} \\ P^{l}(A_i) &\le P^{u}(A_i), \ i = 1, \ldots, m. \end{aligned} \quad (11)$$

Step 5. Ranking the alternatives

Considering there is an interval utility function $\left[P^{l}(A_i), P^{u}(A_i)\right]$ for each A_i we must rank all the intervals. The approach (minimax regret approach) is summarized as follow.

$$\begin{aligned} R_i &= max\left[max_{i \ne j}\left(a_j^{u}\right) - a_i^{l}, 0\right] = \\ & max\left[max_{i \ne j}\{m(A_i) + w(A_i)\} - (m(A_i) - w(A_i)), 0\right]; i = 1, \ldots, n. \end{aligned} \quad (12)$$

where, $m(A_i)$ and $w(A_i)$ is the midpoint $\left(m(A_i) = 1/2\left(a_i^{u} + a_i^{l}\right)\right)$ and widths $\left(w(A_i) = 1/2\left(a_i^{u} - a_i^{l}\right)\right)$, respectively.

1.8 Interval Arithmetic

Interval arithmetic (also known as "interval mathematics", "interval analysis", and "interval computation") is an arithmetic defined on sets of real intervals, rather than sets of real numbers [23]. The development of the model theory of interval arithmetic began in 1959 with the technical reports of Ramon Edgar Moore. He developed a number system and arithmetic for real intervals by introducing the so-called range numbers and range arithmetic. Those terms are the first popular synonyms of modern intervals numbers and interval arithmetic. The subsequently developed theory of interval arithmetic has provided a better solution for the problems occurring between computation and applied mathematics when utilizing floating-point arithmetic or traditional numerical approximation methods [24]. Nowadays, interval arithmetic is widely used and has numerous applications in scientific and engineering disciplines that deal intensely with uncertain data (or range data) [23]. The battle with uncertainty has been addressed in many ways especially in the decision-making field and has resulted in more than one method. Those methods are useful and help DMs when they are faced with uncertainty [24]. Several solutions for the above problem are possible (more on this can be found in [25–27]). In this paper, we have adopted the interval arithmetic proposed by Yamanaka and Oishi [28].

According to Yamanaka and Oishi [28], for any $x \in \Re$, an interval is the closed interval represented with the following (Eq. 13):

$$[\underline{x}, \overline{x}] = \{x | \underline{x} \leq x \leq \overline{x}\}, \tag{13}$$

where, $\underline{x} \leq \overline{x} \in \Re$.

The symbols $\underline{x}$ and $\overline{x}$ are referred to as the lower and upper end of the interval, respectively. For simplicity, $[x]$ shall hereinafter represent the interval that satisfies the following (Eq. 14):

$$[x] = [\underline{x}, \overline{x}] \tag{14}$$

Given any two intervals $[x]$, $[y] \in \Re$, then binary operation on them shall be defined as follows [Eqs. (15)–(18)]:

$$[x] + [y] = [\underline{x} + \underline{y}, \overline{x} + \overline{y}], \tag{15}$$

$$[x] - [y] = [\underline{x} - \overline{y}, \overline{x} - \underline{y}], \tag{16}$$

$$[x] \times [y] = [\min(\underline{x}\underline{y}, \overline{x}\overline{y}, \underline{x}\overline{y}, \overline{x}\underline{y}), \max(\underline{x}\underline{y}, \overline{x}\overline{y}, \underline{x}\overline{y}, \overline{x}\underline{y})], \tag{17}$$

$$[x]/[y] = [\min(\underline{x}/\underline{y}, \overline{x}/\overline{y}, \underline{x}/\overline{y}, \overline{x}/\underline{y}), \max(\underline{x}/\underline{y}, \overline{x}/\overline{y}, \underline{x}/\overline{y}, \overline{x}/\underline{y})] \; (0 \notin [y]). \tag{18}$$

2 Prior Work

According to Salehi and Izadikhah [8], SAW is the oldest, most widely known and practically used method. Hence, in this section, to integrate the survey in various aspects, we divide it into two parts: 1. SAW: SAW under uncertain conditions; and interval arithmetic, and 2. Negative values/preferences. For instance, for part 1: (for SAW); [29–41]; (for SAW under uncertain conditions); [1, 8, 21, 42–51]; (for interval arithmetic); [24, 52–55], could be referred to as an example. And for part 2: (for negative values/preferences); our literature research revealed that a limited number of works has been done in terms of negative values/preferences. For instance, Schoner [56] presented an introduction to the use of negative values in the Analytic Hierarchy Process (AHP). Saaty and Ozdemir [17] shown how negative priorities can be defined as relative numbers and used along with positive priorities, in the AHP model.

Millet and Schoner [57] described why and how ratio scaled multi-criteria analysis techniques, such as the AHP, should allow for a subtraction mechanism where by negative preference is combined with positive preference. Othman et al. [58] incorporated negative values in AHP using rule-based scoring methodology for ranking of sustainable chemical process design options. Ersoy [59] generated two different solutions for negative values in the decision matrix. Firstly, negative decision matrix elements are converted to positive using the z-score standardization method. Secondly, different normalization techniques are used instead of vector normalization in the algorithm of the PIV method. In addition, Gorcun and Kuchkonder [60] pointed that traditional CRITIC and the PSI techniques do not work if there are any negative or zero scores in values of indexes. Therefore, this method has been modified by using the z-score technique. So, the CRITIC method, which is a part of the suggested model has been improved, and the negative value transformation technique has been integrated into this method. During our review of the literature, no model was found that addressed the problem of ranking alternatives with interval SAW method in the presence of negative values. This is the aim of this chapter.

3 Research Design

According to Ersoy [59], the existence of negative value in the decision matrix is not frequently encountered in MCDM problems. However, Saaty and Ozdemir [17] believe that, in decision making there are often criteria that are opposite in direction to other criteria as in benefits versus costs, and in opportunities versus risks and sometimes need to be distinguished by using negative numbers. On the other side, according to Hadiono [24], the aim of the operation of decision-making technique is to reduce the uncertainty in the decision outcomes. So, when the degree of uncertainty is high, we can no longer assume a crisp value but report the data in a ranged (or interval) format [61], such as temperature, time, data and energy-related computations [62]. Further, Sayadi et al. [63] believe that interval numbers are the simplest form of representing uncertainty in the decision matrix, and help in reduction of data degeneration and produce more robust results [64]. Therefore, when attempting to solve a problem, all important factors of the problem should be considered. In this paper, we try to see all factors together. So, the proposed model in this study supports the use of the interval SAW method, especially by implementing interval arithmetic into the decision making process in the presence of negative values.

Table 2 Decision matrix with mix interval data

Alternative	Criteria			
	C_1[a]	C_2	...	C_n
A_1	$[-x_{11}^u, -x_{11}^l]$	$[x_{12}^l, x_{12}^u]$	...	$[x_{1n}^l, x_{1n}^u]$
A_2	$[-x_{21}^u, -x_{21}^l]$	$[x_{22}^l, x_{22}^u]$	...	$[x_{2n}^l, x_{2n}^u]$
⋮	⋮	⋮	⋮	⋮
A_m	$[-x_{m1}^u, -x_{m1}^l]$	$[x_{m2}^l, x_{m2}^u]$	...	$[x_{mn}^l, x_{mn}^u]$

[a]For instance, C_1 is the cost-type criteria

4 Proposed Procedure

This study is conducted with the aims to develop a new interval SAW model, particularly; by implementing interval arithmetic into the decision making process, for MCDM problems with negative values. Nevertheless, the proposed procedure can be expressed as follows.

4.1 Decision Matrix with Mix Interval Data

Decision matrix based on the mix interval data is shown below (Table 2). So, you can have in the decision matrix a mix of positive and negative values.

Where;

A_i: is m possible alternatives,

C_j: is n criteria,

X_{ij}: is the rating of alternative A_i with respect to criterion C_j, so $x_{ij} \in [x_{ij}^l, x_{ij}^u]$, and $w_j = ([w_1^l, w_1^u]; [w_2^l, w_2^u]; \ldots; [w_n^l, w_n^u])$ is the weight of criterion C_j.

Notice; in the proposed model, there is no need for negative weights for any cost-type criteria due to the structure of the model.

4.2 Identifying the Mix Weight of Criteria

According to Odu [65], determining criteria weights is a problem that arises frequently in many MCDM techniques. So, results of the MCDM method largely depend on such weights [66]. In addition, they classified weighting methods into subjective or objective weighting methods. So, in the subjective weighting methods, criteria weights are derived from the DMs judgment on criteria. While, in the objective weighting methods, preferences of DM on multiple criteria are not involved and the criteria weights are obtained from mathematical algorithms or models.

On the other side, Amiri et al. [67] believe that, it is more natural and logical than an interval comparison matrix should give an interval weight estimate rather than an exact point estimate. Hence, the subjective method is preferred in this paper [i.e., $w_j^{Subjective} = ([w_1^l, w_1^u]; [w_2^l, w_2^u]; \ldots; [w_n^l, w_n^u])$].

4.3 Construct the Mix Interval Normalized Decision Matrix

Construct the mix interval normalized decision matrix, by using Eqs. (7) [for cost-type criteria], (8) [for benefit-type criteria], and (18) [for both cost/benefit type criteria]).

4.4 Calculate the Utility Function for Each Index

Calculate the utility function for each index, by using Eqs. (11), (12) and (17).

Notice; here, $\sum_{j=1}^{n} (w_j^l + w_j^u) = 1$.

4.5 Rank the Preference Order of All Alternatives According to the Acceptability Index

- **Acceptability approach**

Let < be an extended relation between the intervals $A = [a_L, a_R]$ and $B = [b_L, b_R]$ on the real line $\Re$, then for $m(A) \leq m(B)$, we construct a premise $A < B$, which implies that A is inferior to B (or B is superior to A). Here, the term 'inferior to' ('superior to') is analogous to 'less than' ('greater than').

Definition 1. Let I be the set of all closed intervals on the real line $\Re$. Here, we further define an acceptability function $A : I \times I \to [0, \infty)$ such that (Eq. 19):

$$A(A \prec B) = \frac{m(B) - m(A)}{w(B) + w(A)}, \tag{19}$$

where $w(B) + w(A) \neq 0$. $A(A \prec B)$ may be interpreted as the grade of acceptability of the 'first interval to be inferior to the second interval'.

The grade of acceptability of $(A \prec B)$ may be classified and interpreted further on the basis of comparative position of mean and width of interval B with respect to those of interval A as follows (Eq. 20):

$$A(A \prec B)\left\{\begin{array}{ll} = 0 & if: m(A) = m(B) \\ \in (0, 1) & if: m(A) < m(B) \text{ } and \text{ } a_R > b_L \\ \geq 1 & if: m(A) < m(B) \text{ } and \text{ } a_R \leq b_L \end{array}\right\} \quad (20)$$

The classification of the acceptability grades is interpreted as follows:

i. If $A(A \prec B) = 0$, then the premise 'A is inferior to B' is not accepted.
ii. If $0 < A(A \prec B) < 1$, then the interpreter accepts the premise $(A \prec B)$ with different grades of satisfaction ranging from zero to one (excluding zero and one).
iii. If $A(A \prec B) \geq 1$, the interpreter is absolutely satisfied with the premise $(A \prec B)$ or, in other words, he accepts that $(A \prec B)$ is true [68, 69].

5 Numerical Examples

In this section, we present two numerical examples to illustrate how the proposed method can be used.

5.1 *Example 1*

Suppose that, $x = \{x_1, x_2, ..., x_6\}$ be a set of 6 project managers, which contains four attributes (criteria) that are; "C_1 = Salary", "C_2 = Poor time management", "C_3 = Integrating and trustworthiness", and "C_4 = Experience in managing projects", (In addition, C_1 and C_2 are cost-type and C_3 and C_4 are benefit-type criteria, respectively), see Table 3.

By using step 1:
Construct the mix interval decision matrix, based on Tables 2 and 3, as follows (Table 4).

By using step 2:
Assignment of weights to the criteria by subjective method, as follows.
$w_j = ([0.105, 0.145], [0.100, 0.150], [0.075, 0.175], [0.120, 0.130])$.

By using step 3:
Construct the mix interval normalized decision matrix, by Eqs. (7) and (18) [for cost-type criteria], (8) and (18) [for benefit-type criteria]), as follows (Table 5).

Notice; here (i.e., for cost-type criteria), $[a, b], [c, d] \rightarrow [a, b] = \left[\frac{1}{x_{ij}^U}, \frac{1}{x_{ij}^L}\right]$ and $[c, d] = \left[\frac{1}{\sum x_{ij}^U}, \frac{1}{\sum x_{ij}^L}\right]$.

Table 3 Decision matrix

Alternative	Criteria							
	C_1[a]		C_2[a]		C_3		C_4	
	L	U	L	U	L	U	L	U
PM_1[b]	1356	1356	2331	3012	30	52	123	167
PM_2	746	746	3541	4356	52	70	452	541
PM_3	1025	1025	3110	3682	41	49	154	167
PM_4	450	450	5106	6234	70	83	338	415
PM_5	236	236	3504	4501	61	71	123	169
PM_6	384	384	4682	5762	66	88	344	444

[a]Cost-type criteria
[b]Project manager number 1

Table 4 The mix interval decision matrix

Alternative	Criteria							
	C_1		C_2		C_3		C_4	
	L	U	L	U	L	U	L	U
PM1	−1356	−1356	−3012	−2331	30	52	123	167
PM2	−746	−746	−4356	−3541	52	70	452	541
PM3	−1025	−1025	−3682	−3110	41	49	154	167
PM4	−450	−450	−6234	−5106	70	83	338	415
PM5	−236	−236	−4501	−3504	61	71	123	169
PM6	−384	−384	−5762	−4682	66	88	344	444

Table 5 The mix interval normalized decision matrix

Alternative	Criteria							
	C_1		C_2		C_3		C_4	
	L	U	L	U	L	U	L	U
PM_1	0.030	0.030	0.107	0.138	0.041	0.071	0.036	0.049
PM_2	0.055	0.055	0.074	0.091	0.071	0.095	0.132	0.157
PM_3	0.040	0.040	0.083	0.104	0.056	0.067	0.045	0.049
PM_4	0.092	0.092	0.052	0.063	0.095	0.113	0.098	0.121
PM_5	0.175	0.175	0.072	0.092	0.083	0.097	0.036	0.049
PM_6	0.107	0.107	0.056	0.069	0.090	0.120	0.100	0.129

By using step 4:
Calculate the utility function for each index, by using Eqs. (11), (12), and (17); notice: here $\sum_{j=1}^{n}\left(w_j^l + w_j^u = 1\right)$, as follows (Table 6).

Table 6 The interval utility

Alternative	Criteria	
	Utility index	
	L	U
PM_1	0.00307	0.02073
PM_2	0.00532	0.02046
PM_3	0.00420	0.01554
PM_4	0.00517	0.01982
PM_5	0.00429	0.02535
PM_6	0.00559	0.02101

By using step 5:
Rank the preference order of all alternatives (or project managers), by using Eqs. (19) and (20), as follows.

$$PM_5 \succ PM_6 \succ PM_2 \succ PM_4 \succ PM_1 \succ PM_3$$

5.2 *Example 2*

Suppose that, $x = \{x_1, x_2, ..., x_6\}$ be a set of 6 project managers, which contains four attributes (criteria) that are; "C_1 = Accepting an unrealistic schedule", "C_2 = Choosing the wrong sponsor", "C_3 = Cool under pressure", and "C_4 = Expertise in technical matters", (In addition, C_1 and C_2 are cost-type and C_3 and C_4 are benefit-type criteria, respectively), see Table 7.

Table 7 Decision matrix

Alternative	Criteria							
	C_1[a]		C_2[a]		C_3		C_4	
	L	U	L	U	L	U	L	U
PM_1[b]	1100	1200	5000	5500	40	50	120	180
PM_2	923	1100	4320	5000	61	71	352	542
PM_3	1000	1086	3887	4326	43	56	198	236
PM_4	620	790	4896	5423	68	84	334	512
PM_5[c]	750	900	4000	4450	55	70	240	390
PM_6[d]	400	600	3000	3400	70	90	360	600

[a]Cost-type criteria
[b]The worst alternative
[c]The average alternative
[d]The best alternative

Table 8 The interval utility

Alternative	Criteria	
	Utility index	
	L	U
PM_1	0.00161	0.01527
PM_2	0.00277	0.03007
PM_3	0.00208	0.01176
PM_4	0.00465	0.04250
PM_5	0.00612	0.01050
PM_6	0.00505	0.02541

In continuation, with no intention to describe the whole procedure, we will only point to the final results (Table 8), as follows.

Repeating the above process, we fondly get the ranking order of the six project managers as:

$$PM_6 \succ PM_2 \succ PM_4 \succ PM_5 \succ PM_3 \succ PM_1$$

According to the above result (explanations marked with asterisk in Table 7 and obtained ranking), the best alternative is PM6, since it is superior to all the other alternatives. Meanwhile, PM5 and PM1 have average and bad performance. In sum, these analyses show that the proposed approach is efficient, and the result is consistent. Here, more studies have been done. To compare this result with the conventional interval SAW approach results, we will use the same numerical example (Tables 3 and 7; for example 1 and example 2, respectively). Also, in order to compare our method with the original interval SAW methods, we will use the same numerical example given by Salehi and Izadikhah [8]. In this section, with no intention to describe the whole procedure, we will only point to the final results (Table 9), as follows.

In sum, according to the results of Table 9, we can find the priority is different from that of the original method. Because the contained negative numbers are considered into the proposed model.

6 Discussion

According to the results, we can find the priority is $PM_5 \succ PM_6 \succ PM_2 \succ PM_4 \succ PM_1 \succ PM_3$ and $PM_6 \succ PM_2 \succ PM_4 \succ PM_5 \succ PM_3 \succ PM_1$ for example 1 and example 2, respectively.

This priority is different from that of the original method (interval SAW method) [$PM_6 \succ PM_5 \succ PM_2 \succ PM_4 \succ PM_1 \succ PM_3$ and $PM_6 \succ PM_4 \succ PM_2 \succ PM_5 \succ PM_3 \succ PM_1$, respectively]. Because the contained negative values in decision matrix are considered into the proposed model, in this situation, PM5 and PM6 will be the

Table 9 The comparison result

	Priority	
	Conventional method	Proposed method
Example 1	$PM_6 \succ PM_5 \succ PM_2 \succ PM_4 \succ PM_1 \succ PM_3$	$PM_5 \succ PM_6 \succ PM_2 \succ PM_4 \succ PM_1 \succ PM_3$
Example 2	$PM_6 \succ PM_4 \succ PM_2 \succ PM_5 \succ PM_3 \succ PM_1$	$PM_6 \succ PM_2 \succ PM_4 \succ PM_5 \succ PM_3 \succ PM_1$
Example given by Salehi and Izadikhah [8][a,b]	$PM_4 \succ PM_2 \succ PM_6 \succ PM_5 \succ PM_1 \succ PM_3$	$PM_4 \succ PM_2 \succ PM_6 \succ PM_1 \succ PM_5 \succ PM_3$

[a]The example has been re-solved, because the original article on data normalization mistakenly treated negative criteria as positive criteria
[b]In the original article, the minimax regret approach (MRA) was used in the ranking step, while here, for comparison with the proposed model, we used the acceptability approach

suitable project manager rather than PM6 and PM6 for example1 and example 2, respectively. In addition, in the example of Salehi and Izadikhah [8], the fourth and fifth ranks were reversed (i.e., from: $PM_4 \succ PM_2 \succ PM_6 \succ PM_1 \succ PM_5 \succ PM_3$ to $PM_4 \succ PM_2 \succ PM_6 \succ PM_5 \succ PM_1 \succ PM_3$).

7 Conclusion

While the Interval Simple Additive Weighting (ISAW) method is widely recognized for its efficiency, a significant limitation lies in its inability to account for negative values, which are crucial in expressing preferences. This chapter addresses this gap by focusing on the integration of negative values into the ISAW method and comparing the proposed approach with the conventional ISAW. The results demonstrate the feasibility and validity of the modified approach, showing that the type of data—whether positive or negative—can significantly influence decision outcomes. The proposed method proves to be robust, allowing for the use of mixed data (both positive and negative), grounded in solid mathematical principles (interval arithmetic), and enhancing decision-making quality. Additionally, it optimizes resource usage and offers reliable outcomes. Overall, the findings confirm the effectiveness of the proposed approach, with potential applications in future studies.

References

1. Wang, Y.J.: Interval-valued fuzzy multi-criteria decision making based on simple additive weighting and relative preference relation. Inf. Sci. **503**(2019), 319–335 (2019)
2. Uzun, B., Ozsahin, I., Agbor, V.O., Ozsahin, D.U.: Theoretical aspects of multi-criteria decision making (MCDM) methods. In: Ozsahin, I., Ozsahin, D.U., Uzun, B. (eds.) A Applications of Multi-Criteria Decision Making Theories in Healthcare and Biomedical Engineering. Elsevier Inc. (2021)
3. Zavadskas, E.K., Antucheviciene, J., Turskis, Z., Adeli, H.: Hybrid multi-criteria decision making methods: a review of applications in engineering. Scientia Iranica **23**(1), 1–20 (2016)
4. Roszkowska, E.: Rank ordering criteria weighting methods—a comparative overview. Optimum. Studia Ekonomiczne NR **5**(65), 14–33 (2013)
5. Azadfallah, M.: A new MCDM approach for ranking of candidates in voting systems. Int. J. Soc. Syst. Sci. **11**(2), 119–133 (2019)
6. Sitorus, F., Cilliers, J.J., Parada, P.R.B.: Multi-criteria decision making for the choice problem in mining and mineral processing: applications and trends. Expert Syst. Appl. **121**(2019), 393–417 (2019)
7. Olson, D.L.: Decision Aids for Selection Problems. Springer, New-York, Inc. (1996)
8. Salehi, A., Izadikhah, M.: A novel method to extend SAW for decision making problems with interval data. Decis. Sci. Lett. **3**(2014), 225–236 (2014)
9. Tzeng, G.H., Huang, J.J.: Multiple Attribute Decision Making: Methods and Applications. CRC Press (2011)
10. Triantaphyllou, E., Shu, B., Sanchez, S.N., Ray, T.: Multi-criteria decision making: an operations research approach. In: Webster, J.G. (ed.) Encyclopedia of Electrical and Electronics Engineering, vol. 15, pp. 175–186. Wiley & Sons, New York, NY (1998)

11. Rao, R.V.: Decision Making in the Manufacturing Environment, Using Graph Theory and Fuzzy Multi-Attribute Decision Making methods, p. 2007. Springer, London (2007)
12. Urosevic, K., Gligoric, Z., Miljanovic, I., Beljic, C., Gligoric, M.: Novel methods in multi-criteria decision making process (MCRAT and RAPS)—application in the mining industry. Mathematics **9**(1980) (2021). https://doi.org/10.3390/math9161980
13. Liu, S.: Novel approaches to multi-criteria decision making with incomplete information system. COMPUSoFT Int. J. Adv. Comput. Technol. **2**(5), 114–120 (2013)
14. Yue, Z.: Group decision making with multi attribute interval data. Inform. Fusion **14**(2013), 551–561 (2013)
15. Jahanshahloo, G.R., Lotfi, F.H., Izadikhah, M.: An algorithmic method to extend TOPSIS for decision making problems with interval data. Appl. Math. Comput. **175**(2006), 1375–1384 (2006)
16. Zhang, Y., Xu, J.: The ELECTRE method based on interval numbers and its application to the selection of leather manufacture alternatives. World J. Model. Simul. **2**(2), 119–128 (2006)
17. Saaty, T.L., Ozdemir, M.: Negative priorities in the analytic hierarchy process. Math. Comput. Model. **37**(9–10), 1063–1075 (2003)
18. Rogers, L.: The History of Negative Numbers. University of Cambridge, Published 2008, Revised 2009 (2008). https://nrich.maths.org/5961. Accessed 23 Dec 2023
19. Alinezhad, A., Khalili, J.: New Methods and Applications in Multiple Attribute Decision Making (MADM). Springer Nature Switzerland AG (2019).
20. Hwang, C.L., Yoon, K.: Multiple Attribute Decision Making. Springer (1981)
21. Azadfallah, M.: Supplier selection with interval SAW for a group of decision makers when a group cannot reach to consensus. J. Supply Chain Manage. Syst. **6**(3), 25–34 (2017)
22. Azadfallah, M.: Supplier selection using MADM method under uncertainty. J. Supply Chain Manage. Syst. **5**(3), 1–8 (2016)
23. Dawood, H.: Theories of Interval Arithmetic: Mathematical Foundations and Applications. LAP Lambert Academic Publishing (2011). ISBN 978-3-8465-154-2
24. Hadiono, K.: A practical implementation of interval arithmetic in AHP. IJAHP **8**(2) (2016). ISSN 1936-6744256-27
25. Benhamou, F., Older, W.J.: Applying interval arithmetic to reach integer, and Boolean constraints. J. Logic Program. **32**(1), 1–24 (1997)
26. Hickey, T., Ju, Q., Emden, M.H.V.: Interval arithmetic: from principles to implementation. J. ACM **48**(5), 1038–1068 (2001)
27. Moore, R.E., Kearfott, R.B., Cloud, M.J.: Introduction to interval analysis. In: Society for Industrial and applied Mathematics, 1st edn., Philadelphia (2009)
28. Yamanaka, N., Oishi, S.: Interval arithmetic and its implementations. RIMS Kokyuroku Bessatsu **B54**(2015), 71–98 (2015)
29. Afshari, A.R., Mojahed, M., Yusuff, R.M.: Simple additive weighting approach to personnel selection problem. Int. J. Innov. Manage. Technol. **1**(5), 511–515 (2010)
30. Agustyaningrum, C.I., Alamsyah, D.P., Susanti, L., Ramdhani, Y., Prasetio, R.T.: Decision support system application with simple additive weighting. In: Proceedings of the International Conference on Industrial Engineering and Operations Management. Monterrey, Mexico, November 3–5, 2021 (2021).
31. Arsyah, U.I., Jalinus, N., Syahril, A., Arsyah, R.H., Pratiwi, M.: Analysis of the simple additive weighting method in educational aid decision making. Turk. J. Comput. Math. Educ. **12**(14), 2389–2396 (2021)
32. Aulawi, H., Nuraeni, F., Setiawan, R., Rianto, W.F., Pratama, A.S., Manlana, H.: Simple additive weighting in the development of a decision support system for the selection of house construction projects. In: International Conference on Computer Science, Information Technology and Engineering, 16 February 2023. IEEE Jakarta, Indonesia (2023). https://doi.org/10.1109/ICCoSiTE57641.2023.1.127813
33. Biswas, T.K., Chaki, S.: Applications of modified simple additive weighting method in manufacturing environment. IJE Trans. A Basics **35**(04), 830–836 (2022)

34. Kaliszewski, I., Podkopaev, D.: Simple additive weighting—a meta model for multi-criteria decision analysis methods. Expert Syst. Appl. **54**(2016), 155–161 (2016)
35. Khoiry, I.A., Amelia, D.R.: Exploring simple additive weighting (SAW) for decision making. J.l Inovtek Polbeng-Seri. Informatika **8**(02), 281–290 (2023)
36. Limbong, T., Sitorus, L., Purba, D., Simarmata, J.: Implementation of simple additive weighting method in teachers teaching assessment quality. In: Proceedings of the 1st Untimed International Conference on Economic Education and Social Science (U1CEES2018), pp. 347–350 (2018). ISBN 978-989-758-432-9
37. Murad, D.F., Fernando, E., More, L.A.W., Aulia, R.C.: Application engineer selection simple additive weighting method approach. In: 2020 International Conference on Information Management and Technology (ICIMTech), 13–14 August 2020 (2020). https://doi.org/10.1109/ICIMTech508302020.9210945
38. Nawindah, N.: Simple additive weighting (SAW) mathematics method for warehouse disaster location selection in central Jakarta, Indonesia. Int. J. Pure Appl. Math. **117**(15), 795–803 (2017)
39. Purwanto, R. Novia, P.D., Maharran, R.H., Syafirullah, L.: Application of simple additive weighting (SAW) method and decision table in decision support system determines the level of problem student punishment levels. In Proceedings of the 4th International Conference on Applied Science and Technology on Engineering's Science (iCAST-ES2021), pp. 1194–1202 (2021). ISBN 978-989-758-615-6, ISSN 2975-8246
40. Putra, D.W.T., Punggara, A.A.: Comparison analysis of simple additive weighting (SAW) and weighted product (WP) in decision support systems. MATEC Web Conf. **215**, 01003, 1–5 (2018). https://doi.org/10.1651/matecconf/201821501003
41. Simahungkalit, E., Tarigan, J.S., Sari, D.C., Hasibuan, A.S.: Implementation of simple additive weighting (SAW) method in decision support system to determine the best university in Medan. Int. J. Res. Vocat. Stud. **2**(4), 168–172 (2023)
42. Afshari, A.R., Yusuff, R., Derayatifar, A.R.: Project manager selection by fuzzy simple additive weighting method. In: International Conference on Innovation Management and Technology Research, 21–22 May 2012. IEEE (2012). https://doi.org/10.1109/CIMTR.2012.6236429
43. Aliyeva, K., Aliyeva, A., Aliev, R., Ozdeser, M.: Application of fuzzy simple additive weighting method in group decision making for capital investment. Axioms 2023 **12**(8), 797 (2023). https://doi.org/10.3390/axioms12080797
44. Atmojo, R., Subagyo, A., Pardamean, B., Abbas, B.S., Cahyani, A., Manulang, I.D.: Fuzzy simple additive weighting based, decision support system application for alternative confusion reduction strategy in SMART phone purchases. Am. J. Appl. Sci. **11**(4), 668–680 (2014)
45. Chou, S.Y., Chang, Y.H., Shen, C.Y.: A fuzzy simple additive weighting system under group decision making for facility location selection with objective/subjective attributes. Eur. J. Oper. Res. **189**(1), 132–145 (2008)
46. Roszkowska, E., Kacprzak, D.: The fuzzy SAW and fuzzy TOPSIS procedures based on ordered fuzzy numbers. Inf. Sci. **369**(2016), 6564–6584 (2016)
47. Rwanda, R., Oetama, R.S.: Fuzzy multi-attribute decision making and SAW for supplier measurement in furniture business. J. Inf. Syst. Inform. **5**(4), 1432–1448 (2023)
48. Stevic, Z., Durmic, E., Ganjic, M., Pamucar, D., Puska, A.: A novel Multi-criteria decision making model: interval rough SAW method for sustainable supplier selection. Information **10**(10), 292 (2019)
49. Wang, Y.J.: A fuzzy multi-criteria decision making model based on simple additive weighting method and relative preference relation. Appl. Soft Comput. **30**(2015), 412–420 (2015)
50. Wang, P., Zhu, Z., Wang, Y.: A novel hybrid MCDM model combining the SAW, TOPSIS, and GRA methods based on experimental design. Inf. Sci. **345**(2016), 27–45 (2016)
51. Zak, J., Golda, P., Cur, K., Zawisza, T.: Assessment of airside an aerodrome infrastructure by SAW method with weights from Shannon's interval entropy. Arch. Trans. **60**(4), 171–185 (2021)
52. Jafari, H., Ehsanifar, M.: Using interval arithmetic for providing a MADM approach. J. Fuzzy Extension Appl. **1**(1), 60–68 (2020)

53. Michnik, J., Grabowski, A.: Modelling uncertainty in the wings method using interval arithmetic. Int. J. Inf. Technol. Decis. Mak. **19**(01), 221–240 (2020)
54. Sharghi, P., Jabbarova, K., Aliyeva, K.: RDM interval arithmetic based decision making on port selection. Proc. Comput. Sci. **120**(2017), 572–579 (2017)
55. Singh, R.K., Benyoucef, L.: Fuzzy logic and interval arithmetic-based TOPSIS method for multi-criteria Veverse auctions. Serv. Sci. **4**(2), 101–117 (2012)
56. Schoner, B.: The introduction of negative values to the analytic hierarchy process. IJAHP 294–299, Vancouver, Canada (1996)
57. Millet, I., Schoner, B.: Incorporating negative values into the analytic hierarchy process. Comput. Oper. Res. **32**(2005), 3163–3173 (2005)
58. Othman, M.R., Repke, J.U., Wozny, G.: Incorporating negative values in AHP using rule-based scoring methodology for ranking of sustainable chemical process design options. Comput. Aided Chem. Eng. **28**(2010), 1045–1050 (2010)
59. Ersoy, N.: Application of the PIV method in the presence of negative data: an empirical example from a real-world case. Hitit J. Soc. Sci. **14**(2), 318–337 (2021)
60. Gorcun, O.F., Kuchkonder, H.: A novel performance evaluation technique based on integrated weighting approach: a case study in the field of sport management. Decis. Sci. Lett. **10**(2021), 511–524 (2021)
61. Chauhan, A., Vaish, R.: A comparative study on decision making methods with interval data. J. Comput. Eng. **2014**, Article ID: 793074, 1–10 (2014)
62. Amiri, M., Nosratian, N.E., Jamshidi, A., Kazemi, A.: Developing a new ELECTRE method with interval data in multi-attribute decision making problems. J. Appl. Sci. **8**(22), 4017–4028 (2008)
63. Sayadi, M.K., Heydari, M., Shahanaghi, K.: Extension of VIKOR method for decision making problem with interval numbers. Appl. Math. Model. **33**(2009), 2257–2262 (2009)
64. Hafezalkotob, A., Hafezalkotob, A.: Interval multi MOORA method with target values of attributes based on interval distance and preference degree: biomaterials selection. J. Ind. Eng. Int. **13**(2017), 181–198 (2017)
65. Odu, G.: Weighting methods for multi-criteria decision making technique. J. Appl. Sci. Environ. Manag. **23**(8), 1449 (2019). https://doi.org/10.4314/jasem.v23i87
66. Zardari, N.H., Ahmed, K., Shirazi, S.M., Yusop, Z.B.: Weighting Methods and their Effects on Multi-Criteria Decision Making Model Outcomes in Water Resources Management.Springer (2015)
67. Amiri, M., Zandian, M., Vahdani, B., Yazdani, M., Soltani, R.: A hybrid MCDM model with interval weights and data for convention site selection. J. Appl. Sci. **8**(15), 2678–2686 (2008)
68. Sengupta, A., Pal, T.K.: Fuzzy Preference Ordering of Interval Numbers in Decision Problems. Springer, Berlin, Heidelberg (2009)
69. Sengupta, A., Pal, T.K., Chakraborty, D.: Interpretation of inequality constraints involving interval coefficients and a solution to interval linear programming. Fuzzy Sets Syst. **119**(2001), 129–138 (2001)

Human Resource Management in Project Management: Software Engineering Case Study

Kenneth David Strang

Abstract This chapter presents an in-depth analysis of best practices in project human resource management (PM HRM) through examples derived from a case study of a multinational software engineering organization. This chapter explores the foundational literature supporting contemporary PM HRM, highlighting significant insights and identifying limitations within the ISO and PMI bodies of knowledge. A comparative review of organizational human resource management (OHRM) literature and PM HRM literature is provided to underscore the differences and intersections between these fields. A detailed examination of the software staffing database, a critical artifact in PM HRM, is presented using a case study example to demonstrate its practical application. Furthermore, essential PM HRM functions, including recruitment, selection, leadership, team development, and performance evaluation, are thoroughly discussed. The research question (RQ) is addressed by illustrating effective PM HRM best practices, with empirical evidence drawn from the multinational software engineering corporation case study.

Keywords Project human resource management · Performance evaluation · Project management · ISO 21502 · Project management body of knowledge · PMBOK · Staffing pool database

1 Introduction

This chapter discusses human resource management (HRM) theory and its application in the context of the project. A project management HRM (PM HRM) approach is developed based on literature and empirical experience. A case study in the fintech

K. D. Strang (✉)
W3-Research, New York, NY 11041, US
e-mail: professor@kennethstrang.com

University of the Cumberlands, Williamsburg, KY 40769, US

RMIT University, Melbourne, NSW 3001, Australia

K. D. Strang and N. R. Vajjhala (eds.), *International Program and Project Management—Best Practices in Selected Industries*, Information Systems Engineering and Management 31, https://doi.org/10.1007/978-3-031-80275-1_13

industry is used to provide examples. We present empirical research analysis to show the important HRM literature and how PM HRM differs. PM HRM processes are explained first according to the PMBOK and ISO literature, and then we interpret those findings to develop a simplified PM HRM approach. This chapter will explain best practices using examples from a multinational software engineering case study organization.

1.1 Rationale for Better Human Resource Management in Projects

There is no specific evidence that PM HRM is failing or succeeding in projects, but the project failure rate is too high at 50%. HRM and PM HRM methods are well-documented in the PMBOK, but how PM HRM is applied to projects differs from the literature. There appears to be a misunderstanding of how PM HRM techniques differ from organizational HRM theory. These issues need to be reconciled, and better PM HRM best practices are needed by practitioners.

1.2 Research Question and Methods

The research question (RQ) underlying this chapter is what effective best practices for human resource management in project management are, focusing on the fintext industry as a case study example. An abductive literature review strategy was applied, which means only selected papers were selected for discussion [1]. A pragmatic ideology was taken, meaning customized techniques were used to collect, analyze, and interpret the data and conduct the single case study [1, 2].

2 Human Resource Management: Organization Versus Project Contexts

This chapter contrasts organizational versus project HRM and then identifies how HRM is typically conducted in projects. Figure 1 conceptually contrasts the differences explained in the following subsections.

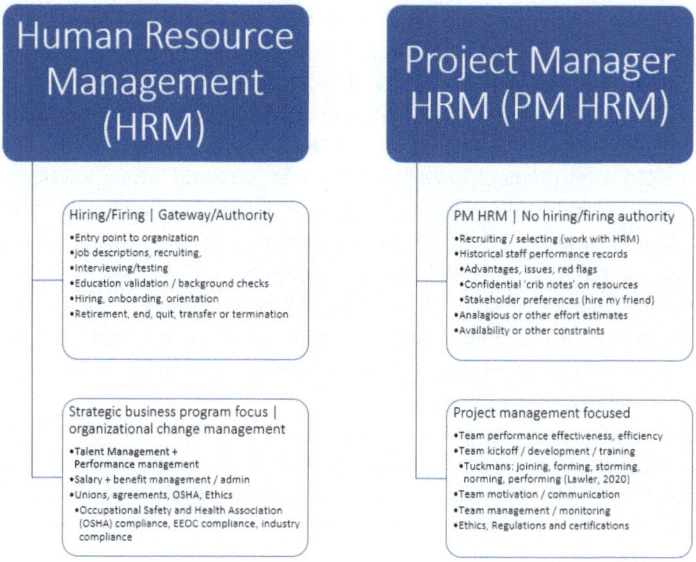

Fig. 1 Organization versus project human resource management

2.1 Organizational HRM

Organizational HRM (OHRM) theory is old and mature compared to PM HRM. The former was developed in the Western world, primarily by academic scholars working in USA-based Midwest universities during the industrial era of the 1960–1980's [3].

As depicted on the left side of Fig. 1, OHRM is described as the overall human resource function across the business entity, the gateway or the hiring/firing entry point to the company, regardless of whether it is a multinational conglomerate or a small business, in any industry [3, 4]. OHRM is an always-present function, although it may be outsourced. OHRM is also considered the cradle-to-grave process for employees, where they are hired, employed, and then eventually, there is either retirement, quitting, transferring, termination, or, in a few cases, passing away before any of the above.

OHRM is considered a strategic function of the business because human resources are generally considered the most important asset in the company [3, 4]. OHRM handles high-level corporate program initiatives such as talent management, organizational change management, and performance management, including salary/benefits administration, company-wide corporate training, and regulatory compliance. In some industries and for certain organizations, unions may be involved; if so, this becomes another function OHRM handles.

From the company standpoint, as shown on the left side of Fig. 1, OHRM could be considered the ethical supervisors of the company, guiding ethical behavior internally and externally, such as social good [1]. The OHRM function is usually responsible for designing the job, writing the job description, or approving what the functional manager provides [3, 4]. The OHRM department also takes responsibility for adherence to regulations, laws, and industry-standard practices. One generic area that OHRM is responsible for is occupational safety and health—which is regulated in the U.S. by the federal Occupational Safety and Health Association (OSHA). Additionally, the OHRM department ensures internal procedures and hiring practices comply with the U.S. Federal Employee Equal Opportunity Commission laws [1]. There are numerous federal laws that the OHRM department must administer, ranging from the deduction of mandatory taxes from wages before employees are paid to remitting the taxes to the government (e.g., the Internal Revenue Agency in the U.S.). OHRM is also responsible for explaining employee retirement savings and, in many cases, providing a pension system—if a system is provided, OHRM administers that as well (or occasionally, they outsource it).

From the employee perspective, OHRM starts with staff planning, writing job descriptions for positions, and advertising to recruit candidates. Next in line for the OHRM function is interviewing, including asking behavioral questions, possibly involving testing, and then the selection decision—hire or move on to the next candidate.

Before hiring, the OHRM department will verify the selected candidate's credentials, academic education, past jobs, certification, and public records to determine if there has been any criminal behavior [1]. OHRM checks for criminal and unethical behavior to identify unfavorable attributes of employees. Federal regulations provide some discrimination protection to candidates, so organizations generally do not use background checks as part of the initial screening. Instead, these become conditions within the hiring offer and agreement.

The principle for using background checks is that in psychology, we know that past behavior tends to repeat, so OHRM departments are usually unwilling to hire employees with unethical or criminal behavior records [1]. In most organizations, candidates will be rejected if they have even minor violations of laws such as speeding, misdemeanor offenses, and lying on the employment application (e.g., faking education or certifications). With a higher supply versus demand for jobs, organizations can apply strict hiring criteria.

Once the candidate accepts the hiring offer, they begin onboarding. Onboarding is a form of training customized and created by the organization based on its standard operating procedures. Onboarding will also involve meeting key people, understanding the company, and hands-on orientation sessions to use the technology, facilities, and software.

2.2 Project Manager HRM (PM HRM)

We call project manager HRM (PM HRM) because the PM conducts HRM in projects, not the OHRM department. The PM is responsible and accountable. As summarized on the right side of Fig. 1, PM HRM involves several of the same functions as OHRM: recruiting, selecting, training/development (to some extent), and performance evaluation. PM HRM does not involve hiring, firing, salary administration, or the federal regulation compliance oversight functions that ORHM oversees. OHRM and PM HRM differ because the latter lack authority or scope. The PM HRM scope is strictly within the project context, which practitioners recognize is a unique endeavor with a specific end date [4].

Some of the HRM tasks a PM will perform include reviewing and sometimes updating the historical staff performance records as part of the team performance evaluations. Performance evaluation is one of the most important and strategic functions in PM HRM. The PM must objectively evaluate all team resources during the project (to avoid forgetting good and bad work results) and at the end of the closing phase. The PM needs to make notes, sometimes we call that data confidential 'crib notes' (from the card game era), on each resource, such as soft skills, hard (functional/technical) skills [4], interests, behavior red flags, and specific words to capture what was done well or poorly on assigned WBS work package activities. This allows the PM to transform project performance into the OHRM performance history so that anyone in the organization can use it as though the PM was a department. In some organizations, a project department is called the Project Management Office (PMO), based on the PMBOK [5].

The PM is usually responsible for team development, as illustrated on the right side of Fig. 1. Since a project is a limited-time endeavor, resources are often combined with an organizational imperative to perform well as a team, like hitting the ground running. The unique requirement in projects to have resources rapidly become productive together has led to several team development theories, including the best-known model of Tuckmans: joining, forming, storming, norming, performing [3, 6]. The PM motivates team members and ensures compliance with organizational standard operating procedures, including ethics, regulations, and external standards. Interestingly and importantly, PM HRM does not mean or include stakeholder management functions. The difference is that PM HRM focuses on motivating team members, using the same engagement techniques used with key stakeholders. Still, the objective in PM HRM is to convince the resource to do the work, as compared to the other domain where the PM tries to obtain resources from stakeholders or to help them understand and accept project WBS activities and deliverables that will impact the organization.

Training is usually not included in PM HRM (compared to OHRM) unless the PM determines a resource needs specific assistance. Often, a PM will not have training within the project budget, so the PM faces a dilemma if a resource is not performing well. In that situation, the PM could assign a mentor to help the underperforming resource, persuade the resource to take corporate training on their own time or replace the resource with another where the above do not work. The PM does not have the authority to terminate a project team resource like this is done in OHRM. Still, the PM can remove a resource from the project schedule, taking them off their assigned WBS work package activities and replacing that resource with another. This punishes the underperforming resource because other team members and the organization (later on) will observe this as an implied failure by that resource. Therefore, removing a team member from the work should be done only if no other feasible alternatives exist. Ultimately, the PM is accountable for the project deliverables, so the PM must make often difficult choices.

Another significant difference between PM HRM and OHRM is that PMs usually do not have control over the billing rates of resources as set by OHRM and the functional manager. The PM commonly borrows resources from functional managers (the key stakeholders). Therefore, the function manager controls the billing rate, usually the salary divided by the working hours. Hence, it becomes an hourly or daily rate that the PM can use in project for cost management. Nevertheless, the PM can impact the salary of a project team member through the performance evaluation results, which go back to the functional manager and OHRM department at the of the project in the closing phase.

3 PM HRM Based on ISO and PMBOK

Neither of the ISO guidelines mentions HRM, human resources, or even the word 'human', other than to note that resources are provided to a project by the sponsor [7, 8]. The PMBOK version 7 from PM [6] does not contain much information about applying PM HRM—there is more data about managing physical resources. PM HRM in the PMI PMBOK should not be confused with procurement, where the latter involves administering the outsourcing of goods (usually), but also external human resources [6]. The above guidelines assume that HRM is already well documented in the literature and ought to apply in PM HRM. However, we argue this is not true—PM HRM differs from OHRM, as contrasted in the previous section. Procurement is beyond the scope of this chapter.

The older PMBOK version from PMI provides high-level advice about PM HRM [5, 9]. PMI mentions the keyword 'staffing plan,' which they defined as a component of the resource management plan, to describe when and how team members will be acquired, along with the length of time they will be needed [5]. The Resource (staffing) management plan is described as the document identifying where the resources in the staffing plan are obtained, how resources will be deployed, evaluated, and managed, along with everything already identified in the staffing plan revised for the current project [9]. The mention of a staffing plan is quite significant since, in actual practice, we use a staffing pool database (sometimes known by other names) to record all the essential information about every project team member. This key PM HRM artifact will be discussed in more detail later.

The older versions of the PMBOK describe PM HRM as:

- 9.1 Plan Resource Management (in the planning phase of a project, describes the approach a PM will use for PM HRM—perhaps this means recruiting?);
- 9.3 Acquire Resources (clearly this refers to recruiting in PM HRM);
- 9.4 Develop Team (implies using the Tuckman model cited above);

- 9.5 Manage Team (this is PM leadership and ongoing performance management in PM HRM);
- 9.6 Control Resources (implies PM leadership and performance evaluation);
- Adapted from PMI PMBOK [6, 9].

What is missing from the PMI PMBOK and ISO guides is the discussion about how to perform the recruiting, ongoing leadership/performance monitoring, and resource performance evaluations in the closing phase. Another confusing aspect of the PMBOK is that estimating as part of time and cost management refers to assigning resources to WBS work package activities, which is correct. Still, there is no explanation about how to do that. Assigning resources is partially a PM HRM function because the PM must review the resource's soft and hard skills [4, 10].

4 PM HRM Based on ISO and PMBOK

Table 1 is a proposed best practice for PM HRM. It is based on what could be salvaged from the PMI and ISO PMBOK. We consider it a simplified best practice for PM HRM, superimposed on the universal project management matrix developed in the first chapters.

As highlighted in Table 1, the PM HRM function starts at the beginning of a project, in the initiation phase, when the sponsor has asked the PM to manage a new project. At this point, the PM is developing the charter. Along with identifying stakeholders, the PM is either given resources by the sponsor to use or the PM will identify the types of resources likely to be needed to complete the charter. The PM will create the staffing pool database at the initiation phase, or in many cases if the PM is starting a new project in the same organizational context (e.g., not a consultant PM), the PM will revise a previous staffing pool database.

The staffing pool database is a spreadsheet in simple form, such as MS Excel or Google Sheets, but it may be a table in a relational database within a commercial software product. Either way, the PM must document the resources and use the staffing pool database throughout the project to record changes and ongoing resource performance notes. This key artifact is explained in more detail below.

In the planning phase (see Table 1), PM HRM acquires or selects resources, which may be internal or external (contracted—and not necessarily through procurement). Here, the staffing pool database is the input (but it can be updated as needed). The PM will research the resource, first using the staffing pool database and then looking at attached employee resumes or previous project documents to determine which hard and soft skills best match the WBS in the project. This implies that the WBS must be developed, which is a paradox because the PM will sometimes select resources first to help develop the WBS.

During the planning phase, the PM typically meets with each resource to confirm with the resource they are suited for (and motivated) to perform the proposed WBS work packages (or specific activities), willing to do the work, and available for the

Table 1 Simplified PM HRM applied to the universal project management matrix

PM function	Project phases			
	Initiate	Plan	Execute	Close
Stakeholder	Stakeholder identification > ***Stakeholder register***	Stakeholder analysis (recursive) > ***Stakeholder register***	Stakeholder engagement (ongoing) > ***Stakeholder register***	
Resource (PM HRM—this chapter)	Identify human resources > *Staffing pool database*	Request in-house + contract resources > *Staffing pool database*	Lead/motivate/direct team appraise performance (ongoing) > *Staffing pool database*	Performance evaluations > *Staffing pool database*
Integration	Charter + governance	Finalize deliverable scope		Surveys + lessons-learned
Integration		Project management plan	Formal change control	Final close administration
Scope		Work breakdown structure [WBS]		
Scope		Define WBS tasks for deliverable(s)		
Time		Estimate task durations from WBS		
Time		Sequence tasks from WBS		
Time		Assign resources to tasks		
Risk		Identify + quantify internal risks		
Risk		Brainstorm external risks		
Risk		Develop risk mitigation plan	Risk mitigation/change control	

(continued)

Table 1 (continued)

PM function	Project phases			
	Initiate	Plan	Execute	Close
Quality		Plan quality assurance	Quality control/rework/ signoff	
Procurement		Plan non-human purchase governance	Administer/manage procurements	
Cost		Estimate costs/budget	Formal change control/ earned value	
Communication		Plan communications	Distribute information + progress	

scheduled duration. There is guesswork in this planning stage as other tasks must be completed, such as the WBS, before the staffing pool database is finalized. Once the team is finalized, the PM usually removes or hides resources in the staffing pool database who are not selected for the project—but it will be strategic to have the entire database available later in case an existing resource needs to be replaced, or an additional mentor is needed.

5 PM Staffing Pool Database

One of the most important and seldom documented PM HRM activities for a PM is to keep track of the project team resources, from project to project, so historical performance, attributes, soft skills, and hard skills will be available in the future as input to team member selection and other OHRM activities. The PM uses the staffing pool database from their records or sometimes from the OHRM department, even if they are an external contractor, because clients often re-hire a successful PM for future projects [10, 11].

The key PM HRM artifact is the staffing pool database cited in Table 1. Unfortunately, there is ambiguity in the literature concerning which fields or attributes are needed for the staffing pool database. We present the generally accepted structure of the staffing pool database in Table 2 and explain the fields below.

A staffing pool database should include columns or fields to identify resources, summarize their key soft and key hard skills, and record ongoing performance notes during the project execution and closing phases. The important descriptive fields are the name (first and last), nickname to be used in everyday communication between the PM and the resource, contact information including location, and employee number to correlate back to OHRM (for external contracted resources, the employee number can sometimes be the procurement or contract number).

The hard skills are functional, technical skills associated with the subject matter that the resource is familiar with or has identified competencies in based on education, certification, or self-professed understanding. Hard skills are specific to each discipline. For example, Python full-stack programmers in the computing services software industry will be familiar with frontend design methodologies, coding transaction scripts, and developing the database structure in the back end [10, 12]. The additional resource attributes for the staffing pool database should include years of experience, credentials (degrees/certifications), billing rate (per hour, preferably to facilitate WBS time estimating), and availability date. Many project management software programs store additional contact and operational data, including email and cell phone, to facilitate easy communication. The case study shows that our staffing pool is linked directly to Microsoft Teams for interactions.

Performance evaluations in PM HRM are a special topic. First, the PM must objectively evaluate every team member's ongoing and final performance. The evaluation must be ethical as well as objective. The case study illustrates this. A qualitative comment field records each performance evaluation comment by the PM,

Table 2 Staffing pool database from case study company after closing project phase

Experience (years)	Billing rate	Available (date)	soft skills	Hard skills	Rating	Performance evaluation (Confidential)
3	$20/hr	3/4/2024	Conflict and crisis management, leadership qualities, resourcefulness	Organizational abilities, communication skills, project management proficiency	5	Color highlighted important metrics on the RAM, recent deliverables have been receiving perfect score. Has shown improvements in performance through the weeks
3	$20/hr	3/4/2024	Collaboration, communication, critical thinking, initiative	GAP analysis, data analysis, UI mockup/ wireframe design, usability testing	5	Perfect work on all the deliverables through the semester. Got a shout out from the sponsor in all assignments. Timely submissions
4	$40/hr	3/4/2024	Leadership, communication, strategic thinking	Project management, data analysis, marketing	5	Consistently received a perfect score on all deliverables
8	$30/hr	3/4/2024	Critical thinking, problem solving, negotioation, team work	Account management, financial acumen	5	Perfect work on all deliverables
3	$20/hr	3/4/2024	Communication, empathy, integrity, leadership, conflict and crisis	Environmental policy research, cost management, project planning, technical knowledge	5	Perfect or near-perfect work on all assignments, sent the roster and asked helpful questions

(continued)

Table 2 (continued)

Experience (years)	Billing rate	Available (date)	soft skills	Hard skills	Rating	Performance evaluation (Confidential)
3	$20/hr	3/4/2024	Persuasiveness, professionalism, reliability, integrity	Power systems analysis, familiarity with relevant codes and standards, engineering designing skills and knowledge, knowledge of engineering science and technology	5	Consistently received a 5 out of 5 rating! perfect submissions for all deliverables. Also did a good job engaging with peers
6	$20/hr	3/4/2024	Critical thinking, problem solving, communication, leadership, teamwork	CRM, project planning and scheduling	5	Perfect work on all deliverables
7	$20/hr	3/4/2024	Leadership, problem solving, communication	Resource management, procurement and supplier management, inventory management	5	Exceptional work on all deliverables, shoutouts from the sponsor on additional elemenets like picking a designers who actually designed a course at the university
6	$20/hr	3/4/2024	Effective communication, positive attitude, strong motivational skills	Spatial planning and visualization, technical proficiency, documentation and reporting	4	Started off with 3 out of 5 rating, but showed progress and improvement to eventually achieve near-perfect results
6	$20/hr	3/4/2024	Cognitive skills, individual skills, social skills	Front-end and back-end technologies, web application, Agile and SCRUM	4	Good job on RAM and staffing pool, strugged with some deliverables

(continued)

Table 2 (continued)

Experience (years)	Billing rate	Available (date)	soft skills	Hard skills	Rating	Performance evaluation (Confidential)
2	$20/hr	3/4/2024	Communication, teamwork, problem-solving	Data analysis, statistical modeling, database management	4	Perfect work on RAM and stakeholder communication plan. Received a perfect score in almost all assignments except kickoff agenda
8	$20/hr	3/4/2024	Effective communication, collaboration, adaptability and flexibility, critical thinking	Automotive design and engineering, prototyping and testing, quality control and assurance	4	Exceptional stakeholder communications plan, has received perfect rating, some initial deliverables were poor but has shown progress and improvement through the weeks
5	$20/hr	3/4/2024	Communication, leadership, teamwork, adaptability, time management, decision making	Quality management	4	Good work on RAM and staffing pool, near perfect work on all assignments, struggled a little bit on the communications plan
3	$20/hr	3/4/2024	Effective communication, adaptability and problem-solving capabilities	Technical skills, communication skills and problem-solving abilities	4	Good job on stakeholder register and stakeholder engagement plan. Struggled with RAM

(continued)

Table 2 (continued)

Experience (years)	Billing rate	Available (date)	soft skills	Hard skills	Rating	Performance evaluation (Confidential)
4	$20/hr	3/4/2024	Communication, leadership, collaboration	Data analysis, reporting, dash boarding	4	Near perfect work on the RAM, performed well on all deliverables except communications plan
3	$20/hr	3/4/2024	Leadership, communication, teamwork, adaptability, conflict resolution	System architecture, analysis and optimization	4	Good work on RAM and staffing pool, slightly struggled with a few other deliverables
4	$20/hr	3/4/2024	Leadership, empathy, adaptability	Resource management	4	Early work, followed complete instructions. Consistently received 4/5 rating on all deliverables. Called out by professor in all evaluations for near perfect results
2	$20/hr	3/4/2024	Communication, empathy, collaboration	Training advice, student support, program coordination	3	Missed minor details during deliverables. Good & timely work
5	$20/hr	3/4/2024	Problem solving abilities, leadership, communication	Performance management, quality management, technical tools	3	Early submissions, followed complete instructions
3	$20/hr	3/4/2024	Resourcefulness, leadership	Data analysis	3	Did not follow complete instructions for most deliverables

(continued)

Table 2 (continued)

Experience (years)	Billing rate	Available (date)	soft skills	Hard skills	Rating	Performance evaluation (Confidential)
4	$20/hr	3/4/2024	Resourcefulness, leadership, communication, teamwork	Performance monitoring, decision making, project lifecycle management	3	Near perfect work on some assignments, 2/5 on a few
2	$20/hr	3/4/2024	Communication, adaptability and problem-solving abilities	Proficiency, a strong foundation and effective communication skills	3	Consistently received a 3 out of 5 rating; did a good job on stakeholder register
4	$20/hr	3/4/2024	Emotional intelligence, adaptability, team collaboration	Data analysis, communication of insights, data visualization	3	Near perfect work on deliverables 5–6. Some deliverables had incomplete details
3	$20/hr	3/4/2024	Effective communication skills, problem-solving abilities and attention to detail	Technical proficiency, data analysis and interpersonal communication skills	3	3 out of 5 rating on average across deliverables
7	$20/hr	3/4/2024	Communication, criticial thinking, ethics, awareness	Examination, evaluation, investigation	3	Consistently received 3 out of 5 rating, but did a good job on the RAM
3	$20/hr	3/4/2024	Communication, interpersonal skills, strategic thinking	Analytical abilities, probject planning	2	Incomplete submissions of deliverables
4	$20/hr	3/4/2024	Leadership, communication, collaboration, adaptability	Technical tools, risk assessment	2	2 out of 5 rating on RAM. could not find submissions for most deliverables

(continued)

Table 2 (continued)

Experience (years)	Billing rate	Available (date)	soft skills	Hard skills	Rating	Performance evaluation (Confidential)
4	$20/hr	3/4/2024	Leadership, communication, problem-solving	Strategic planning, resource allocation, risk management	2	Consistently received 2 or 3 out of 5 rating. Did not follow complete instructions for the deliverables
4	$20/hr	3/4/2024	Communication, critical thinking, problem solving	Advanced data analysis and interpretation, strong programming and statistical modeling skills and domain knowledge	1	Incomplete deliverables, did not follow instructions

usually stored by date to be reviewed historically for trends. Accompanying each performance evaluation will be a quantitative score, an objective rating (1–5 where five is highest), or another metric scale indicating how the resource performed the WBS-assigned work identified in the evaluation comment.

The comments must be objective, factual, and polite—they are used to discuss the performance with the specific resource, to obtain their agreement or disagreement (which should also be recorded), and to share the results with OHRM and the functional manager of the resource. Any special codes or abbreviations should be explained in the footnotes on the performance evaluation report. Still, it is best to avoid special coding since it may not be carried over from the staffing pool database.

A performance evaluation done by the PM should be considered confidential, secured, and never provided to anyone except for, as mentioned above (the OHRM department, functional managers, and the resource should get to see their performance evaluation during the project closing phase). Other than the above, no one should be given anyone's performance evaluation results unless the supervisor or a Court order authorizes it. Consider this confidential information in the same classification of protected and privileged information that a functional human resource manager would store about every employee.

There is a wealth of literature documenting leadership approaches for a PM when performing HRM. We advise you to consult the literature. The critical aspect of PM HRM emphasized in this chapter was to formally document the resources available to the project in the staffing pool database (e.g., using technology, not writing notes on restaurant napkins), understand the abilities of each resource (particularly their soft and hard skills), and to use the data to allocate the resources to WBS activities on a best-fit and availability basis, then objectively monitor performance on an ongoing basis. Finally, we noted the importance of performing a final performance evaluation.

6 PM Staffing Pool Database Applied in a Software Engineering Case Study

The case study company is a multinational corporation valued at over $14 billion USD, based in India, an offshore provider of expertise and technology resources for cloud infrastructure management, system automation scripting, and supply chain load balancing with security cadence [12–14]. The cloud infrastructure services include managing platforms like AWS, Azure, and GCP and provisioning, configuring, and optimizing cloud resources to support the operational needs of applications and services. The automation scripting uses languages like Python, Shell, and PowerShell to streamline operational tasks and improve efficiency in cloud environments. The security protection and firewall are accomplished using Kubernetes approaches to balance and protect containerized workloads, manage clusters, and implement best practices for container deployment, scaling, and networking.

In the initiation phase, the sponsor appointed the PM as a specialist contracted to a reputable Indian company because the client was based in the U.S. and had requested this assignment. The stakeholder register and staffing pool database were developed without any history, as this was a new project in a new context.

Table 2 shows an excerpt from the staffing pool database with confidential identifying attributes removed from the team members. This illustrates how the soft and hard skills were categorized to be leveraged later by the PM for WBS work package activity allocations in the resource allocation matrix. In Table 2, we can observe that the PM sorted the staffing pool database according to performance evaluation rating, in descending order, and then by resource surname (not shown). This artifact was completed in the closing phase of the project so we cannot illustrate the in-progress results. However, it can be emphasized that the author was the PM, and the progress of each resource was periodically documented in the staffing pool database based on each milestone, as the PM had time to do so.

In summary, the staffing pool database was the focal point of PM HRM. There was much less effort put into managing the team members, leading them, or communicating with them. In fact, the author can emphasize that by establishing the staffing pool database, sharing the non-confidential rows during the project kickoff meeting, and sharing the WBS work package assignments part of the time and cost management (not shown in this chapter—beyond PM HRM scope), a solid working and trustful relationship was established with all team members.

We can mention advice for communicating with international resources, and practices supporting EEOC. It is recommended to follow the advice given at the outset to record the resource nickname and verify this with the resource, to avoid using surnames instead of first names for international resources and migrants to any country. Additionally, and not shown in the chapter, it is recommended the PM make an effort to understand how to pronounce each resource's name, either in English or in the native tongue, to respect the individual. The PM in the case study made notes in the staffing pool database to indicate phonetic pronunciations of the nickname, and sometimes the full name, in order to hold meaningful and respected conversations with each resource. As a reminder, literature proves that human resources are the most important assets of a corporation, and the same can be hypothesized about a project. The author found that it was well worth the extra effort of making individual notes about each resource to improve the quality of interactions, including establishing initial credibility, building trust, and maintaining an understanding of the final performance evaluations.

7 Conclusion

This chapter has summarized the most important and relevant literature underpinning contemporary project human resource management (PM HRM). While the ISO and PMI bodies of knowledge were consulted, they lacked any relevant advice beyond the general PM HRM nomenclature. The organizational (OHRM) literature was reviewed and contrasted with the PM HRM literature. A misunderstanding was found in the literature concerning how PM HRM techniques differed from organizational HRM theory and how PM HRM should be applied to projects. A comparative model was discussed so practitioners could understand the differences in these two levels of analysis (organization versus project).

Interestingly, the literature review revealed that neither of the ISO guidelines mentioned HRM, human resources, or even the word 'human,' other than to note that resources are provided to a project by the sponsor [4, 5]. The PMBOK version 7 from PM [6] did not contain information about how to apply PM HRM—there were more instructions about managing physical resources like equipment. The older PMBOK version from PMI provided only fleeting advice about PM HRM [7, 8], mentioning the keyword 'staffing plan,' in an older 2017 document which they defined as a component of the resource management plan, to describe when and how team members will be acquired, along with the length of time they will be needed [7]. Additional problems identified in the PMBOK included describing PM HRM as planning resources, controlling resources, acquiring resources, and developing the team—but without any details about how a PM ought to perform those functions!

This chapter provided the missing details to close the gap in the literature. The RQ was answered, demonstrating effective PM HRM best practices based on a multinational software engineering corporation case study. As part of these best practices, a critical and essential PM HRM artifact was discussed in great detail, namely the staffing pool database. The specific fields or attributes required in this artifact were explained, namely resource attributes, soft skills, functional hard skills, and performance evaluation data. The chapter attempted to show how an experienced PM applied PM HRM particularly to evaluate team resources. Nonetheless, other

PM HRM functions such as leadership, team development, and the basic recruiting/ selecting activities were discussed.

The limitations of this chapter is that the intended audience is emerging or inexperienced PMs. The recommendations were simplified. Nevertheless, alternative theories, techniques, and models were explained so that experienced practitioners and researchers could enhance this study. The research design was also explained in sufficient detail so as to allow researchers to understand and extend this study.

The second significant limitation of this chapter concerned the case study—which was focused on a successful multinational software engineering company based in India. This industry is much different than many other manufacturing industries and also different from practicing project management in other highly-populated industries such as education and public administration. This constraint must be kept in mind for practitioners outside the software engineering industry. Even within the software engineering industry, there are many different subsectors such as software language programming and functional applications such as ERP or HRIS.

Additionally, since a successful project was used for the case study examples, the negative events were not encountered to fully illustrate how the PM would need to compensate for failures in the project plan. All resources were moderately successful (as shown by the performance evaluation in Table 2) but no resource was terminated, left, or had to be replaced. Even though one resource received a 1 rating out of 5 (where 5 was highest and 1 was lowest on the scale), noting that most other resources received ratings from 3 to 5. Consequently, the limitation is that poor-performing resources were not in the sample. Other researchers ought to experiment with more extreme conditions such as resources rated very low.

Finally, although international literature and empirical research studies from Europe were favored, much of the PM HRM project management body of knowledge was developed in the USA and the case study organization was based in the USA. Consequently, other researchers ought to study project stakeholder management best practices by examining empirical studies and case study organizations, from outside the USA. Furthermore, there was some reliance on European literature. Additional global regions ought to be targeted for empirical evidence and research.

References

1. Strang, K.D.: Cybercrime risk found in employee behavior big data using semi-supervised machine learning with personality theories. Big Data Cogn. Comput. **8**(4), Art. no. 4 (2024). https://doi.org/10.3390/bdcc8040037
2. Strang, K.D. (ed.): Handbook of Research Design in Business and Management, 1st edn. In: Research Methods. Palgrave Macmillan US, New York (2015). https://doi.org/10.1057/9781137484956
3. Lawler, E.: Effective Human Resource Management: A Global Analysis. Stanford University Press, Harford, CT (2020). https://doi.org/10.1515/9780804782685
4. Cerezo-Narváez, A., Ceca, M.J.B., Blanco, J.L.Y.: Traceability of intra- and interpersonal skills: from education to labor market. In: Otero-Mateo, M., Pastor-Fernandez, A. (eds.), Human

Capital and Competences in Project Management. IntechOpen, UK (2018). https://www.intechopen.com/chapters/57566
5. PMI.: PMI Lexicon. In: PMI Lexicon of Project Management Terms—Version 3.2 [free public version]. Project Management Institute (PMI), PA, USA (2017). https://www.pmi.org/pmbok-guide-standards/lexicon. Accessed 2 Jan 2024
6. PMI.: The Standard for Project Management and A guide to the Project Management Body of Knowledge (PMBOK Guide), 7th edn. Project Management Institute (PMI), PA, USA (2021)
7. ISO.: BS ISO 21500:2021 Project, Programme and Portfolio Management. Context and Concepts (2023). https://www.en-standard.eu/bs-iso-21500-2021-project-programme-and-portfolio-management-context-and-concepts/. Accessed 24 Nov 2023
8. ISO.: ISO 21502:2020(en), Project, Programme and Portfolio Management—Guidance on Project Management (2012). https://www.iso.org/obp/ui/en/#iso:std:iso:21502:ed-1:v1:en. Accessed 23 Nov 2023
9. PMI.: The Standard for Program Management, 4th edn. Project Management Institute (PMI), PA, USA (2017). https://www.pmi.org/pmbok-guide-standards/foundational/program-management. Accessed 2 Jan 2024
10. Strang, K.D.: Resistance, diseconomies, and abnormal AI behavior in HRM: a real-time big data action research experiment at a pharmaceutical. Hum.-Intell. Syst. Integr. **4**(3–4), 35–52 (2022). https://doi.org/10.1007/s42454-022-00046-6
11. Strang, K.D., Sun, Z.: ERP Staff Versus AI Recruitment with Employment Real-Time Big Data. Discov. Artif. Intell. **2**(1), 21 (2022). https://doi.org/10.1007/s44163-022-00037-1
12. Strang, K.D.: How effective is business education in the workplace: structural equation model of soft and hard skill competencies. SN Bus. Econ. **3**(1), 28 (2022). https://doi.org/10.1007/s43546-022-00404-1
13. Che, F.N., Vajjhala, N.R., Strang, K.D.: Perceived student learning, attitudes and experiential learning in software engineering education. In: 2021 IEEE World Conference on Engineering Education (EDUNINE), pp. 1–6. IEEE, Guatemala City, Guatemala (2021). https://doi.org/10.1109/EDUNINE51952.2021.9429095
14. Strang, K.D., Vajjhala, N.R.: Analysis of university engineering program skill effectiveness in research & development companies. In: 2023 IEEE World Engineering Education Conference (EDUNINE), pp. 1–5 (2023). https://doi.org/10.1109/EDUNINE57531.2023.10102861

The manufacturer's authorised representative in the EU is Springer Nature Customer Service Centre GmbH, Europaplatz 3, 69115 Heidelberg, Germany. If you have any concerns regarding our products, please contact ProductSafety@springernature.com

Printed and bound by CPI Group (UK) Ltd, Croydon, CR0 4YY
10/02/2026
02051042-0008